AF732505

handbook of nanoindentation with biological applications

handbook of
nanoindentation
with biological applications

editor
michelle l. oyen
University of Cambridge, UK

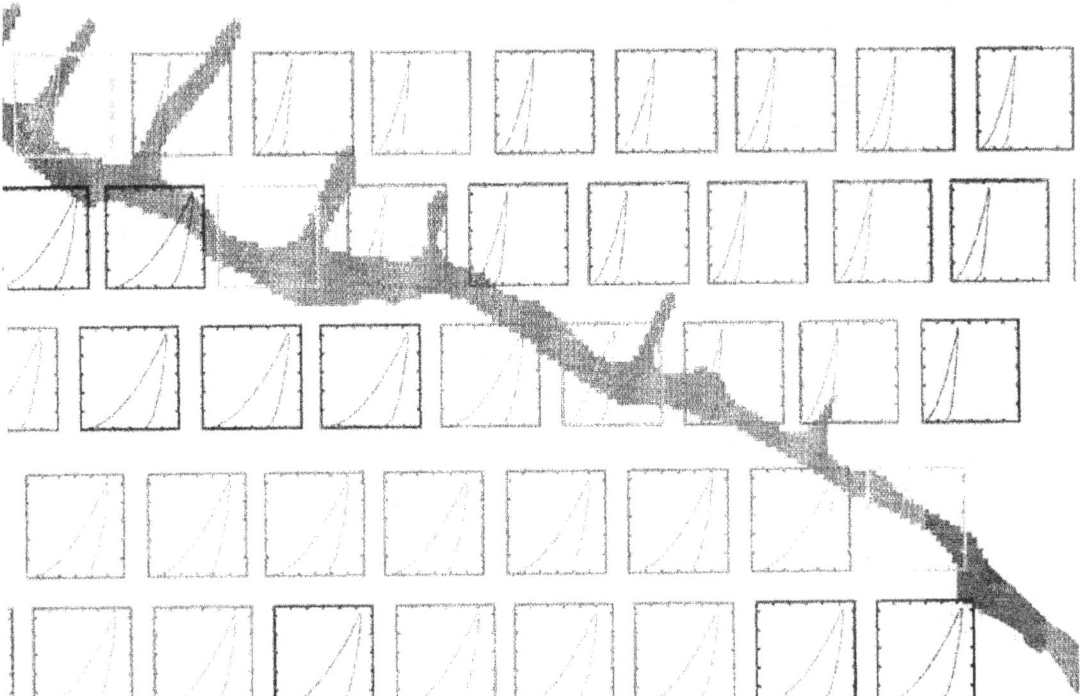

CRC Press
Taylor & Francis Group
Boca Raton London New York

CRC Press is an imprint of the
Taylor & Francis Group, an **informa** business

HANDBOOK OF NANOINDENTATION WITH BIOLOGICAL APPLICATIONS

First published 2011 by Pan Stanford Publishing Pte. Ltd.

Published 2019 by CRC Press
Taylor & Francis Group
6000 Broken Sound Parkway NW, Suite 300
Boca Raton, FL 33487-2742

© 2011 by Taylor & Francis Group, LLC
CRC Press is an imprint of the Taylor & Francis Group, an informa business

No claim to original U.S. Government works

ISBN-13: 978-981-4241-89-2 (hbk)

This book contains information obtained from authentic and highly regarded sources. While all reasonable efforts have been made to publish reliable data and information, neither the author[s] nor the publisher can accept any legal responsibility or liability for any errors or omissions that may be made. The publishers wish to make clear that any views or opinions expressed in this book by individual editors, authors or contributors are personal to them and do not necessarily reflect the views/opinions of the publishers. The information or guidance contained in this book is intended for use by medical, scientific or health-care professionals and is provided strictly as a supplement to the medical or other professional's own judgement, their knowledge of the patient's medical history, relevant manufacturer's instructions and the appropriate best practice guidelines. Because of the rapid advances in medical science, any information or advice on dosages, procedures or diagnoses should be independently verified. The reader is strongly urged to consult the relevant national drug formulary and the drug companies' and device or material manufacturers' printed instructions, and their websites, before administering or utilizing any of the drugs, devices or materials mentioned in this book. This book does not indicate whether a particular treatment is appropriate or suitable for a particular individual. Ultimately it is the sole responsibility of the medical professional to make his or her own professional judgements, so as to advise and treat patients appropriately. The authors and publishers have also attempted to trace the copyright holders of all material reproduced in this publication and apologize to copyright holders if permission to publish in this form has not been obtained. If any copyright material has not been acknowledged please write and let us know so we may rectify in any future reprint.

Except as permitted under U.S. Copyright Law, no part of this book may be reprinted, reproduced, transmitted, or utilized in any form by any electronic, mechanical, or other means, now known or hereafter invented, including photocopying, microfilming, and recording, or in any information storage or retrieval system, without written permission from the publishers.

For permission to photocopy or use material electronically from this work, please access www.copyright.com (http://www.copyright.com/) or contact the Copyright Clearance Center, Inc. (CCC), 222 Rosewood Drive, Danvers, MA 01923, 978-750-8400. CCC is a not-for-profit organization that provides licenses and registration for a variety of users. For organizations that have been granted a photocopy license by the CCC, a separate system of payment has been arranged.

Trademark Notice: Product or corporate names may be trademarks or registered trademarks, and are used only for identification and explanation without intent to infringe.

Visit the Taylor & Francis Web site at
http://www.taylorandfrancis.com

and the CRC Press Web site at
http://www.crcpress.com

British Library Cataloguing-in-Publication Data
A catalogue record for this book is available from the British Library.

Cover designed in consultation with Gavron Designs

Preface

This book represents a first attempt to aggregate the recent literature concerning nanoindentation testing of biological materials. The topic has grown quickly in the last decade, and has been the subject of a number of recent conference symposia and journal issues; many of these included contributions by one or more of the chapter authors represented here. The field is developing rapidly, and it is inevitable that by the time this book is published there will have been many (if not hundreds) of new technical papers on this subject.

For this reason, this volume has been focussed on general principles, instead of specific results, wherever possible. Following an introduction, the volume is divided into two parts. In the first part, Chapters 2-6, fundamental contact mechanics principles underlying the nanoindentation technique are explored in the context of biological materials. Although biological materials are the focus here, the content of these chapters is equally applicable in the context of polymers, hydrogels and other non-biological systems with mechanical properties in the range of those found in natural systems.

The second part of the book, Chapters 7-10, examines examples of specific biological systems in which nanoindentation testing has been a key tool in explicating nanomechanical structure-properties relationships. These chapters are not intended to be comprehensive reviews of the literature on any one topic. Instead, they are meant to be instructive examples, demonstrating the potential for nanoindentation methods in these and other biological systems, while investigating the challenges associated with application of the technique to such systems.

I am grateful to the chapter authors for their contributions and for their overall enthusiasm with, and participation in, the process of bringing this volume together. With the fast-paced growth of nanoindentation in recent years, this effort represents only a first broad view of the field. I am inspired by having had the opportunity to work on this project with some true pioneers of nanomechanics.

Michelle L. Oyen
Cambridge University
Engineering Department
Cambridge, UK
2 June, 2010

Contents

CHAPTER 1

INTRODUCTION

Tamaryn A.V. Shean, Michelle L. Oyen and Michael F. Ashby

Department of Engineering, Cambridge University
E-mail: tavs2@cam.ac.uk

This introductory chapter provides background to motivate the study of biological materials with nanoindentation techniques. The reasoning behind investigations of the mechanical behavior of biological materials is introduced, as are the particular reasons for conducting these investigations at small physical length-scales. A taxonomy for biological materials is presented, emphasizing a top-down hierarchical system for classification, from organs to tissues and down in scale, including the molecular building blocks for both cells and extracellular matrix material. Typical implementation of nanoindentation testing is described and limitations in applying the technique to biological systems are presented. This overview motivates the content contained in the remainder of this volume, including investigation of fundamental contact mechanics principles appropriate to biological systems and application of nanoindentation techniques to the characterization of biological materials.

1. Introduction

There are many naturally occurring materials with fundamental mechanical functions. Both people and plants have structural skeletons; many shells and some bones (*i.e.* the ribcage and skull) provide protection for delicate organisms or internal organs. There has been recent growth in the measurement of mechanical properties of natural materials with structural functions,[1-4] in the context of trying to understand how these materials respond to mechanical forces and

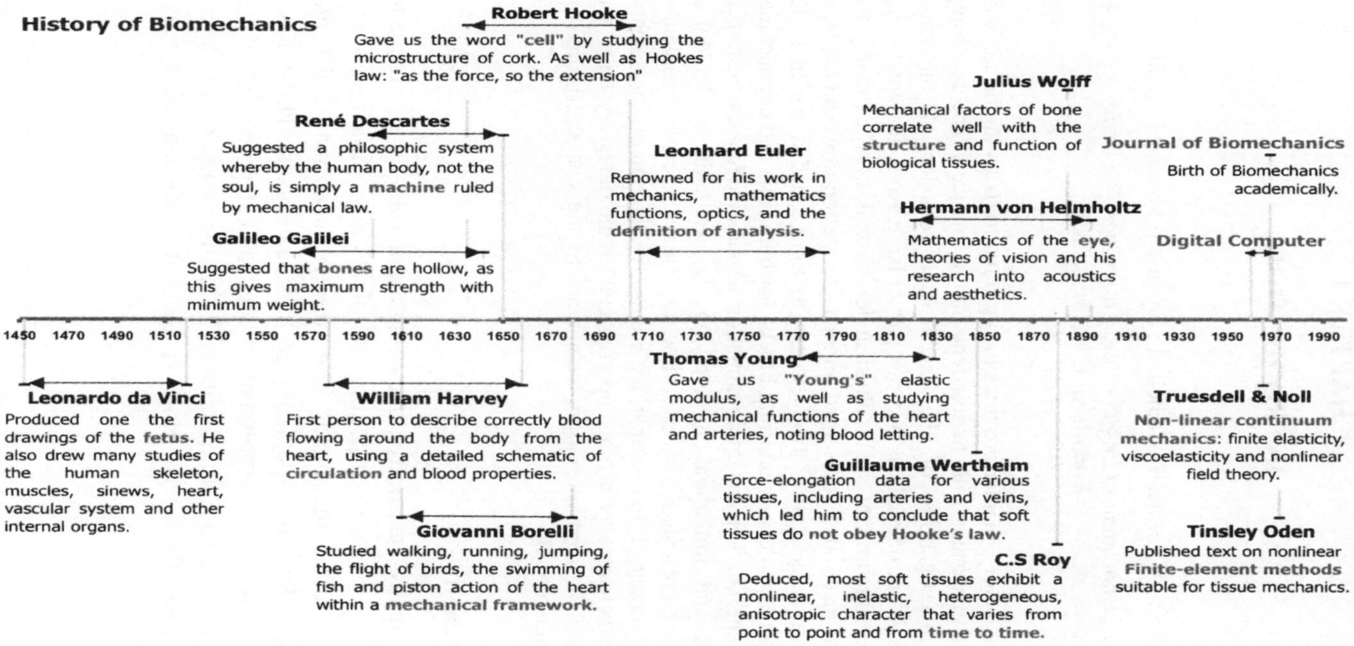

Figure 1-1. The history of biomechanics, prior to the 1960s boom, in which Professor Y.C. Fung and others helped establish the field as an autonomous scientific discipline.

deformations. In addition to trying to gain fundamental physical understanding, there are direct applications in fields such as medicine, an example being the use of mechanical assays for diagnosis of disease.

In the context of human and medical applications, research on mechanical functioning of biological materials is often labeled "biomechanics". The recent history of biomechanics extends back to 1452, with Leonardo da Vinci (Fig. 1-1) in work that included investigations at dominantly macroscopic length-scales, including the description of whole-body motion, experiments on the mechanical responses of organs and tissues, and the development of continuum-based constitutive laws to describe tissue behavior. The pioneering work of "the Father of Biomechanics," Professor Y.C. Fung,[5,6] helped establish the field as an autonomous scientific discipline within the broader field of bioengineering.

Investigation of mechanical behavior of natural materials *not* associated with human medicine is a field that has largely developed in parallel to traditional (medically-oriented) biomechanics. These distinct fields, however, have begun to merge as research has increasingly focused on fundamental length-scales, as shown in Fig. 1-2. The study of natural materials—either in a medical or a basic science context—is expanding with the application of modern materials characterization techniques, many developed for probing materials and many for imaging at the atomic and molecular scale, to the study of biological systems (Fig. 1-2). The introduction of materials science techniques to the study of biomechanics has also been associated with the application of materials science philosophy to biological problems: focus has shifted subtly from purely mechanical explorations to investigations concerned with structure-property relationships. New and developing subspecialties have appeared including multi-scale theoretical modeling of hierarchical biological structures, nanoindentation and nanotribology of biological materials, nanomechanics of individual biological constituents, single cell mechanics, and examination of bioadhesion and biolubrication.

The recent interest in studying mechanics of natural materials is combined with a growing interest in biomimetic materials synthesis for the creation of robust, and in many cases mechanically-functional,

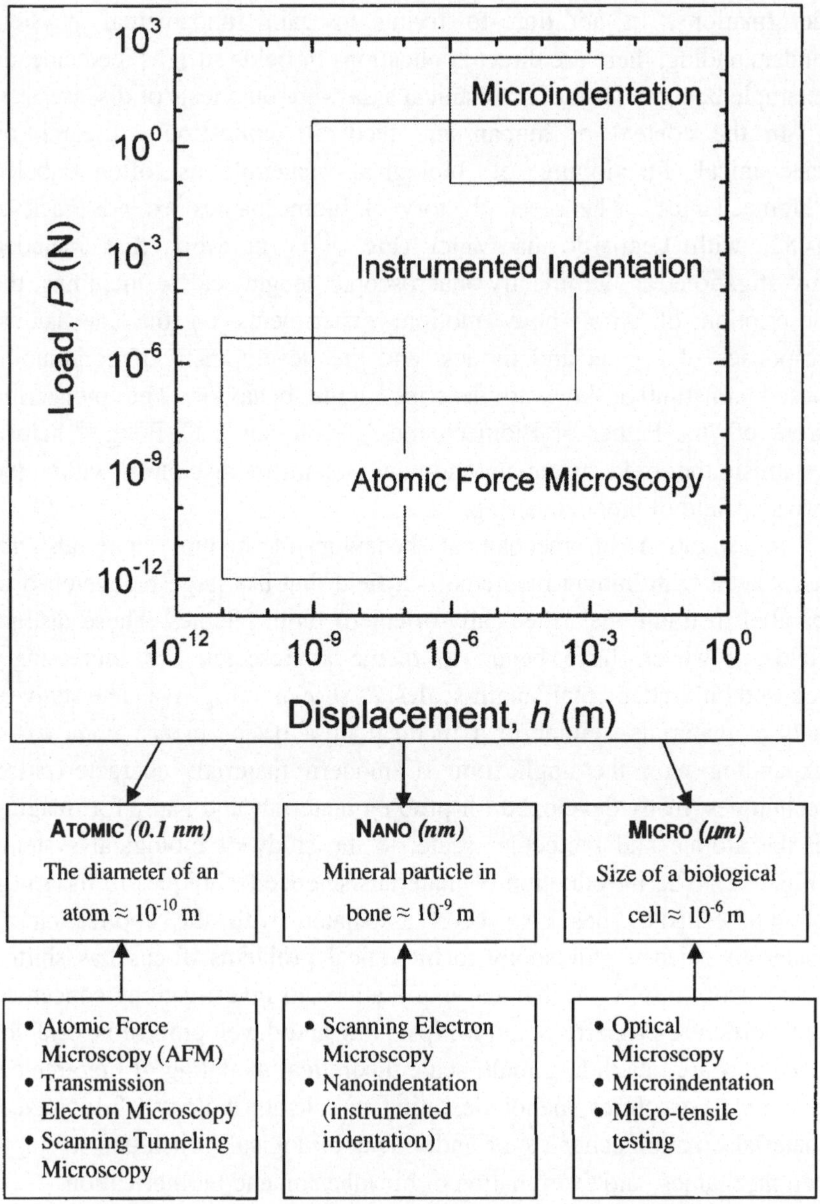

Figure 1-2. Mechanical characterization diagram, explaining the relevant length scales involved in biological mechanical testing and the types of testing available.

materials based on the "design principles" active in nature.[7-9] For mimicking the structure and mechanical properties of biological composites, detailed investigation of the mechanical properties of the native materials is required. Biomimesis also includes exciting developments involving novel materials processing techniques aimed at the generation of hierarchical and complicated structures.

Nanoindentation has emerged as a leading technique for characterizing the mechanical behavior of engineering materials, and is emerging as an increasingly popular technique for the study of biological and biomimetic materials. Traditional mechanical testing, such as a tensile test, involves the gripping of a sample with an approximately constant cross-sectional area (A_0) and pulling (or pushing in compression) in a direction perpendicular to the cross-section. A large volume V of material is loaded during this type of test, equal to the product of the sample gage length (l_0) and A_0; V is similar in magnitude to the total sample volume V_0 (less only the gripped region). In contrast,

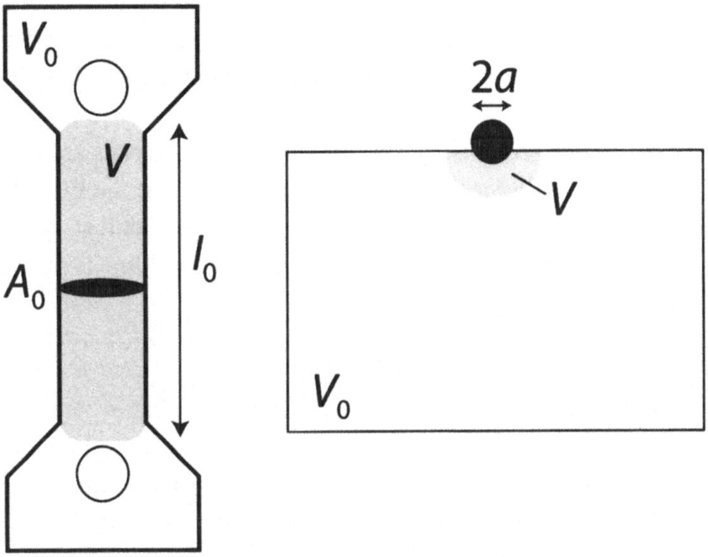

Figure 1-3. A schematic illustration of the mechanically tested volume (V) relative to the total sample volume (V_0). Other variables are defined in the text.

in contact mechanical testing a probe with known or calibrated shape is brought down onto a flat surface of material. The total sample volume V_0 can be substantially larger than the volume of material mechanically tested, V (Fig. 1-3). Thus, while a tensile test averages the mechanical response over the whole sample, a contact test measures local behavior over a contact diameter of $2a$. Contact mechanical testing has become popular for examining local variations in mechanical behavior because of this small ratio of V/V_0. Indentation testing at very small length scales has been frequently referred to as "nanoindentation" although in reality it ranges from true nanoindentation to microindentation (Fig. 1-2).

Due in part to the hierarchical and multiscale organization of natural materials, the mechanical response can vary depending on the length-scale of observation. Especially at fundamental length-scales, there is dramatic structural heterogeneity in natural materials due to distributions in shape, size, composition, and mechanical properties of the constituents. Nanoindentation is thus ideal for probing local heterogeneities in natural materials, and can be used to map out local heterogeneity, and great scope exists both for experiments at very small (nm) length scales and for multi-scale investigations incorporating nanomechanical results with higher levels of organization (length-scales from nm to μm).

In the following sections, classes of biological materials are reviewed. Nanoindentation testing techniques are introduced, and the limitations of traditional nanoindentation are discussed in the context of natural materials. The "roadmap" for the remainder of this volume is presented.

2. Biological Materials Taxonomy

The natural world is broad and diverse. There are existing mechanisms for classifying biological organisms based on an eight-level hierarchy: domain, kingdom, phylum, class, order, family, genus, species. The two-name "scientific" designation according to genus-species is the standard identification mechanism for biological organisms, for example *Escherichia coli* (*E. coli*, a bacterium) or *Homo sapiens* (humans). These classification schemes are fundamentally cell-based. Biological organisms can be unicellular or multicellular, where the cell is the smallest individual unit of living matter, the presence of which signals a difference between a living tissue and a nonliving entity. Cells are typically micrometers in diameter, and the products of cells range from molecules sized in the nanometers, to multimolecular assemblies larger than the cells that produced them.

The subject of this work is biological materials, where it is recognized that this could include non-living materials that were manufactured by biological entities. In addition, materials *per se* can be sub-units of an organism, and similar tissues (or materials) can exist across many different species. As such, a slightly different taxonomy is required to categorize the natural materials considered here, as opposed to the organisms and species considered by classical Linnaen taxonomy. This overall structure—starting with the kingdom designation from traditional taxonomy—is shown in Fig. 1-4. Within the individual kingdoms, there are large structures called organs, which are found both in animals (bones, kidneys, brain, skin) and in plants (root, leaf, flower). In general organs are a high-order, complicated structure that is made up of several individual tissues, where tissues contain cell types for performing specific functions. Since the same tissues can be found in multiple organs, organs are not considered as a separate level in the taxonomy. There are four animal tissues (epithelium, nervous tissue, muscle tissue, connective tissue) and three plant tissues (epidermis, ground tissue, and vascular tissue). Tissues are themselves composed of both cells and extracellular matrix (ECM). While many tissues are primarily cellular, the connective tissues in animals and the sclerenchyma in plants (part of the ground tissue) are dominantly extracellular matrix and this is where the majority of structural and mechanical functions are found.

Biological tissues are not continuous media, but are composites consisting of living cells in the non-living ECM that is built and remodeled by the cells. Biological materials are dynamic—there exists great potential for changes in the tissue hierarchy with growth and development or with disease because of the action of live cells. The complexity of natural materials can be quite astonishing, with many levels of hierarchical structure over a range of length-scales. The ECM can itself be a composite material, with multiple phases including any number of organic and inorganic constituents. It is therefore useful to break things down according to the molecular building blocks from which they are made. Cells and tissues are constructed from four classes of organic macromolecules: proteins, saccharides (carbohydrates), lipids and nucleic acids (including DNA and RNA).[10] The lipids and nucleic acids are dominantly cell-associated in the body, while proteins and saccharides are important in cells but also make up the organic component of the ECM. In addition, there are inorganic biominerals present in many tissue types. Specific examples of each of these molecular "building blocks" will follow in Section 2.1, with examples of specific tissues composed of these building blocks summarized in Section 2.2.

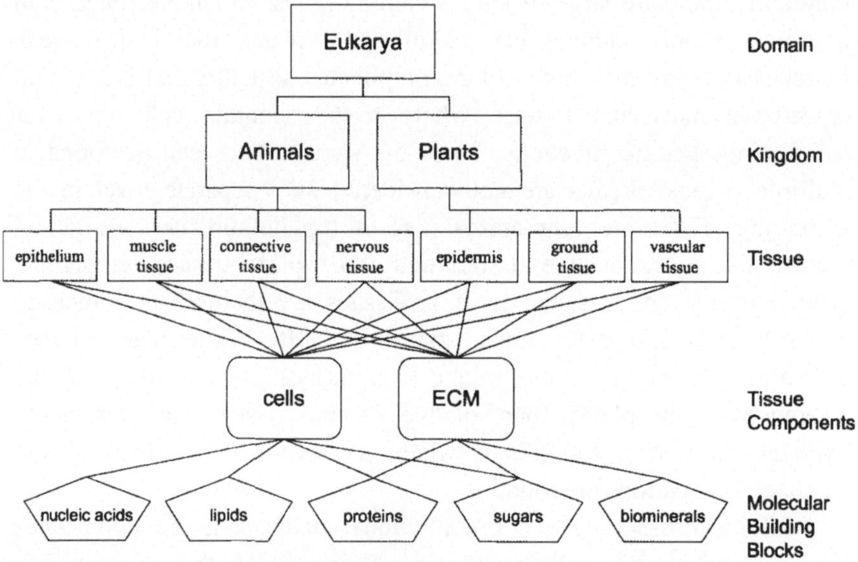

Figure 1-4. A new taxonomy for the categorization of biological materials. The taxonomy initiates with the domain and kingdom designations from typical botanical and zoological taxonomy but then focuses on structures relevant from a materials—instead of species—perspective.

2.1. *Molecular building blocks*

Here, briefly, the building blocks of biological materials are discussed including the cell-associated components, proteins, sugars, biominerals and water.

2.1.1. *Cell-associated components*

Cells are made up of carbon, hydrogen, nitrogen and oxygen—these four atoms make up 99% of the cell by weight, including 70% water. The four atoms make up four classes of small molecules: simple sugars (monosaccharides), fatty acids (saturated and unsaturated), amino acids and nucleotides. Macromolecules are then formed from these small molecules: monosaccharides into polysaccharides, amino acids polymerize into proteins, and nucleotides polymerize into nucleic acids. The lipids in cells are fatty acids that form phospholipids, the major component of bilayer cell membranes. Fatty acids are found in many places and it is therefore not easy to categorize one main set of macromolecules into which they polymerize. Nucleic acids are based on the sugars ribose and deoxyribose, forming RNA and DNA, respectively.

2.1.2. *Proteins*

Proteins are classed as either primarily structural—a category dominated by the fibrillar proteins including the ubiquitous macromolecule collagen—or mainly associated with non-mechanical functions, such as catalysis by enzymes.

Protein chains are all synthesized from different linear combinations of 20 amino acids. An amino acid structure is centered on an alpha carbon atom –**C**; off this atom is an amino group $-\mathbf{NH_2}$, a carboxyl group $-\mathbf{CO_2H}$, a hydrogen atom –**H** and a side chain $-\mathbf{R}_i$, as shown in Fig. 1-5. These side chains can range from a single hydrogen atom to a very complex long chain aromatic group,[10] each side chain creates a different amino acid, which has different physicochemical properties: amino acids can be basic or acidic, uncharged polar or nonpolar. A polypeptide chain is a chain of amino acids, linked together by peptide bonds. Peptides bonds are inflexible and therefore chain rotation can only occur on the bonds surrounding these peptide bonds. While the primary structure of proteins arises from covalent peptide linkages, proteins have secondary, tertiary and quaternary structures that arise from the folding of long polypeptide chains (frequently more than one chain per protein) into compact, three-dimensional strcutres due to the actions of ionic bonds plus "weak" interactions, including hydrogen bonding, van der Waals bonds, and hydrophobic interactions. These higher-order structures form spontaneously, as the folded structure is the lowest energy conformation for that particular sequence of amino acids, and arises due to interactions between the side-chains (and based on their classification as basic, acidic, uncharged polar, or nonpolar).

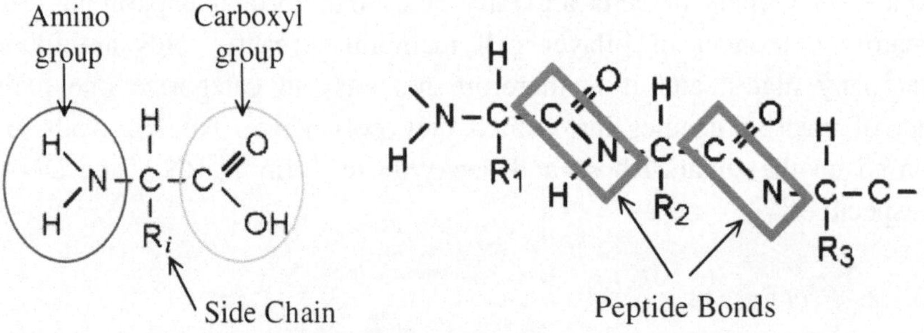

Figure 1-5. An amino acid on the left and a protein chain, made by polymerization of amino acids, showing the peptide bonds, on the right.

2.1.3. *Sugars*

Monosaccharides, or simple sugars, are the basic units of carbohydrates. Glucose, fructose, galactose, xylose and ribose are all examples of monosaccharides (Fig. 1-6). Most have the chemical composition $C_X(H_2O)_Y$, although there are a few exceptions, including deoxyribose from DNA. Monosaccharides are the building blocks for disaccharides, such as sucrose.

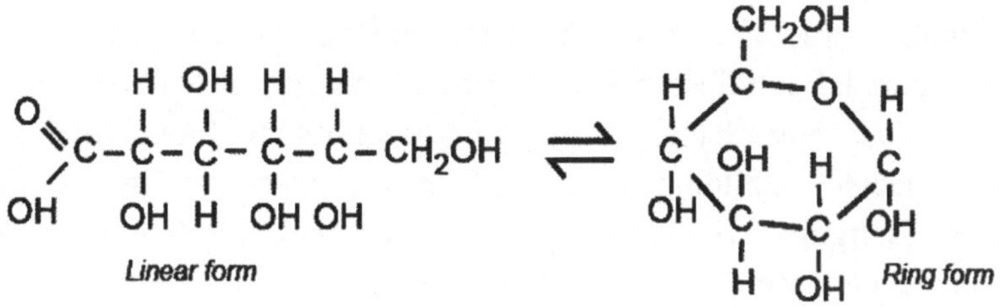

Figure 1-6. Glucose, a monosaccharide, in both its linear and ring form.

Monosaccharide subunits are linked together with glycosidic bonds to form long polymers, polysaccharides, which are complex carbohydrates (Fig. 1-7). These polysaccharides perform two essential functions: structural and storage. Glycogen and starch store energy in a way that can be retrieved when the organism needs it. Structural saccharides, such as cellulose from the plant kingdom and chitin in invertebrate arthropods and insects, act as a mechanical reinforcement phase in a weak organic "matrix". Mono- and polysaccharides exist in two structural forms: linear molecules or as a ring. Monosaccharides in the ring form have the ability to polymerize, via a condensation reaction (releasing water), to give stiff and straight chains of polysaccharides. Figure 1-7 demonstrates this reaction, which is reversible by hydrolysis (addition of a water molecule).

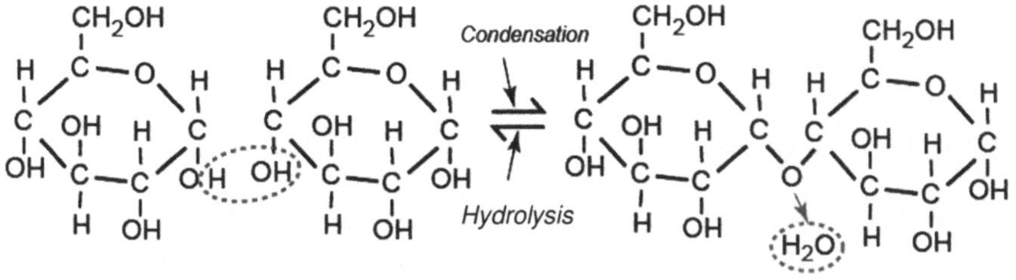

Figure1-7. Monosaccharides reacting by condensation to form polysaccharides, and by undergoing hydrolysis polysaccharides return to monosaccharides.

The building block of primary interest in the plant kingdom is cellulose, which can have thousands of monosaccharide units, arranged such that the bonding oxygen atoms are staggered (Fig. 1-8). Because of this regular structure, there are large numbers of hydrogen bonds between parallel cellulose chains, forming a regular crystalline structure with outstanding mechanical properties for an organic biopolymer. (The other molecule "building block" of most importance in the plant kingdom is lignin, which is a poorly-understood ring-based complex macromolecule formed from alcohols, and which forms an integral part of the secondary cell walls of plants.)

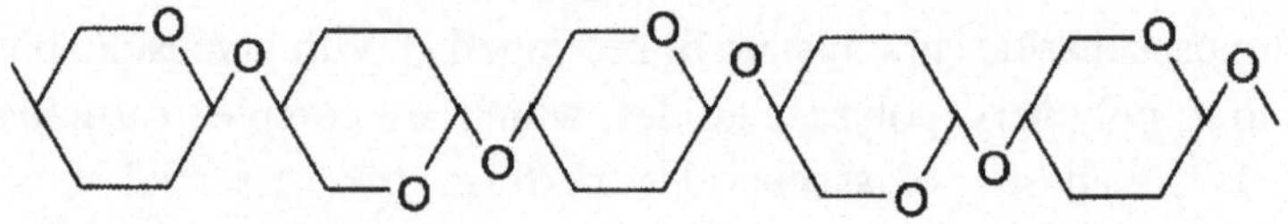

Figure 1-8. Cellulose molecule in short-hand, where each hexagon represents a carbon ring and in which only the oxygen atoms are identified.

The saccharides in the vertebrate ECM are mainly glycosaminoglycans, which are long, unbranched, highly charged molecules formed of disaccharide repeats. There are two important categories of protein-sugar complexes, glycoproteins and proteoglycans. Glycoproteins are proteins with covalently linked oligosaccharide chains, while proteoglycans are a subset of glycoproteins that are heavily glycosylated, in which many glycosaminoglycans are linked to a single protein core. These glycoproteins and proteoglycans assist in protein folding and improve the protein stability; as well as frequently binding water and controlling the local water concentration.

2.1.4. *Biominerals*

Biominerals are the ceramics of the biological world. There are a small number of bioceramics with dominantly structural functions; the biominerals of primary interest here are calcite, aragonite, hydroxyapatite and bio-silica. It is these biominerals that give stiffness, hardness and strength to many natural materials, including bone, teeth and shells. Figure 1-9 shows the formulae and shape of the unit cell that describes the structures of calcite, aragonite and hydroxyapatite; bio-silica is an amorphous glass. Biominerals are analogs of geological minerals but tend to be impure, highly substituted, and are typically not fully dense.

Biomineralization is the process by which living tissues control the deposition of minerals within their structure, either intra- or extra-cellularly, creating composite materials containing both organic and inorganic components. This process is equally useful in both the plant kingdom and animal kingdom, for example, bio-silica gives the nettle its sting, the cactus its spines and bamboo its ability to grow tall yet remain slender. Hydroxyapatite gives teeth and bones their stiffness. There are many reasons for studying the process of biomineralization; biominerals are formed at body temperature and pressure, in stark contrast to most ceramic processing techniques. Mimicking this process introduces energy savings compared with typical ceramic processing, which is advantageous both for the environment and for the price of production.

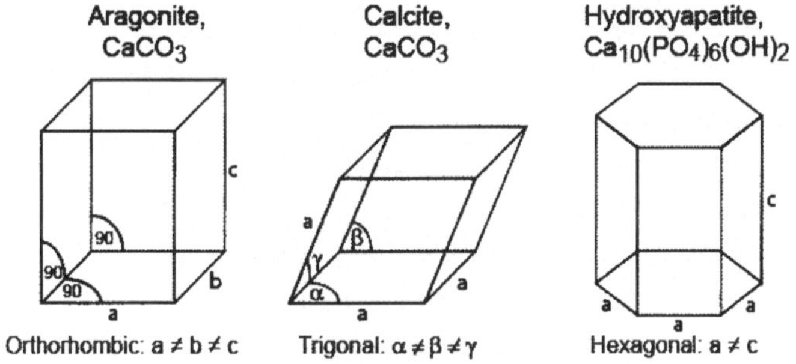

Figure 1-9. Three of the most important biominerals in the natural world, aragonite, calcite, and hydroxyapatite, with and their structural formulae.

2.1.5. *Water*

Biological macromolecules have a chemical structure that is highly dependent on hydrogen bonds, hence the importance of water in proteins, tissues and minerals. Although not explicitly listed in Fig. 1-4 as a building block, water makes up 70% of the human body, both intra- and extra-cellular, and is responsible for many aspects of chemistry that make life possible. Water is used to tune the mechanical properties of chitin-based composites, allowing for variation in elastic modulus over many orders of magnitude.[11] Water acts as a plasticizer for protein molecules such as collagen.[12] Water is bound by the charged glycosaminoglycans molecules in proteoglycans, and electrostatic interaction form as much as 50% of the compressive resistance of proteoglycan-rich tissues such as articular cartilage.[13]

2.2. *Natural tissues*

Biological materials are formed from these primary building blocks—proteins, sugars, minerals and water—and a wide range of composition options are represented in both the plant and animal kingdoms. Next we briefly summarize exemplar biological materials with mechanical functions, including examples in each kingdom.

2.2.1. *Animal kingdom*

There are many interesting structural materials in the animal kingdom, including chitin-based materials, such as insect cuticle, nacre (a chitin-aragonite composite), and silks. In addition, from the vertebrate subphylum are mineralized (hard) tissues with hydroxyapatite reinforcement, such as bones and tooth tissues, non-mineralized (soft) tissues such as ligaments and cartilage, keratin-based tissues, such as nails and hair, and dominantly cellular tissues such as brain and liver.

Mineralized tissues are composites of protein and hydroxyapatite, where the protein is collagen in bone and dentine. The structure of bone will be considered extensively in Chapter 7 of this volume, and teeth will be examined in Chapter 8. Soft tissues are composites of protein and glycosaminoglycans; these will be examined in further detail in Chapter 9 along with silks and nacre. Cellular tissues are cell-dominated, but there is an extracellular matrix component that provides a framework on which or in which the cells reside; to date these materials have been the subject of less nanoindentation research than the ECM-dominated materials while the indentation of individual cells has been more common; cell indentation will be considered in Chapter 10 of this volume.

2.2.2. *Plant kingdom*

In the plant kingdom, tissues are roughly divided by their being "woody" (wood, bark, cork, cone) or "non-woody" (leaf, sisal, pitcher plant, fruit, coir, seed, cotton, bast fiber, hemp, jute, flax, straw). All wood and wood-like materials rely on cellulose for their mechanical stiffness and strength. They are light, parallel to the grain, stiff, strong and tough—as good per unit weight as many man-made materials, except carbon fiber reinforced plastic (CFRP, a very strong, lightweight and expensive fiber reinforced polymer used in engineering).

All tree-woods have broadly the same chemical composition: 40-50 wt% crystalline cellulose fibers in a matrix consisting of 20-25 wt% partly-crystalline hemi-cellulose and another 25-30 wt% amorphous lignin; in addition there are 0-10 wt% oily extractives that give wood its color and smell. Three main features characterize the microstructure—elongated cells that make up the bulk of the wood, called *tracheids* in softwoods and *fibers* in hard woods; *rays*, radial channels connecting the inner part of the trunk to the thin outer region or *cambium; sap channels*, which are enlarged cells with thin walls and large pore space but conduct fluids up the tree, driven by osmotic pressure.

Palms are monocots, more closely related to grasses and ferns than to trees. Bamboos are grasses, mineralized with bio-silica. Many trees have a thin layer of cork just below the outer bark. Structurally it is low density, polymeric closed-cell foam. Nutshells are made up of a network of fiber tissues with a composition close to that of wood, but with an approximately random 3-D fiber lay-up. Materials from the plant world are briefly examined in Chapter 9 along with soft tissues, as nanoindentation testing of these materials is a rapidly growing area of current research.

2.3. *Biomimetic and derived materials*

In addition to naturally-occurring materials, there are a number of quite important materials that are naturally-derived. For example, engineered wood is a composite made of real wood with a polymer binder, and leather is animal skin that has been artificially treated with chemical cross-linkers. In addition, there has been significant recent interest in materials that are purified from natural components. Gel materials are quite common in this arena, including agarose gel, fibrin gel, and gelatin (denatured collagen) gels.

Finally, biomimetic composite materials are being engineered to resemble natural materials. Examples include bone-like materials made from gelatin and hydroxyapatite[14] and nacre-like materials made from polyelectrolytes and clays, producing the organic and inorganic layers similar to those in the nacre.[8] Issues associated with the nanoindentation of biomimetic materials are similar to those encountered in studies of the material being mimicked, and as such there is no separate treatment of these classes of biomimetic materials in this volume.

3. Nanoindentation

In nanoindentation testing, a probe is pressed into a material surface, typically under load-control, and the load and displacement are monitored during the full loading-unloading contact cycle. Following the publication of the landmark paper by Oliver and Pharr in 1992,[15] a straightforward mechanism has existed for the deconvolution of elastic modulus (E) and hardness (H) of an elastic-plastic half-space based on the indentation load-displacement (P-h) data. The Oliver-Pharr algorithm is incorporated directly into the software shipping with most commercial nanoindentation instruments such that automated, high-throughput mechanical testing is facilitated. This actually introduces a problem, in that the Oliver-Pharr procedure is not appropriate for materials that are not characterized as an elastic-plastic half-space,[16] including essentially all polymers and biological materials. Further complications arise when there is substantial adhesion between the sample and the indenter tip, as is often the case with "softer" materials including polymers and soft tissues. As such, the values for E and H that are ejected automatically from a commercial nanoindenter can be meaningless[16] in the context of the materials of interest in this volume.

An additional issue arises with the interpretation of "hardness" data from Vickers or nanoindentation tests of biological materials. Clearly the parameter H has meaning, but it is in this context quite unlike in traditional metal testing, where there can be large hardness variations at constant elastic modulus due to local microstructural effects associated with dislocation motion. Hardening mechanisms such as grain size refinement due to mechanical work are simply unavailable in the non-metal materials found in the natural world. In the context of instrumented indentation testing, the hardness is the mean supported contact stress,[17] a measure of the total resistance to deformation, regardless whether the deformation is elastic, plastic,[18] viscous,[19] or due to crushing fracture.[20] As such, on average, over a wide range of biological materials, the hardness is not an independent material property but depends directly on elastic modulus[21] (Fig. 1-10).

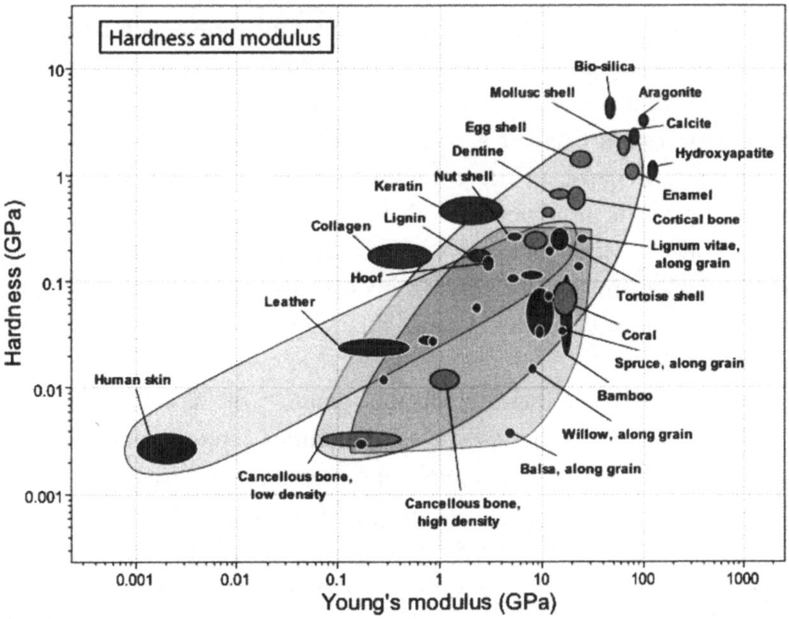

Figure 1-10. The relationship between elastic modulus and "hardness" in biological materials.

Although biological materials often do not exhibit elastic-plastic responses, there are reasons to consider nanoindentation testing as particularly useful for their mechanical characterization. Nanoindentation testing is well-suited for studying the composite nature of biological materials. Biological materials can be functionally graded, where the mechanical properties vary from point to point for reasons to do with the biological function.[22] These property variations result from local variations in tissue composition and microstructure, and the variations are frequently associated with the length-scales of ECM (*i.e.* length scales of tens of nanometers to micrometers) and of cell activity (*i.e.* length scales of micrometers to tens of micrometers). There is no question that elastic modulus is a useful parameter to measure in biological materials, and in many cases the viscoelastic response may also be relevant for interpretation of the elastic response. The "hardness" may still be a useful measure, although with the caveats as noted above. The conditions in which Oliver-Pharr[15] analysis is appropriate, and the alternatives for analysis of indentation data, will form a common theme throughout the remainder of this work, with detailed consideration of different contact mechanics models for data analysis including tip-sample adhesion in Chapter 4 and in the context of viscous deformation in Chapter 5. Because of the complexity of both the material response and the geometric nonlinearities that arise in contact problems, much recent work in this field has used computational models of the contact problem instead of relying on closed-form analytical solutions for simpler contact problems, and this will be considered further in Chapter 6.

Another challenge arising in the context of nanoindentation of biological materials is the large elastic modulus range found, especially compared with the glass, metal and ceramic materials for which nanoindentation testing was originally designed (Fig. 1-11). Nanoindentation instrumentation was optimized for the narrow range of engineering materials in the GPa to hundreds of GPa range, where typical experiments are conducted in mN and nm ranges. In contrast, for soft tissues, with kPa-MPa modulus values, the corresponding working ranges might be μN-μm or even μN-mm. There are thus significant instrumentation challenges associated with the adaptation of indentation instrumentation to the full range of elastic modulus values that arise when considering biological materials; these issues will be examined in detail in Chapters 2 and 3 of this volume.

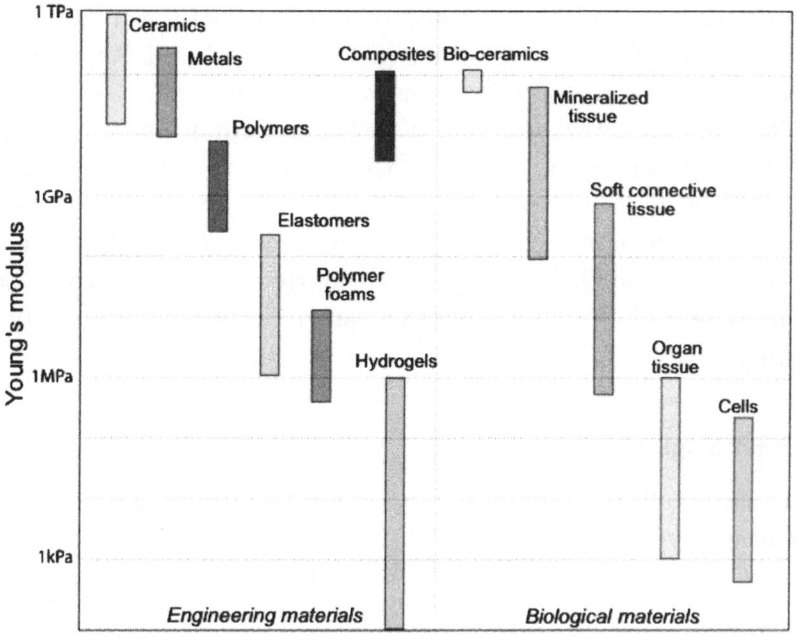

Figure 1-11. A schematic illustration of the elastic modulus ranges for engineering materials and biological materials. The elastic modulus range typically associated with traditional nanoindentation testing is 1 GPa and greater. The majority of non-mineralized biological materials have much smaller modulus values, and for these materials there are significant challenges in adapting standard nanoindentation techniques.

4. Overview of this Volume

The remainder of this volume is divided into two parts. Part I considers the practical principles that underlie the two primary nanoindentation test methods, beginning with the distinction between instrumented indentation and atomic force microscope (AFM)-based testing (Chapter 2) and experimental considerations for each (Chapter 3). Next, fundamental contact mechanics principles are presented; these influence the mode of data analysis in cases of adhesion between the indenter and sample (Chapter 4) and different deformation modes that occur under indentation, *i.e.* elastic, plastic, viscous (Chapter 5). The section concludes with considerations for computational modeling of contact problems (Chapter 6). Having established the basic experimental framework for nanoindentation and the fundamental contact mechanics principles on which nanoindentation data analysis is based, Part II of this volume contains reviews of nanoindentation testing results in chapters divided by material groups: bone (Chapter 7), tooth tissues (Chapter 8), soft (non-mineralized) tissues and other biological materials (Chapter 9). The indentation of biological cells and cell membranes is also considered, including the introduction of new analytical frameworks for membrane problems (Chapter 10). Finally, the volume concludes with an outlook for future research and developments in this field (Chapter 11).

Acknowledgement

The authors thank Robert F. Cook, NIST, Gaithersburg, MD, for assistance with Fig. 1-2.

References

1. C. Viney, K. Katti, F.-J. Ulm and C. Hellmich (editors), Mechanical properties of bioinspired and biological materials, *Mater. Res. Soc. Symp. Proc.* 844, Warrendale, PA (2005).
2. P. Fratzl, W. J. Landis, R. Wang and F. H. Silver (editors), Structure and mechanical behavior of biological materials, *Mater. Res. Soc. Symp. Proc.* 874, Warrendale, PA (2005).
3. A. J. Bushby, V. L. Ferguson, C.-C. Ko and M. L. Oyen (editors), Mechanical behavior of biological and biomimetic systems, *Mater. Res. Soc. Symp. Proc.* 898E, Warrendale, PA (2006).
4. M. L. Oyen, A. J. Bushby, A. Mann and C. Ortiz, *J. Mater. Res.* **21**, 1869 (2006).
5. Y. C. Fung, *Biomechanics: Motion, Flow, Stress and Growth*, Springer-Verlag (1990).
6. Y. C. Fung, *Biomechanics: Mechanical Properties of Living Tissues, 2nd edition*, Springer-Verlag (1993).
7. J. F. Vincent, O. A. Bogatyreva, N. R. Bogatyrev, A. Bowyer and A. K. Pahl, *J. R. Soc. Interface* **3**, 471 (2006).
8. Z. Tang, N. A. Kotov, S. Magonov and B. Ozturk, *Nature Mater.* **2**, 413 (2003).
9. M. L. Oyen, *Mater. Res. Soc. Bull.* **33**, 49 (2008).
10. B. Alberts, A. Johnson, J. Lewis, M. Raff, K. Roberts and P. Walter, *Molecular Biology of the Cell, 4th ed.*, Garland Science, Taylor and Francis (2002).
11. J. F. V. Vincent and U. G. K. Wegst, *Arthropod. Struc. Dev.* **33**, 187 (2004).

12. H. Saito and M. Yokoi, *J. Biochem.* (Tokyo) **111**, 376 (1992).
13. J. Seog J, D. Dean, B. Rolauffs, T. Wu, J. Genzer, A. H. Plaas, A. J. Grodzinsky and C. Ortiz, *J. Biomech.* **38**, 1789 (2005).
14. C.-C. Ko, M. L. Oyen, A. M. Fallgatter, J.-H. Kim, W. H. Douglas, J. Fricton and W.-S. Hu, *J. Mater. Res.* **21**, 3090 (2006).
15. W. C. Oliver and G. M. Pharr, *J. Mater. Res.* **7**, 1564 (1992).
16. M. L. Oyen and R. F. Cook, *J. Mater. Res.* **18**, 139 (2003).
17. K. L. Johnson, *Contact Mechanics*, Cambridge University Press (1996).
18. M. Oyen, *J. Biomech.*, **39**, 2699 (2006).
19. M. L. Oyen and C.-C. Ko, *J. Mater. Sci. Mater. Med.* **18**, 623 (2007).
20. A. M. Fallgatter and C.-C. Ko, *J. Mater. Res.* **21**, 2052 (2006).
21. M. Sakai, *J. Mater. Res.* **14**, 3630 (1999).
22. J. L. Cuy, A. B. Mann, K. J. Livi, M. F. Teaford and T. P. Weihs, *Arch. Oral. Biol.* **47**, 281 (2002).

CHAPTER 2

INDENTATION AND ADHESION AT SMALL LENGTH SCALES

Robert F. Cook

Ceramics Division, National Institute of Standards and Technology
E-mail: robert.cook@nist.gov

Indentation and adhesion are considered in the context of the two experimental methods predominantly used to measure contact behavior at small length scales, instrumented indentation testing and atomic force microscopy. The typical loads and displacements applied and sensed by each measurement technique, both within the (probe + surface) contact system and exterior to it, are considered in turn. The loads and displacements are placed in the context of the atomic-scale parameters quantifying material indentation and adhesion behaviors, atomic bond stiffness and rupture force, respectively. The central role played by the probe stiffness in determining whether a contact is likely to be more indentation- or adhesion-like—or something in between—is highlighted, with an emphasis on quasi-static contact. Dynamic behavior is also considered, along with the likelihood of measurable adhesion effects appearing in typical indentation measurements of biological materials using atomic force microscopes.

1. Introduction

Adhesion and indentation are obviously intimately connected: If two bodies are brought into close enough contact such that the forces between their surfaces can interact to form a single connected entity—adhesion—then it is entirely possible, if not probable, that one body might deform the other locally at the contact—indentation.

Two extremes of behavior might be imagined. If the forces between the surfaces of the two bodies are negligible and the forces exerted in deforming the bodies (or at least one of them) dominate the load-displacement response measured exterior to the bodies, then this is an indentation contact. If the two bodies are extremely stiff (or even rigid, undeformable) and the forces exerted between their surfaces dominate the load-displacement response, then this is an adhesive contact. Clearly, real life will lie somewhere between these extremes, but experience suggests that at large length scales contacts are typically indentation-like and load-displacement responses are determined by the resistance of bodies to bulk deformation (reversible or irreversible). At small length scales contacts are typically adhesion-like and load-displacement responses are determined by the interaction of bodies via surface forces (attractive or repulsive). The load-displacement behavior of real contacts, adhesive-indentation contacts, will thus be determined by the relative forces associated with bulk deformation and surface interactions.

Attention here is restricted to interactions between solid bodies and thus characteristic quantities describing bulk deformation and surface interactions may be associated with these at the atomic scale. The characteristic quantity describing deformation at the atomic scale is the stiffness of a single interatomic bond, k_{bond}, which for longitudinal deformation may be expressed in terms of the bulk Young's modulus, E, and an interatomic spacing, d, by

$$k_{\mathrm{bond}} \simeq Ed \tag{2-1}$$

For a material with a Young's modulus of 100 GPa and an interatomic spacing of 0.5 nm, this gives a characteristic bond stiffness of $k_{\mathrm{bond}} \simeq 50\,\mathrm{N\,m^{-1}}$. The characteristic quantity describing surface interactions at the atomic scale is the maximum force that a bond can support (or that one atom can exert on another), $F_{\mathrm{max,\,bond}}$, which may be expressed in terms of the modulus and interatomic spacing as

$$F_{\mathrm{max,\,bond}} \simeq (c_1)Ed^2 \tag{2-2}$$

where c_1 is a proportionality factor accounting for the details of the interatomic interaction. Taking the values for E and d above and $c_1 = 0.1$, typical for many solids, gives $F_{\mathrm{max,\,bond}} = 2.5\,\mathrm{nN}$. These characteristic atomic-scale properties are very useful in nano-scale contact mechanics and in separating the limits of indentation and adhesive contacts; the characteristic values calculated here will be used below.

2. Contact Behavior Sensed by Instrumented Indentation

Let us consider the extremes of indentation and adhesive contact in the context of the typical experimental ranges of the two instruments used to measure load-displacement responses of contacting bodies: Nanoindenters and atomic force microscopes. Instrumented indentation testing, IIT, sometimes referred to as depth-sensing indentation (with acronym DSI), enables the displacement of a probe to be measured along with the load during contact with a test surface. The resistance of the probe to elastic and plastic deformation is typically much greater than that of the surface and IIT, as the name implies, is typically used to measure indentation behavior. Indentation displacements < 1 μm, measured with < 1 nm resolution, are typical and thus the technique is well suited to measure contact responses at very small length scales and is often referred to as "nanoindentation."

A schematic diagram of a typical IIT device is shown in Fig. 2-1. The indenter consists of a very stiff frame supporting the probe and surrounding a platform for the test surface. The probe contacts the surface through an indentation column that defines an indentation axis. The indentation apparatus controls the position of the probe along the axis or the load applied through the column. The load and, more importantly, the position of the probe, are measured remotely from the contact, which occurs between the tip of the probe and the surface. Hence, although the indentation load experienced by the tip-surface contact is identical to that measured by the remote load cell (as the cell and the contact are elements in series connected by the indentation column), the displacement of the tip relative to the surface is not the same as that imposed by the indenter on the probe as a whole and measured by the remote displacement gauge. The imposed probe displacement and the tip reaction displacement are related by the stiffness of the indentation column between the tip and the gauge and the tip-surface interaction stiffness. It is convenient to view the indentation column between the tip and the gauge as a linear-elastic element with a given spring stiffness, such that the probe consists of the (tip + spring) in series.

From a thermodynamic perspective, the (probe + surface) comprise the "system" under consideration, in this case a contact system, and the indenter is the "environment." Work parameters at the boundary of the system are specified, here the load measured or imposed by the indenter and the measured or imposed conjugate load-point displacement. We will take the boundaries of the system to be the interface between the sample surface and the indenter platform and the interface between the probe spring and the indenter column. The size of the system is quantified by

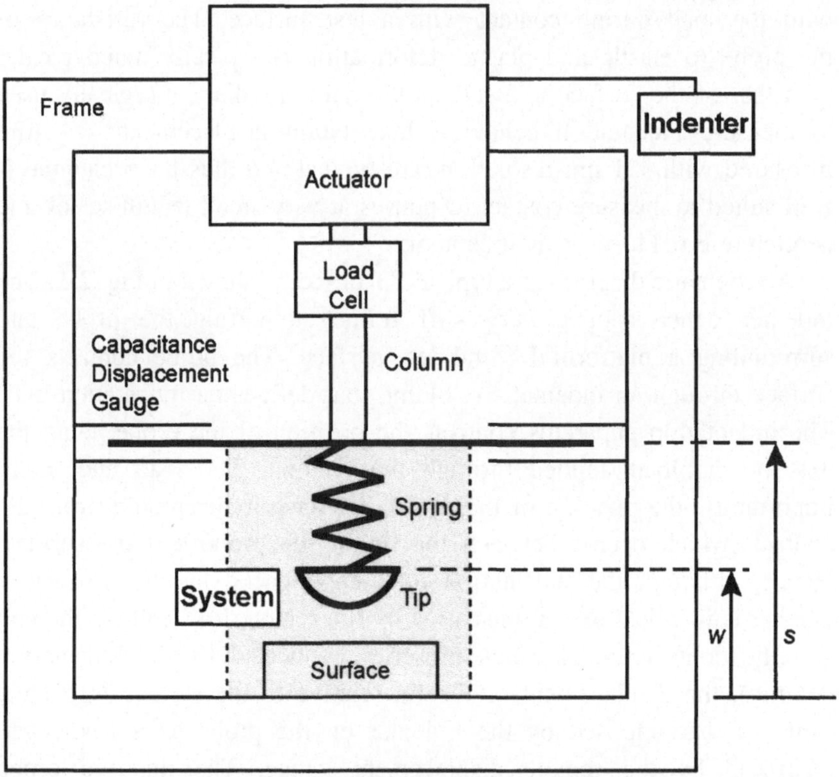

Figure 2-1. A schematic diagram of an instrumented indentation testing apparatus. The spring and tip comprise a probe, which together with a sample test surface comprise a simple mechanical thermodynamic system. An indenter environment imposes and measures load, displacement, and temperature exterior to the system.

the distance between these two interfaces, s, and a change in s measured by the displacement gauge is recognized as the load-point displacement conjugate to the load, F, measured by the load cell. The indenter (environment) imposes one of s or F, and this is then the independent variable in a test, and the system reacts leading to a change in F or s, and this latter is then the dependent test variable. Hidden from view (from a thermodynamic perspective) are the changes in the system and its internal variables leading to the reaction. In particular, adhesive and indentation interactions between the tip and the surface, quantified by the internal tip-surface separation variable, w, and deformation responses of the spring, quantified by the internal spring extension, $(s - w)$, determine the relationship between the changes in the independent and dependent variables, and whether in fact an equilibrium (F, s) configuration exists or can be attained.

The whole point of the above considerations is that while it is F and s that are measured and it is the (F, s) relationship that determines the engineering performance of bodies in contact, it is the underlying (F, w) contact relationship mediated by the $(F, s - w)$ spring relationship that determines what is observed and that provides scientific insight into the bulk and surface properties of the body materials.

Let us consider a typical indentation contact as measured by a displacement-controlled (s is the independent variable) IIT device to examine these interrelationships. As a specific example, we will consider a contact between a cylindrical flat-punch tip of radius a and a flat surface. Both tip and surface are linearly elastic materials and have no surface force interactions. Force and position will be taken as positive when directed along the outward normal to the surface. An indentation cycle begins with s large enough such that the tip and the surface are out of contact and F is thus zero. If s is reduced so that the tip contacts (indents) the surface, the spring is compressed, the tip contacts the surface with a fixed circular contact perimeter, and the force required to maintain the system in equilibrium, $F < 0$. The relative displacements of the spring and tip-surface contact comprising the value of s at this force depend on the relative stiffness of the two components, k_S and k_a, respectively. The stiffness of the indentation system, k, after contact is given by

$$k = \frac{\mathrm{d}F}{\mathrm{d}s} \tag{2-3}$$

and is related, via elements in series, to the spring and contact stiffness values by

$$\frac{1}{k} = \frac{1}{k_S} + \frac{1}{k_a} \tag{2-4}$$

The stiffness of the probe spring, k_S, is given by

$$k_S = \frac{\mathrm{d}F}{\mathrm{d}(s-w)} \tag{2-5}$$

and may be estimated by considering the spring as a simple column, for which

$$k_S = E_S A_S / L_S \tag{2-6}$$

where E_S is the Young's modulus of the column material and A_S and L_S are the cross-sectional area and length of the column, respectively. Taking the column modulus as E_S = 200 GPa (steel) and the column to be a 50 mm long circular rod of radius 2.5 mm gives $k_S \simeq 8\times10^7$ N m^{-1}. The stiffness of the contact, k_a, is given by

$$k_a = \frac{\mathrm{d}F}{\mathrm{d}w} \tag{2-7}$$

and for a circular contact is related to the contact radius by

$$k_a = 2aM \tag{2-8}$$

where M is the indentation modulus of the tip-surface contact. Similar to an elements-in-series concept, the indentation modulus takes into account deformation in both the tip and the surface during contact and is given by

$$\frac{1}{M} = \frac{1}{M_{\text{tip}}} + \frac{1}{M_{\text{surface}}} \tag{2-9}$$

where M_{tip} and M_{surface} are the indentation moduli of the tip and surface, respectively. For elastically isotropic materials, the isotropic indentation modulus, M_{iso}, is related to the material Young's modulus, E, and Poisson's ratio, ν, by

$$M_{iso} = \frac{E}{1-\nu^2} \tag{2-10}$$

which may be recognized as the isotropic "plane-strain" modulus. (For elastically anisotropic materials, calculations involving all the elastic constants are required.) For a typical (cubic crystal) diamond indenter used in nanoindentation, $M_{tip} \simeq 1150\,\text{GPa}$. Taking the surface as isotropic and the material Young's modulus and Poisson's ratio as E = 100 GPa and ν = 0.25 gives $M_{surface} \simeq 107\,\text{GPa}$. The indentation modulus in this case is thus $M \simeq 98\,\text{GPa}$. For a typical nanoindentation contact radius of a = 100 nm, this gives the contact stiffness as $k_a \simeq 2\times10^4\ \text{N m}^{-1}$ (this is about as stiff as a diving board at a swimming pool).

The above result reflects the usual conditions during nanoindentation: The contact stiffness is much less than the spring stiffness, $k_a \ll k_S$, and hence most of the displacement imposed on the system is taken up by the contact, with negligible deformation of the spring, *i.e.*, $w \gg s$. Similarly, within the contact, the surface indentation modulus is much less than the probe indentation modulus, $M_{surface} \ll M_{tip}$, and most of the contact displacement appears as surface deformation. The nature of this surface deformation may be inferred from comparison of the bulk-like contact stiffness just calculated, k_a, and the interatomic bond stiffness, k_{bond}, estimated above. If the contact stiffness were purely determined by the deformation of bonds in the circular area directly beneath the contact (in the manner of a Winkler foundation), the contact stiffness would be given by

$$k_{a,\,Winkler} \simeq k_{bond}(d/3a)(a/d)^2 \tag{2-11}$$

The first term in parentheses accounts for the bonds in series under the contact; such strings of bonds are taken to extend about three times the contact radius, a typical extent for an indentation deformation field. The second term accounts for the number of such strings in parallel within the contact area. Using the values given above for k_{bond}, a, and, d

gives, $k_{\mathrm{a,\,Winkler}} \simeq 3\times10^{3}\ \mathrm{N\,m^{-1}}$, which is much less than the bulk k_{a} calculated. The implication is that the contact deformation must be distributed transversely beyond the contact area, and in fact this is the case: There is considerable deformation in material adjacent to the contact outside the contact area, giving rise to an elements-in-parallel stiffening.

The contact displacement is given by

$$w = \frac{F}{k_{\mathrm{a}}} \tag{2-12}$$

and taking a (small) nanoindentation load of $F = 1$ mN gives $w = 50$ nm. In contrast, the deformation of the indentation column under this load is about 0.01 nm, negligible by comparison with the contact displacement.

3. Contact Behavior Sensed by an Atomic Force Microscope

Atomic force microscopy (AFM) typically operates at the other extreme of partitioning the imposed displacement of the probe between the contact and the spring. Similar to IIT, AFM enables the displacement of a probe to be measured along with the load during loading onto a test surface. However, in AFM the resistance of the tip to elastic and plastic deformation is not always much greater than that of the surface (often the tip and surface are the same material), and AFM, as the name implies, is often used to measure adhesive force behavior. AFM displacements < 100 nm, measured with < 0.1 nm resolution, are typical and thus the technique is well suited to measure contact responses at ultra-small length scales and contact areas and thus the "atomic" in AFM. If these forces are used in a closed feedback loop as part of a lateral scanning system, images of the surface may be generated and thus the "microscope" part of AFM. (AFM is one of the many scanned-probe microscopy, acronym SPM, techniques in which attention is focused on the force interactions between the tip and the surface, as opposed to the electrical or magnetic interactions that are also commonly used to image surfaces.) Such imaging is not considered here, but instead the operation of an AFM with open force feedback is considered; contact measurements performed in this manner are known as force spectroscopy and are closely analogous to IIT measurements in that they generate "force-distance" curves. The great difference between AFM and IIT, which enables force spectroscopy with AFM to measure adhesive interactions, is the stiffness of the probe spring.

A schematic diagram of a typical AFM is shown in Fig. 2-2. The microscope consists of a very stiff frame supporting the probe and surrounding a platform for the test surface. The probe in this case consists of a cantilever beam that is built in (clamped) at one end to a displacement actuator controlled by the AFM (and the AFM is thus the "indenter" environment in this case). Attached to the other end of the cantilever is the probe tip that interacts with the surface. Such interactions impose a force on the cantilever through an indentation axis that passes through the tip but not the actuator and as a consequence a bending moment is imposed on the cantilever. This moment causes the

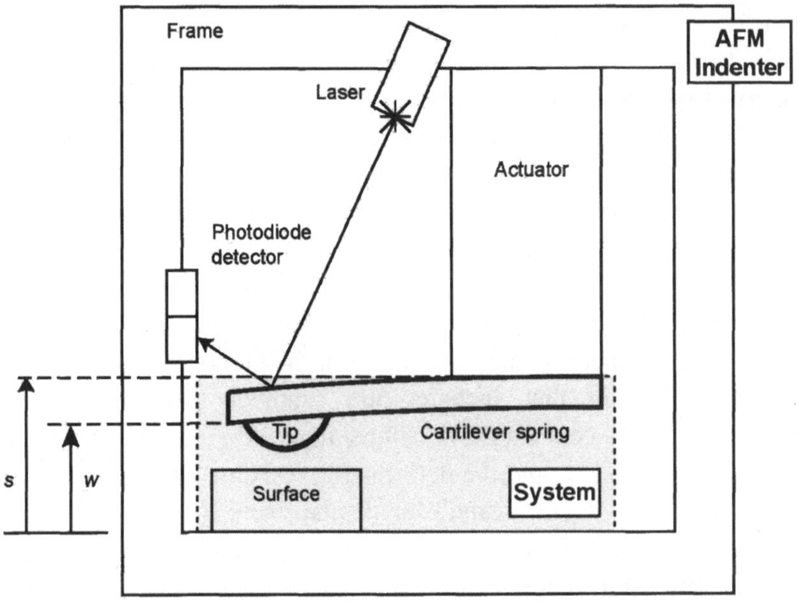

Figure 2-2. A schematic diagram of an AFM contact testing apparatus. The cantilever spring and tip comprise a probe, which together with a sample test surface comprise a simple mechanical thermodynamic system. The AFM indenter environment imposes and measures load, displacement, and temperature exterior to the system.

tip end of the cantilever to displace and deflect relative to the actuator end. A beam of light from a laser source is reflected from the back of the cantilever and collected by a split photodiode detector; deflection of the cantilever by tip-surface interaction-induced bending moments causes movement of the beam spot at the detector and thus changes in photodiode output. Geometrical optics allows the cantilever deflection to be determined from the photodiode output and cantilever mechanics enable the deflection to be related to the tip-end displacement, and thus, from the cantilever stiffness, the force between the tip and surface. Once again, although the indentation apparatus controls the position of the probe at the built-in actuator end, the displacement of the tip relative to the surface is not the same as that imposed by the indenter on the probe as a whole. The imposed probe displacement and the tip reaction displacement are related by the stiffness of the cantilever between the tip and the actuator and the tip-surface interaction stiffness. Mechanically, however, the IIT and AFM geometries are equivalent: The probe consists of the tip + spring in series.

As before, from a thermodynamic perspective, the probe + surface comprises the system under consideration and the microscope indenter is the environment. Here we will take the boundaries of the system to be the interface between the sample surface and the microscope platform and the interface between the cantilever and the actuator. The size of the system, quantified by distance between these two interfaces, s, is imposed by the actuator, which, along with the microscope frame, is well approximated as rigid for typical AFMs. The load, F, conjugate to the actuator displacement, is that inferred from the photodetector "load cell" after a calibration that includes the cantilever stiffness and the microscope optics geometry. As before, there is an internal tip-surface separation variable, w, and the deformation response of the cantilever is quantified by the internal cantilever displacement, $(s - w)$. There is a much greater recognition in AFM studies that the underlying (F, w) contact relationship is significantly mediated by the $(F, s - w)$ cantilever spring relationship to determine what is observed in terms of (F, s). This is because there is a much greater range of probe spring constants available for AFM studies through selection of the cantilever material and geometry.

Let us consider a typical indentation contact as measured by a displacement-controlled AFM device. As before, we will consider contact between a cylindrical tip and a flat surface. In this case, more typical of AFM experiments, we will take the tip and the surface to the same (isotropic) material with Young's modulus of 100 GPa and Poisson's ratio of 0.25. In this case the indentation modulus is $M \simeq 53\,\text{GPa}$ and, for the same assumed contact radius of 100 nm, the contact stiffness is $k_a \simeq 1\times10^4\ \text{N m}^{-1}$. The stiffness of the probe spring in this case may be estimated by considering the spring as a simple rectangular (clamped-fixed) cantilever, for which

$$k_S = b_S t_S^3 E_S^* / 4L_S^3 \tag{2-13}$$

where E_S^* is the plane-strain modulus of the cantilever material along its length, and b_S, t_S, and L_S are the width, thickness, and length of the cantilever, respectively. (The flexibility in tuning AFM probe spring stiffness even for this simple cantilever geometry is obvious given the number of parameters in the above expression.) Taking $E_S^* = 170$ GPa ([110] silicon) and the cantilever to be 30 μm wide, 10 μm thick and 500 μm long gives $k_S \simeq 10\ \text{N m}^{-1}$ (which is at the stiffer end of the spectrum of AFM cantilever stiffness values, 0.002 N m^{-1} to 200 N m^{-1}, but would be typical of that used for AFM indentation).

The above result reflects the usual conditions during AFM indentation, the contact stiffness is much greater than the spring stiffness, $k_a \gg k_S$, and hence most of the displacement imposed on the system is taken up by the cantilever spring, with negligible relative deformation of the contact, *i.e.*, $w \ll s$. Within the contact, the comparable (in this case, identical) surface and probe indentation moduli, $M_{\text{surface}} \simeq M_{\text{tip}}$, lead to contact displacement appearing equally as surface and probe deformation. Within this limit, the cantilever displacement is given by

$$s \simeq \frac{F}{k_S} \tag{2-14}$$

and taking a (moderate) AFM contact load of $F = 10\ \mu\text{N}$ gives $s = 1\ \mu\text{m}$; typical cantilever displacements are very small relative to the cantilever thickness (and this enables a linear relationship to be approximated between cantilever spatial displacement and angular deflection; the photodiode senses deflection, but the linear relation allows for a simple relation between diode output and load). The contact displacement is still given by

$$w = \frac{F}{k_a} \tag{2-15}$$

and for the load and contact stiffness here gives $w = 1$ nm.

The very small, atomic-scale contact displacement above ($w \approx d$ here) and the similarity of the cantilever spring stiffness to the characteristic bond stiffness ($k_S \approx k_{bond}$ in AFM) suggest that other small-scale effects may influence AFM contacts. That is, the effects of interatomic forces may play a role in determining AFM contact behavior in addition to, perhaps even dominating, the effects of interatomic stiffness; the small scales may introduce a significant adhesion component into the nature of the contact as distinct from the purely indentation behavior considered above. To examine this, consider the surface interaction force between the surface and the probe at the characteristic maximum bond force

$$F_{a,\,surface} \simeq F_{max,\,bond}(a/d)^2 \tag{2-16}$$

Using the values of $F_{max,\,bond}$, a, and d above gives $F_{a,\,surface} \simeq 100\ \mu N$, which is much greater than the above contact force of $F = 10\ \mu$N. This calculation suggests that forces associated with surface interactions will definitely be comparable to forces associated with bulk deformation at AFM contacts; AFM contacts will have strong adhesion character.

It might be argued that the above calculation assumes ideal solid-like (primary) bonding between the materials forming the contact and that, even if the materials are identical as assumed here, this is unlikely to occur between typical uncleaned surfaces in air at typical contact time scales (≈ 1 s). This is true, although such a calculation certainly pertains for cleaned surfaces in ultra-high vacuum. However, surface interactions arising from secondary bonding between the materials will also lead to adhesive effects at small contacts. Secondary bonding influences, particularly those arising from van der Waals interactions, are pervasive between surfaces in close proximity (separations of $\approx d$). Such influences are typically an order of magnitude weaker than those of primary bonding and thus also lead to significant adhesive effects at small contacts; for the case above such secondary bonding influences would lead to $F_{\text{a, surface}} \simeq 10\ \mu\text{N}$, still comparable to the deformation forces. Consideration of other interactions between the tip and the surface, such as those arising from electrostatic effects or capillary meniscus effects, lead to the same conclusion: Contact measurements using AFMs will be significantly influenced, if not dominated, by surface force effects and such contacts will be more adhesion-like than indentation-like.

A key feature of surface forces is that their influence can extend beyond the surface and hence they can deform the probe spring prior to contact (as opposed to the forces associated with material deformation, which can only affect the spring once contact has been made). For the very stiff nanoindentation probe springs, this effect is negligible, but for the much more compliant AFM cantilever probe springs this effect, as noted above, can be significant. Hence, just as in a predominantly indentation-like contact formed by a nanoindenter, the force imposed by the indenter is balanced against those associated with deformation of the body, in a predominantly adhesion-like contact formed by an AFM the surface forces are balanced against those associated with deformation of the cantilever.

4. Dynamic Behavior

The difference in the stiffness values of the probe springs for nanoindentation and AFM contact instrumentation is also reflected in the dynamic properties of the spring. In the absence of other influences (such as contact forces) the tip behaves as a simple harmonic oscillator with motion about the imposed equilibrium position given by

$$w = s + a_{\text{S}} \sin(\omega_{\text{S}} t) \qquad (2\text{-}17)$$

where a_{S} is the free amplitude of vibration, t is time, and ω_{S} is the characteristic free resonant frequency of oscillation. The resonant frequency is related to the probe stiffness by

$$\omega_{\text{S}}^2 = k_{\text{S}} / m^* \qquad (2\text{-}18)$$

where m^* is the effective mass of the oscillating element. (For closely integrated tip-spring systems as in nanoindenters and AFMs, the effective mass reflects the mass of the larger oscillating spring element much more than the mass of the smaller tip; for AFM cantilevers, $m^* \simeq m_{\text{cantilever}} / 4$, for example.) In the presence of contact forces, the tip still behaves as a simple harmonic oscillator but with motion about a new equilibrium position determined by the force balances discussed above and with a resonant frequency determined by the effects of the probe spring and the contact stiffness field. Typically, the behavior of the contact stiffness field is not measured directly, but indirectly by the changes in the resonant frequency and amplitude of vibration of the tip as the probe position is changed. Such a determination can be made by a dynamic measurement in which the tip is forced into oscillation by an imposed oscillatory motion of the probe by the indenter and the frequency of maximum vibration amplitude measured.

Determination can also be made by a "thermal" measurement, in which the tip oscillates in response to imposed thermal equilibrium with the indenter at temperature, T. The classical equipartition theorem prescribes the average free amplitude of oscillation of the tip as a function of temperature:

$$k_S \langle a_S^2 \rangle = \boldsymbol{k}T \tag{2-19}$$

where $\boldsymbol{k}$ is Boltzmann's constant and $\langle\ \rangle$ indicates the time-averaged value of the quantity. Thus the free amplitude and frequency of oscillation may be estimated from

$$a_S = (\boldsymbol{k}T / k_S)^{1/2} \tag{2-20}$$

and

$$\omega_S = (k_S / m^*)^{1/2} \tag{2-21}$$

Taking T = 300 K and using the density of silicon to estimate the effective mass of the $k_S \simeq 10\ \text{N m}^{-1}$ AFM cantilever considered above gives $a_S \simeq 0.02\ \text{nm}$ and $\omega_S \simeq 340\ \text{kHz}$, an oscillatory behavior easily measurable by good AFM photodiode detection systems. If the thickness of the cantilever is reduced such that $k_S \simeq 0.01\ \text{N m}^{-1}$ (as might be used for indentation contact of very soft or compliant materials), the oscillations become larger and slower, $a_S \approx 0.6$ nm and $\omega_S \approx 34$ kHz, and very easily measured. If the stiffness of the probe spring is increased, the amplitude of oscillation clearly decreases, and this is the case with typical nanoindentation systems; the amplitude of tip oscillation is negligible and usually undetectable. (The larger probe stiffness is compensated by the larger probe mass to give comparable resonant frequencies; for the steel probe considered above the probe resonant frequency is about 140 kHz. Dynamic properties of nanoindenters, however, are typically limited by the smaller resonant frequencies of the instrument as a whole, typically much less than 1 kHz).

Variations in tip-resonance oscillation amplitude and frequency will also occur if the stiffness experienced by the tip is changed not by change in probe spring stiffness but by the superposition of a contact stiffness field. For indentation-like contacts, in which the tip is deforming the surface, the total stiffness experienced by the tip is greater than the probe stiffness and the resonant frequency will be greater than the free resonance; this is the basis for contact-resonance AFM methods of modulus mapping of surfaces. For adhesive-like contacts the total stiffness experienced by the tip can decrease significantly in attractive surface force fields, leading to large decreases and increases in resonant frequency and thermal amplitude, respectively; this is the basis for dynamic stiffness field (and, by integration, force field) AFM force spectroscopy. In many cases, the stiffness field experienced by the tip can vanish completely and such instability points are the basis for adhesion measurements.

5. Contact Responses of Biological Materials

It is important to note that many, if not most, biological materials will have Young's moduli significantly less than the 100 GPa material considered here as an example (in fact this value approaches that for the mineral apatite, the stiff constituent of bones and teeth). Taking 100 MPa as a representative value gives a contact stiffness for a 100 nm radius contact of about 20 N m^{-1} and thus 1 μN is required to create 50 nm of indentation deformation. These values are much more amenable to measurement with AFM and nanoindenters designed explicitly for such small load testing; in both cases the probe spring is much more compliant than that of a typical nanoindenter and the loads required are comparable to those exerted by surface forces. Adhesion effects are thus expected to be significant for many indentation measurements of mechanical properties of biological materials. If 100 kPa is taken as a representative modulus value, measurement by AFM is clearly required and adhesion effects will be almost inescapable; very stiff cantilevers may be used to minimize deflection from surface force effects, but such cantilevers will also exhibit very little deflection during indentation contact with the very compliant material. Hence a framework is required to incorporate adhesion effects into indentation measurements of biological materials. It is clear from the discussion above that such a framework should focus heavily on the stiffness of the measurement probe relative to the stiffness of the surface force and bulk deformation fields. Such a framework will be developed in Chapter 4, but first Chapter 3 will provide a survey of the experimental issues and challenges associated with measurements of contact responses of biological materials.

CHAPTER 3

INSTRUMENTATION AND EXPERIMENTATION

Krystyn J. Van Vliet

Department of Materials Science and Engineering,
Department of Biological Engineering,
Massachussets Institute of Technology
E-mail: krystyn@mit.edu

In this chapter, practical experimental issues associated with the mechanical testing of biological materials are considered. Nanoindenters and atomic force microscope devices are considered in separate sections, with best practice recommendations for testing materials of different stiffnesses in each configuration. Technical aspects of commercial instrumentation are examined, including load and displacement resolution, load and displacement ranges, stability of the instrument signals and compliance of the load-frame. Significant emphasis is placed on mechanisms for achieving environmental control necessary for biological materials testing, including temperature and humidity effects, specimen preparation and stability of the biological samples themselves.

1. Introduction

There are several differences between the experimental set-ups for instrumented indentation (sometimes termed 'nanoindentation') and atomic force microscope (AFM)-enabled indentation, as discussed at length in Chapter 2. There are therefore slightly different experimental considerations for the two types of testing, and one or the other technique may be more suited for the experiment being planned. There are also differences based on material class, especially related to material stiffness. A summary of experimental design variables for a range of materials (with different stiffnesses) is presented in Table 3-1; in the

Table 3-1. Chief concerns in experimental design for nanoindentation of biological materials.

	Displacement		**Load**		**Signal**	**Frame**	**Environmental**
Material	**Resolution**	**Range**	**Resolution**	**Range**	**Stability**	**Compliance**	**Control**
Cell	X			X	X		X
Extracellular protein (film)	X	X	X		X	X	X
Hydrogel (film)	X		X				X
Rubber (bulk)	X	X		X			
Soft organ tissue	X	X		X	X		X
Muscle tissue	X	X		X	X	X	X
Cellulose composite	X	X	X	X		X	
Connective tissue		X		X	X	X	
Mineralized bone tissue		X	X			X	

remainder of this chapter, these considerations will be examined in more detail and for both nanoindentation and AFM testing configurations.

2. Instrumented Indentation

There currently exist several commercially available instrumented indenters with the force and displacement resolution required for contact loading of biological materials. Certain of these instruments are also capable of the larger displacements and aqueous-solvent immersion that is required for nanomechanical analysis of compliant, hydrated tissues. The operating principles, modes of signal acquisition, and signal stability of such commercial or customized instrumented indenters vary, such that specific instruments may be better suited than others for particular biological materials and mechanical phenomena of interest.

2.1. *Instrumentation*

The first requirement of instrumented indentation is that the output signals of the experiment, namely force and displacement, are calibrated in terms of the machine signal outputs which are typically sensor voltages. In most commercial instrumented indenters, transducer devices are chosen such that there is a linear relationship between the signal voltage and output value (load or displacement).

2.1.1. *Resolution and range required for testing biological materials*

As commercial instrumented indenters were originally designed to enable mechanical characterization of metal and ceramic films that were typically of submicrometer-scale surface roughness in the as-deposited state, it is worthwhile to consider whether the resolution and range of force and displacement required of biological material characterization may differ from the original instrument requirements. Indeed, some mineralized biological materials (*e.g.*, trabecular bone) contain a significant volume fraction of mechanically stiff hard mineral, and can exhibit volume-averaged stiffness and mean contact pressures on the order of 1–10 GPa.[1-3] In such mineralized biomaterials, the load and depth resolution and range of standard commercial instrumentation is typically sufficient to extract repeatable data. However, in such structurally heterogeneous samples, the volume of the deformed material volume as compared to the size of the composite constituents (*e.g.*, the diameter and aspect ratio of mineralized phase) requires particular attention if accurate calculation of phase and macroscale mechanical properties is desired.

In more compliant biological materials, ranging from muscle and brain tissue to extracellular matrices to individual cells, the load and depth resolution and range of commercial instrumentation is tested severely. First, let us consider the load resolution required to deform a model extracellular matrix, a polymeric hydrogel exhibiting macroscale elastic modulus E on the order of 10 kPa. Let us assume that this hydrogel remains fully hydrated and structurally stable at room temperature, exhibits minimal adhesion to the probe material, and deforms within the linear elastic regime under the applied stresses and strain rates imposed by a spherical probe of radius R = 1 mm. Let us also simplify the effects of finite thickness to assume that we can neglect mechanical contributions of the underlying substrate for indentation depths h as large as 1 μm, which is approximately the maximum displacement in several commercial indenters. If the probe material is made from a comparably rigid material such as diamond, glass, or a glassy polymer such as polystyrene, the elastic properties of the probe can be reasonably neglected, such that the elastic modulus of the indenter-material contact (Eq. 2-9) equals the indentation (plane strain) elastic modulus of the material. The maximum load P_{max} corresponding to this desired maximum indentation depth can be calculated as $P_{max} = 4/3\ E_r\ R^{1/2}\ h^{3/2}$ (see Chapter 5) and is only 420 nN. This P_{max} competes with the actual load resolution of most commercial instrumented indenters. In contrast, the maximum load required to deform an idealized, 60% mineralized bone of E = 10 GPa[1] for the same h_{max} and indenter probe geometry would be 420 mN, well within the normal operating range of the same instrumented indenters.

From this simple example, several points regarding resolution and range are apparent. First, the load resolution required of the most compliant biological samples is not a common feature of commercial instrumented indenters; this requirement differs from the electronic noise floor of the force transducers that may be quoted by manufacturers, and is most easily verified through direct experimental testing on well-characterized samples. This calculation demonstrates the increasingly reported use of structurally compliant, cantilevered scanning probe or atomic force microscope probes to infer mechanical properties of tissues and cells; the associated challenges and advantages of that approach are detailed in Section 3. Second, the depth range required to deform materials to loads within the calibrated range of instrumented indenters can easily exceed the range defined by the displacement transducer devices, which is typically on the micrometer scale. Third, the depth resolution and the load range available on commercial instrumented indenters are sufficient for mechanical characterization of biological materials, owing to the comparably lower stiffness but greater toughness of such natural composites. Thus, the challenges in designing informative experiments are in the judicious choice of (1) probe geometry and maximum loads or depths required to infer mechanical response that are attributable to individual phases, interphases, or macroscale volumes as desired by the user; and (2) loading times and physicochemical environments that reflect the extent of viscoelastoplasticity and energy dissipation of the material, which are necessarily a function of these conditions. Finally, the depth range over which many biological materials are probed and analyzed is not at the nanoscale, according to International Organization for Standardization (ISO) 14577 definitions specified originally for metallic materials ($h <$ 200 nm).

Note that stability of these signals over time is also a valid concern, particularly for experiments of extended duration such as contact creep compliance measurements. The signal stability (in terms of thermal drift [N/s] or [m/s]) stated by the vendors is typically at the level of electronic noise. The actual signal stability also depends on operating environment, so this should be measured via comparable experiments on a rigid material for each new experiment.

2.1.2. *Frame compliance*

All instrumentation that applies loads to a sample within a load train has the potential to dissipate some of that applied energy through the load frame itself. The magnitude of this dissipation during sample loading is quantified as the frame compliance C_f in units of [m/N], the distance over which the frame deflects per Newton of applied load during quasistatic, load-controlled extension or compression.

The same concerns apply to instrumented indentation of biological materials. In fact, when stiff biological materials are indented on universal testing frames at the macroscopic scale ($P \sim$ N and $h \sim$ mm), C_f is inferred from a reference sample that is subject to the same loading conditions, and is thus typically a hardness standard for which both E and H are independently verified. For commercial instrumented indenters, two approaches are possible: indirect estimation of C_f from micro- or nanoscale indentations on a reference sample[4] or direct measurement of C_f via compression of the indenter holder against the rigid indenter stage.[5] The former approach was included as an iterative procedure in the seminal paper of Oliver and Pharr,[4] but this iteration (coupled to estimation of the indenter contact area $A_c(h_c)$) can be challenging to implement and relies on several assumptions and idealizations regarding the reference material's response to contact loading. Alternatively, direct measurement of C_f can be achieved via fully elastic contact between a flat punch version of the indenter and the stage surface, provided that normal contact can be achieved to eliminate air gaps and compression of surface asperities at the indenter-stage interface. The frame compliance will depend, in part, on the details of the instrumented indenter mount and other load train components. As expected, the significance of the frame compliance to inferred mechanical properties is nonlinear and depends on the stiffness of the material and maximum loads applied. For a hydrogel case above, the additional frame deflection is clearly negligible (10^{-7} nm); for the bone the additional deflection due to frame compliance can be more substantial.

2.2. *Adapations required of environmental control*

One of the major differences between instrumented indentation of synthetic vs. biological materials is that the latter are typically stable only in non-ambient environments. For example, hydrogels require immersion in water to maintain the structure that confers the low elastic moduli of these materials; tissues rapidly denature if not maintained at a near-neutral pH in aqueous media that is devoid of degrading proteolytic enzymes; and cells do not maintain the molecular adhesions to underlying surfaces unless specific levels of CO_2 and metal ions in the surrounding media are maintained at approximately body temperature (37°C). In all of these cases, the physical environment and signal stability of instrumented indenters require special consideration.

2.2.1. *Temperature and associated thermal drift*

For biomaterials that exist or will be used at body temperature, it is relatively straightforward to elevate the testing or sample temperature. This is achieved differently in various commercial indenters, either by heating the entire indenter enclosure to achieve a stable operating temperature at which *P-h* data are collected, by heating only the sample via devices such as a thermoelectric or Peltier stage, or by resistively heating both the sample and the indenter to eliminate any quenching at the probe-sample interface.

In all cases, it is critical that the increased temperature does not result in significantly increased rate of change of the output signals (*i.e.*, thermal drift), and that this thermal drift is known. As in the case of contact creep compliance, these drift rates can be measured via sustained contact loading of rigid surfaces. However, it is always preferable to modify the instrument or the operating condition to minimize drift in such signals by minimizing heat differentials and fluxes on the electronics, rather than to subtract a supposedly constant drift function $h_{\mathrm{drift}}(t)$ or $P_{\mathrm{drift}}(t)$ from the experimental data. Theoretically, this can be achieved via heating of the entire apparatus to a uniform temperature, but this can be difficult due to the size of these instruments. Alternatively, the electronic components can be physically shielded from the heated components (*e.g.*, the sample stage). As long is care is taken to avoid construction of effective chimneys that lead to thermal drafts around the instrument, this solution has routinely demonstrated electronic drift levels that are comparable to room temperature signal stability.

Expansion of all heated components would be expected due to finite thermal expansion coefficients of these materials, so the time required to achieve a stable testing temperature is considerably greater than for ambient experiments. Typically, one allows 12–24 hrs to achieve thermal stability of the system inclusive of the sample. This pause can be reduced if load train components are machined from materials of low thermal expansion coefficients. When the sample must also be maintained in a hydrated state and may be prone to proteolytic degradation, this extended pause requires special care to maintain biomaterial integrity. Finally, the elevated temperatures of testing are necessarily transmitted through the indenter probe, such that attachment of the probe to the indenter holder within the load train requires either mechanical fixture or high-temperature adhesives.

2.2.2. *Controlled gaseous atmosphere*

As the structure and mechanical properties of many biomaterials can be sensitive to %CO_2, oxygen tension, and ambient air composition, it may also be important to regulate the gaseous environment. This can be achieved readily via gas inlets and outlets in all commercial indenters. However, care must be taken to either seal the sample chamber well or to monitor the actual gas content of the chamber if gas content is an important variable. As a rule, the smaller the environmental chamber surrounding the sample and indenter, the easier it is to maintain known environmental parameters. In the particular case of gaseous content, the gas flow must be sufficiently slow and laminar that acoustic and mechanical noise are not created within the chamber. As a result, stabilization of a gaseous environment also typically requires several hours of lead time prior to data acquisition.

2.2.3. *Fluid cells for hydrated biological materials*

Many biomaterials require immersion in aqueous environments (or at least high humidity environments) to maintain structure and concurrent mechanical properties. Chambers that maintain sample submersion can be readily constructed, but peculiarities of the instrumented indenter load train must be considered carefully. For example, a typical instrumented indenter displaces the tip vertically into the sample surface; if the indenter shaft were to plunge through an air-water interface to contact a submerged sample, a meniscus will form and move readily along the shaft perimeter. The associated surface tension of the water will deflect the indenter shaft, indicating a load that is due not to sample stiffness but to this moving meniscus. If properly calibrated and minimized compared with the indentation loads of interest, such artifacts can be accounted for during data interpretation. Alternatively, a horizontal orientation of the pendulum-based system eliminates the moving meniscus, as the extended indenter can remain fully submerged during sample contact and indentation loading (Fig. 3-1). Constantinides *et al.* have shown that such a horizontal fluid cell design enables measurement of mechanical properties for hydrogels and animal tissues over durations of observation on the order of 24 hrs.

Note that immersion of the biomaterial also requires consideration of fixation to the underlying substrate, such that the biomaterial should not float (if it is oriented horizontally) or sag (if it is oriented vertically).

Standard adhesives such as cyanocrylate may not be stable in submerged, saline environments typical of tissue analysis, but biomedically specialized adhesives such as Cell-Tak™ can be employed. In the case of vertically oriented, hydrated samples, both continuum simulations and experimental observations indicate that the buoyancy of such samples minimizes any surface deflections over immersion for at least 24 hrs (Fig. 3-1).

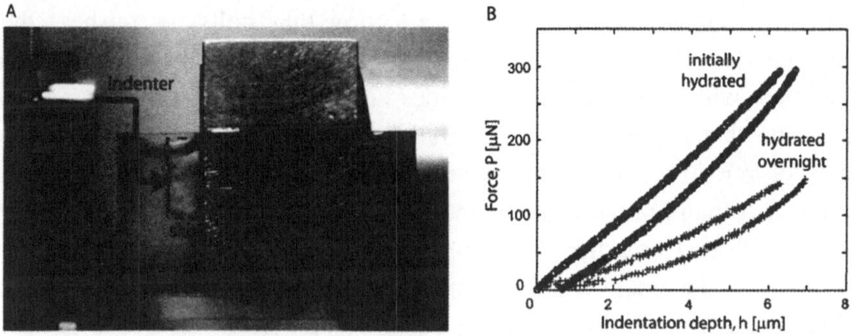

Figure 3-1. (A) Indentation of porcine skin within aqueous fluid cell. Probe is maintained at a constant submersion depth during indentation. Meniscus shown for reference of fluid level. (B) Quasistatic force-depth response of skin changes appreciably over several hours of sustained hydration, due ostensibly to tissue degradation. Courtesy J. F. Smith, G. Constantinides, Z. I. Kalcioglu, and K. J. Van Vliet.

2.3. *Indenter probe materials*

Although biological materials tend to be more mechanically compliant than the metals and ceramics for which diamond indenters were originally developed, the choice of indenter material is not as simple as selecting any material which is mechanically stiffer and harder than the biomaterial sample. If that were the case, polystyerene spheres could be used to measure the mechanical response of soft tissues, and glass spheres or cones could be considered in the analysis of mineralized collagen. Rather, the chemical functionality of the probe surface is an increasingly important consideration in the mechanical analysis of biomaterial surfaces, which often exhibit appreciable surface charge and nonspecific adhesion. These interactions can effect unintended probe interactions during approach and retraction from the surface.

2.3.1. *Electrostatic and hydrophobic interactions*

If the probe is strongly attracted to or repulsed from the sample surface, it is increasingly challenging to automatically detect the initiation of true mechanical contact. Especially in the case of proteinaceous or glycosylated materials, electrostatic interactions with the probe can occur to pull a like-charged probe onto the sample surface or push an oppositely charged probe away. The manifestation of these interactions can be a large and rapid probe deflection that pulls (or pushes) the indenter out of a measureable range of actuation. If data acquisition is initiated when the probe overcomes a specific load (voltage), it is possible that the instrument will fail to display this interaction. In the case of attractive interaction, one may then infer the material to be much more stiff than in reality, as the indentation depth corresponding to "near zero load" will be artifactually high. Ideally, one would seek to acquire all load-displacement data from the initial to the final point of load train displacement, in order to easily identify such nefarious probe-sample interaction events that may not register with actual sample loading. With such data, one could manually or semi-automatically identify the actual contact point beyond which mechanical load was increased at the sample surface.

These interactions can be buffered via immersion of the sample and probe in ion-containing, aqueous media; this will decrease the Debye length, or distance over which the probe "feels" an effective charge from the sample surface, for increasing ionic strength of the solution. Alternatively, one can chemically modify the probe to enhance or mitigate interactions with the biomaterial surface. Here it is important to note that certain probe materials are more amenable to chemical modification than others. For example, diamond is a crystalline carbon phase that is resistant to biochemical modification, whereas silica, polystyrene, and gold are more readily functionalized with sequential adsorption steps of molecular moities such as amine or carboxyl groups. Chemically modified instrumented indenters have even been fabricated by mounting AFM-cantilevered borosilicate spheres onto the indenter shaft, such that spheres of μm-scale radii could be used to test hypotheses of probe-surface interaction for synthetic polymers.

2.3.2. *Adhesion*

Actual adhesion of the probe to the biomaterial surface is initiated by attractive interactions, and may be due to specific or nonspecific interactions between the probe and sample surfaces. Under conditions of adhesion, the contact load can be increased and the contact stress essentially reduced due to increased probe-material contact area. Elastic moduli can be computed directly from the adhesion force, according to Johnson-Kendall-Roberts (JKR) theory (see Chapter 4). In synthetic polymers, it has been shown that various means of calculating elastic moduli (including the JKR theory and Oliver-Pharr method) give a wide range of results. This indicates that the extent of adhesion between a probe and a biomaterial remains a valid concern in the analysis of mechanical properties, though the most accurate way to infer such properties for sticky, compliant materials is a matter of current research.

2.4. *Experimental approaches*

2.4.1. *Calibration over appropriate signal range*

For quantitatively accurate measurements, load and depth signals must be calibrated over a range appropriate to the conditions of interest. Although in certain commercial indenters the internal signal calibrations are not readily accessible to the user, such full calibration is recommended on at least an intermittent basis to confirm the stability of such critical conversion factors. If direct calibration is not possible, the next best alternative is a standardized series of experiments on a reference sample. Interestingly, there is no clear standard for mechanical properties of polymers and biological materials; this is in contrast to standards for stiffer and harder materials, which tend to be fused silica (stiff) and aluminum (ductile). Several potential materials are currently under consideration, including relatively stiff, amorphous polymers such as polycarbonate and polystyrene, and relatively compliant elastomers. As noted above, load as [N/V] and displacement as [m/V] are signal transducer conversion factors that can be calibrated by the user via methods such as weighing of calibrated masses placed on the load train, and displacement of the unconstrained indenter via calibrated stage translation, respectively.

Frame compliance can also be measured directly, *e.g.*, via compression of the indenter shaft against the sample stage,[5] or indirectly. Indirect estimation of C_f can be achieved via concurrent modification of C_f and $A(h_c)$ to achieve the independently verified E and H of a standard material using a probe for which the area function is established independently,[4] or via assumption of constant (but unknown) stiffness and hardness of an indented sample, whereby C_f is the linear extrapolation of unloading compliance dh/dP vs. the inverse contact depth $1/h_c$. Both of these indirect methods are particularly sensitive to the accuracy (and paucity) of data obtained at low h_c, and require assumptions of material behavior and contact that are not required of direct frame compliance determination.

Even when these signals are well calibrated, one challenge particular to the analysis of biological materials is identification of the contact point – the indenter displacement beyond which true mechanical contact with the material is established and loading commences. This is due chiefly to the high compliance of biological materials as compared to the stiff materials for which indenters were originally designed (and thus to the load resolution of these instruments). However, this early stage contact can also be complicated by electrostatic and even chemical interactions at the probe-material interface, especially if the load train contains flexures that allow the probe to "snap" into contact with attractive surfaces or to resist contact with repulsive surfaces. As biomaterials are often highly charged, such interfacial complications can be appropriately minimized via analysis in fluid cells filled with ionic aqueous media. Of course, any swelling associated with aqueous immersion further increases sample compliance, and thus load resolution may then challenge accurate contact point identification. Beyond minimizing any user-specified, initial contact loads beyond which data acquisition commences, it is advisable to always collect load-displacement (P-h) data prior to the instrument's assumed identification of mechanical contact initiation ($P = h = 0$ upon loading) and to collect these data beyond the instrument's assumed identification of mechanical contact loss ($P = h = 0$ upon unloading). Post processing of the full P-h

sweep, though more time consuming and peculiar to each sample and loading condition, is preferable to any overestimation of stiffness and neglect of adhesion that would otherwise be undetected by reported *P-h* data.

2.4.2. *Consideration of calibration-standard materials*

If calibrations are to be determined via direct testing of reference samples, we should then consider the optimal reference materials for compliant biomaterials. These reference materials or calibrations standards, should be of comparable stiffness, strength, energy dissipation, and loading time sensitivity as biological materials, so that the machine can be exercised over comparable ranges for the reference and actual samples. Here we consider the requirements of such calibration-standard materials.

2.4.2.1. Mechanical compliance

Polymeric materials exhibit a wide range of mechanical compliance, from the kPa-stiffness of hydrogels like polyacrylamide to the GPa-stiffness of glassy polymers like polystyrene. In the strictest sense of calibration, it would be ideal to identify several materials that represent this wide range of mechanical properties, as well as hydration condition. These materials should behave under linear elastic responses for high strains, such that this simplification could be met in analyses such as creep compliance. Further, these materials should be available at large thicknesses (to minimize effects of the underlying substrate) and in large quantities (so that these are readily available).

2.4.2.2. Homogeneity and chemical reactivity at surface

Additionally, these materials would be structurally homogeneous at least over the length scale of the indentation diameter, and ideally over the diameter of the strained material volume. The chemical reactivity of the biomaterial surface with the probe surface should be minimal, except in instances where the goal is to calibrate probe adhesion.

2.4.2.3. Roughness

As most models of contact loading assume normal contact of the probe with respect to the sample surface, surface roughness of calibration-standard materials would be minimized. One possible exception would be in calibration of experiments for which the biomaterials are inherently rough and this roughness confers the mechanical property of interest, as in the case of adhesion of patterned elastomers.[6]

2.5. *Sample mounting and preparation*

Mounting of nonbiological samples to indentation stages typically involves coarse clamping via vacuum pressure, cyanoacrylate adhesives, or high-temperature cements. In such cases, the goal is to minimize the mechanical compliance due to the adhesive, while maintaining a stationary sample position over the course of testing. This requirement remains in analysis of biological samples, but the presence of hydration and the organic nature of such materials constrains the methods of sample preparation.

2.5.1. *Extent of hydration*

Most biological material samples exist at superambient states of humidity and hydration. To the extent possible, the physiological degree of hydration, salinity, and overall ion content of the biological solutions surrounding the biomaterial *in vivo* should be recapitulated in experiments. This is because the structure of these materials is maintained by such hydration, and is responsible for the mechanical properties of the biomaterial. For biological materials such as soft tissues that are fully hydrated *in vivo*, it is optimal to conduct instrumented indentation experiments within a fluid cell that is filled with solution of appropriate ionic strength and pH. For biological materials that are in humidified but not necessarily submerged environments, it is optimal to maintain relative humidity feedback and avoid condensation onto electronic components.

2.5.2. *Limiting biodegradation rates*

If biological materials are hydrated in full or in part, it is common for these materials to denature or degrade over time. This is due to the presence of enzymes that act on proteins or nucleic acids in natural degradation processes. As these structural changes will modify the mechanical properties of such samples, it is optimal to limit the rate of degradation during mechanical analysis. This can be achieved by lowering the sample temperature below room temperature (4°C), as is often done in other analyses of biological materials and processes, or by limiting the number or activation of proteolytic enzymes. Alternatively and least optimally, one could attempt to minimize degradation simply by analyzing the biological material for short intervals at room temperature, followed by storage in saline conditions at lower temperature (4°C). To minimize degradation of the biomaterial samples over the course of experimental analysis, RNAase and proteinase neutralizers can be added to the bathing solution.

2.5.3. *Compatible adhesives*

Given that secure mounting of biological samples is a requirement of accurate measurement, care must be taken to ensure that any adhesives used are both minimally compliant and also stable under the physical environment of test (e.g., immersion in saline). Cyanoacrylate adhesive meets the first requirement, in that such thermoplastic adhesives of low viscosity (<5 cP) wet and spread easily to contribute a maximum thickness of ~50 μm. Based on the mechanical properties of such adhesives (E = 4 GPa), we can estimate the maximum compliance due to compression of such a linear elastic layer to be 2 nm/N.[5] Although this is a sufficiently low compliance for most testing, this adhesive is not stable over months in submerged aqueous environments. Other adhesives have been developed especially for sustained adhesion of hydrated cells and tissues, such as BD Cell-Tak™ (BD Biosciences, Franklin Lakes, NJ). These adhesives are of similar compliance, but considerably more expensive than cyanoacrylate. The chief advantage of such proprietary formulations is that proteinaceous biomaterials adhere well to including amorphous polymers, glass, metal, and other biological materials, for sustained immersion times.

3. AFM-Enabled Indentation

As compared to industrial, structural materials for which instrumented indentation is now ubiquitous, biomaterials exhibit high compliance and significant structural heterogeneity. For these reasons, there is an increasing number of studies that employ scanning probe or atomic force microscopy to indent and infer mechanical properties of biomaterials. Although both instrumented indenters and AFMs are capable of applying load and acquiring load-displacement traces, there are several important differences and caveats to the use of AFMs as indenters. This section outlines the distinct operating principles and experimental considerations unique to AFM-enabled indentation.

3.1. *Instrumentation*

3.1.1. *Acquisition of calibrated load and displacement signals*

The operating principles of scanning probe microscopes based on optical detection of cantilever deflection, often termed atomic force microscopes due to the picoNewton-scale forces achievable, are responsible for the advantages and limitations of this mode of indentation on biological materials.

Figure 3-2 shows a schematic of this instrument, comprising a silicon-based cantilever of calibrated stiffness k that is typically displaced normal to the sample surface via a calibrated piezoactuator. Deflection of this cantilever upon contact is detected via a Class II (red) laser, which reflects from the free end of the cantilever into a position sensitive detector such as a quadrant photodiode. Signal feedback relates the cantilever base position required to induce or maintain a user-specified probe deflection; this feedback is implemented during scanning (imaging) of a sample surface in AFM contact mode, as well as during indentation. Indentation loading of the sample surface proceeds via user-specified displacement of the cantilever base (or the sample stage, depending on the AFM design) over a user-specified time; unloading occurs at a user-specified maximum base displacement or cantilever deflection. As the feedback loop operates on an optoelectronic signal based in mechanical contact, even the inventors of the first contact-mode AFM recognized the potential to image biological surfaces under ambient and near-physiological conditions[7]: full immersion of the probe and sample in aqueous media (ranging in complexity from a drop of water into which the probe is submerged to a gasketed fluid cell for continuous media exchange) does not degrade imaging resolution. Such fluid cells can be engineered to maintain hydrogels, tissues, and cells at stable hydration, pH, and temperature comparable to *in vitro* culture conditions.

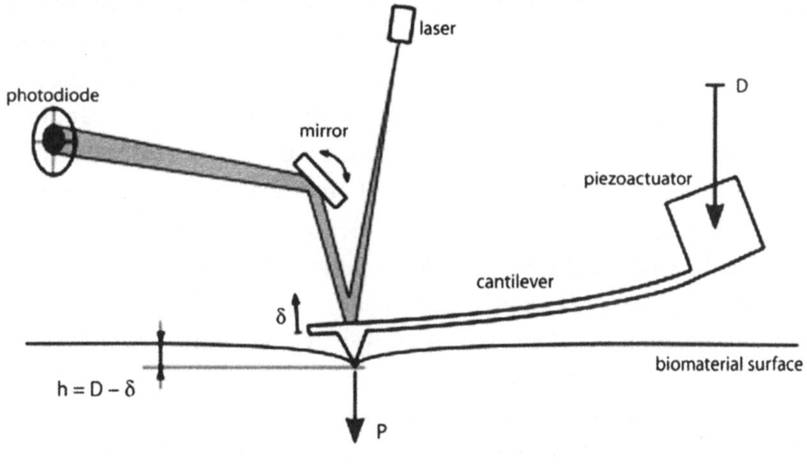

Figure 3-2. Schematic of scanning probe or atomic force microscope feedback required for both imaging and indentation. After Silva and Van Vliet.[8]

3.1.2. *Cantilever as force transducer*

Instrumented indenters are designed to minimize sample compliance and, in many cases, acquire load and displacement signals through independent transducers. In contrast, AFMs rely on cantilevered probes that employ and report contact loads via cantilever deflection. Cantilevers are typically fabricated from silicon or silicon nitride via standard microelectronic fabrication mask lithography, and the cantilever length, width, and thickness are varied to achieve a range of spring constants (k) on the order of 50 N/m < k < 0.001 N/m. The macroscopic shape of these cantilevers is typically rectangular (diving board) or triangular (V-shaped); the latter profile is typical of low-stiffness cantilevers and reduces cantilever torsion during scanning in contact mode. The actual probe that contacts the biomaterial surface is located at the distal end of the cantilever, and typically extends 1 μm normal to the cantilever.

Just as in the case of instrumented indenters, this force transducer must be calibrated prior to use. As these force transducers are microscale, thin, mass-produced devices, there is considerable variation in residual stress and even feature sizes among a wafer of fabricated cantilevers. Thus, the nominal stiffness of an individual cantilever can vary to factors of 2, 5 and even 10 from the nominal spring constant classifications of the vendor. Calibration of the cantilever k prior to indentation experiments is critical to conversion of the cantilever deflection δ to force ($P = k\delta$), and can be conducted in two steps. First, one must establish the sensitivity of the cantilever, in terms of photodiode output voltage per unit cantilever deflection. This is termed the Optical Lever Sensitivity (OLS) in units of [V/m], and is measured simply by displacing the cantilever toward a rigid and hard surface (*e.g.*, glass). When contact occurs, the cantilever free-end will deflect 1 nm for every 1 nm of the calibrated piezoactuator, so the cantilever deflection in photodiode [V] is easily converted to [m] via the slope of the cantilever deflection δ vs. cantilever base displacement D relation. For the relative laser and photodiode positions typical of commercial AFMs, changes in cantilever deflection of 0.1 Å are readily measurable, and precision is limited by thermal fluctuations. Second, one must establish the stiffness of the cantilever by displacing it a (now calibrated) distance under a known force. One facile approach is to measure the deflection of the cantilever as it oscillates in air, acquiring the power spectral density function or fast Fourier transform of these oscillation frequencies and amplitudes. By fitting a simple harmonic oscillator model to the resonance peak of this spectrum, one can quickly and reliably establish the force transducer calibration constant k. Note that although k is a property of the cantilever and it has been shown that the force resolution of such a spring is independent of k magnitude, the OLS is a function of cantilever reflectivity and system properties that may change over time. Thus, it is advisable to measure k of even a repeatedly used cantilever in this way, prior to each experiment. The load resolution of these cantilevers is limited by the OLS and sensitive to thermal fluctuations and acoustical noise; it can be measured directly via in-contact and out-of-contact deflection oscillations of the cantilever, and is on the order of pN for commercially available cantilevers.

Although AFM cantilevers are often depicted as translating normal to the sample surface, for two reasons the cantilever is fact sharply inclined toward the surface (Fig. 3-2). This angle θ varies among AFM designs but is approximately $\theta = 10–15°$. The first reason for this attack angle is the optics by which cantilever deflection is reflected and detected at sufficient distance to maximize accuracy. The second reason is that the probe extending from the cantilever free-end is only μm-scale in height, such that the cantilever of 100s μm length should be inclined to ensure that the probe contacts the sample surface without the rest of the cantilever contacting the surface and impeding deflection. As a result, there are significant tangential forces generated upon contact of the probe with the indented biomaterial, in addition to the intended normal forces of indentation. It remains challenging to quantify the magnitude of these competing shear forces and distinguish the onset of normal loading from the "digging in" to a compliant sample or "sliding along" a stiff sample under such conditions. At present, the most judicious option is to minimize θ to the extent possible via cantilever mounting modifications, and to consider in one's analysis the possible effects of shear forces at early stages of contact loading.

3.1.3. *Piezoactuator as displacement transducer*

The cantilever base is typically translated via a piezoelectric crystal that can actuate by increasing or decreasing in overall length as a function of applied voltage bias. The typical voltage and displacement range of $BaTiO_3$ piezocrystals is 150 V over a 10 μm excursion, such that ~65 nm of displacement can be achieved per volt. These devices remain useful tools because of the rapidity of actuation (> μm/s), but also confer disadvantages related to inherent piezoelectric material limitations. First, there is limited extent of actuation in such crystals (piezostacks moving in staggered *x*, *y*, and *z* steps typically exhibit maximum displacement of 10 μm, and material cost rises prohibitively with physical size and displacement range). Second, as the piezoelectric effect proceeds via motion of atoms to produce a net increase or decrease of the crystal unit cell dimensions, there is a natural inertial effect and apparent nonlinearity in displacement as a function of voltage. For example, piezoactuators continue to displace in the direction of driving force even after the voltage bias has been removed, eventually reestablishing equilibrium. For these reasons, feedback control of actuation and also calculations of indentation depth, the difference between cantilever base displacement and cantilever free-end deflection ($h = D - \delta$) can be inaccurate. This can be addressed in several ways. For example, one commercial instrument actuates via the piezocrystal but conducts feedback and depth calculations based on a tandem, linearized voltage-displacement transducer.

3.1.4. *Limitations of linearized signal acquisition*

Position sensitive detectors (PSD) such as quadrant photodiodes also have a limited and defined linear range of operation. In such PSDs, this range is defined by the position of the reflected laser spot with respect to the photodiode center. If the cantilever bends beyond this linear range, the linearized OLS used to calculate deflection and then force will be in error, creating nonlinear errors in load-displacement (*P*-*h*) responses. If one is constrained to work within the linear regime of the PSD, the maximum force achievable with a cantilever of a particular stiffness will not be realized.

This limitation can be overcome by a straightforward procedure to linearize this nonlinear signal over the full operating (deflecting) range of the cantilever of interest. This approach is detailed,[9] with results summarized in Fig. 3-3. Application of this approach reduces the maximum systematic error on calculated quantities, including dP/dh from which E may be inferred, from 50% to 5%, and at least doubles the range of calibrated force one can apply to the biomaterial surface.[8]

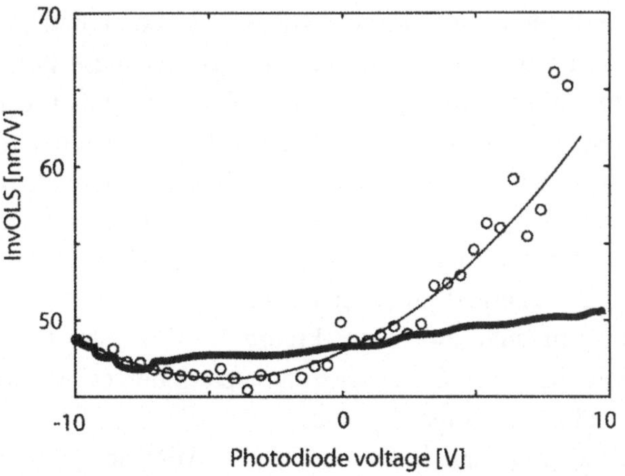

Figure 3-3. Although not obvious to the user, devices that indicate AFM cantilever deflection (and thus force) are not linear over a wide range of operating voltages (circles). This can be linearized through analysis of full nonlinear profile to increase accuracy and range of AFM indentation forces. After Silva and Van Vliet.[8]

3.1.5. *Integration of AFM feedback and instrumented indentation probes*

Most commercially available AFM cantilevers include probes that are readily fabricated in silicon wafer platforms, such as pyramids of apex angle $\alpha \sim 60°$ etched along close-packed planes of the wafer. These probes are quoted to have radii on the order of 25 nm but, due to the nature of this etching and filling process, range widely in size and facet severity. For indentation into compliant surfaces to depths of several μm, this probe imperfection is minimized and one assumes the geometry of a sharp cone. However, there are several cases in which biomaterial indentation requires more well defined probe geometry, including very thin, very compliant films that cannot be probed to μm depths; and analyses that require low imposed strains. In such cases, two distinct approaches have developed.

The first approach is to use colloidal probes, so named because colloidal particles of silica, polystyrene and other materials can be glued to AFM cantilever free-ends to better define the contact area. This approach decreases the imposed strain of indentation, but also increases the strained volume, such that finite thickness effects must be considered.

A further advantage of these probes, as compared to diamond conospherical probes of comparable radii, is that these probe materials are easily modified with biochemical moieties to enhance or mitigate specific interactions between the probe and the biomaterial. These AFM cantilevered probes can even be used within commercial instrumented indenters, simply by mounting the fractured cantilever to the indenter shaft. In this case, the cantilever "legs" are essentially handles to position a well defined, spherical probe and achieve loads and loading profiles typical of instrumented indenters.[9] Figure 3-4 illustrates this application for the particular case of an instrumented indenter mounted with a borosilicate sphere of μm-scale radius.

The second approach is to modify the AFM in order to maintain nanoscale imaging resolution via cantilever scanning, while achieving nanoindentation loading via a separate device more similar to instrumented indenters. For example, some commercial AFMs include an indentation module that actuates diamond indenters within the same mount that holds the AFM cantilever, but is comparably rigid and enables normal approach to the sample surface. The compromise in such designs is that the exquisite load resolution of AFM cantilevers (pN) is sacrificed.

Contact creep compliance and stress relaxation cannot be inferred reliably from AFM contact experiments with biological samples, due to the control variables inherent in AFM contact. For example, contact creep compliance requires that contact load is maintained constant for an appreciable duration while contact depth increases. This condition requires that cantilever free-end deflection is maintained constant and, as the cantilever base position is the signal in feedback, would require an additional calculated feedback loop to attain. Additionally, the signal drift in AFMs can be tens to hundreds of nanometers per second, owing to the sensitivity of these cantilevers to thermal fluctuations; this can exceed the expected range of depth changes during creep or stress relaxation experiments. In short, for such experiments it is strongly preferable to use instrumented indentation capable of either true load or displacement control at the point of contact.

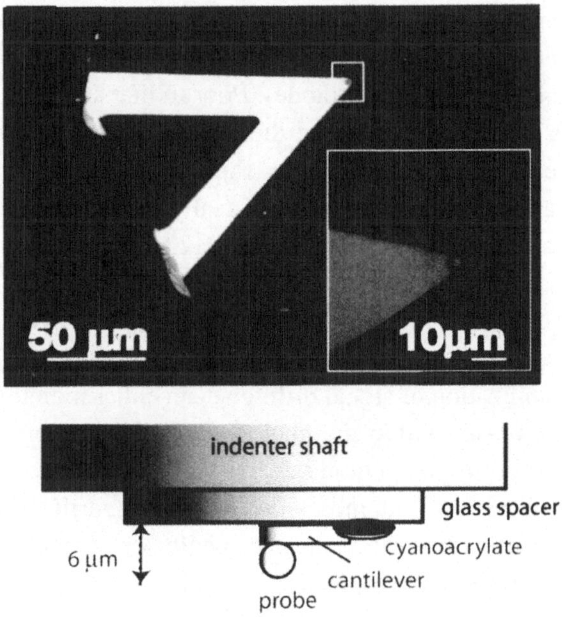

Figure 3-4. Mounting of an AFM-secured spherical probe to an instrumented indenter, leveraging the size and chemical functionalization potential of the AFM cantilever while maintaining the mechanical feedback loop control of the instrumented indenter.[9]

3.2. *Adapations required of environmental control*

As AFMs have been operated in aqueous environments for as long as instrumented indentation has existed, many adaptations required of environmental control for AFM-enabled indentation of biological materials are standard components of these microscopes.

3.2.1. *Temperature and associated thermal drift*

Although typically operated at room temperature, several different resistive heating element stages are compatible with commercial AFMs to enable imaging and indentation at 37°C or higher temperatures. Regardless of the desired imaging temperature, it is important to allow the system to stabilize before acquiring data. For example, if experiments are planned for a hydrated biological tissue in 150 mM phosphate buffered saline at 22°C, at least 30 min will be required for the immersed cantilever to achieve a stable deflection signal (position of the reflected laser on the photodiode). Prior to that stabilization time, the difference in temperature between the air and imaging fluid, as well as any laser-heating of the imaging fluid, will cause the cantilever dimensions and position to fluctuate. Ideally, the imaging fluid will be placed in a sealed container within the AFM overnight to stabilize the imaging fluid temperature and, during the experiment, fluid will be added as required to minimize changes in salinity and pH that will affect both cantilever position stability and sample degradation. To confirm the extent and stabilization of signal drift, one can either monitor the position of the reflected laser within the photodiode and its reported voltage, or equivalently record the photodiode output voltage over time until a threshold drift rate is attained. This threshold drift rate should be considerably less than the displacements or depths expected in mechanical analysis of the biomaterial sample.

3.2.2. *Controlled gaseous atmosphere*

For some commercial AFMs, it is possible to define the gaseous environment around the AFM stage and transducers with gasketed gas inlets. More typically, it is sufficient to control the %CO_2 and pH required for cell survival and tissue stability via use of buffered aqueous solvents such as HEPES or Hank's buffered saline. Whenever solvents are used to fill the instrument or sample chamber, including gases, it is important to confirm compatibility with the instrument materials. One common point of concern for AFM-enabled indentation is chemical stability of the AFM cantilever mount, which braces the chip on which cantilevers are fabricated and also intentionally traps fluid between the cantilever and biomaterial sample.

3.2.3. *Fluid cells for hydrated biological materials*

Several types of fluid cells, or confined chambers that maintain hydration of the biomaterial sample, exist and are proprietary to each AFM. These range from as simple as cantilever mounts fabricated from hydrophobic polymers to o-ring gasketed stages with injection inlets and outlets for media exchange. In addition to the chemical stability of such fluid cells that was noted above, it is also important to confirm the duration over which the fluid cell is impervious to leaks (especially gradual wicking of moisture toward electronic components) and to corrosion (given that saline is often a component of biomaterial imaging fluids).

3.3. *Selection of indenter probes*

As distinct from the selection of instrumented indenter probes, the mechanical stiffness of the AFM cantilevers strongly affects experimental design. This stiffness directly controls the range of contact forces and indentation depths attainable over the linear range of the position sensitive photodiode that monitors cantilever free-end deflection. Consider a typical contact-mode Si_3N_4 V-shaped cantilever of spring constant k = 0.1 N/m and an associated OLS (for a particular AFM) of 100 nm/V. For a photodiode that is maximized at an output voltage of 10 V, the maximum free-end cantilever deflection that can be measured is 1 μm, and thus the maximum force that can be imposed on the biomaterial surface is 0.1 μN or 100 nN. This, of course, assumes that the entire range of the photodiode is linearized.

If one needs to impose more force than allowed by this cantilever, one would choose a stiffer cantilever geometry. As will be discussed in Section 3.5, the choice of cantilever stiffness is further complicated by the fact that indentation with a cantilever that is much stiffer than the biomaterial sample results in penetration without deflection, and with one that is much more compliant results in deflection without penetration.

There is much more limited choice of indenter probe geometry in AFM-enabled indentation. Essentially, there exist the as-fabricated pyramidal cones of Si and Si_3N_4 cantilevers, terminating in radii of ~25 nm; colloidal spheres of μm-scale radius, comprising glass or polystyrene beads that are glued to tipless cantilevers, and diamond probes glued or brazed to stiff, tipless cantilevers. These diamond probes are typically from the same vendors who supply the instrumented indenters, and therefore exhibit semi-apex angles of ~70° for Berkovich and Vickers pyramids, with finite apex radii on the order of 100s nm.

3.4. *Experimental approaches*

3.4.1. *Calibration over appropriate signal range*

To convert cantilever deflection to load-depth responses, three calibration values are key: OLS, cantilever spring constant k, and signal drift rate. These calibrations assume that the user has already confirmed the range of linear voltage-displacement response of the laser-photodiode feedback loop, and is either working within the linear range or has linearized over a larger operating range as discussed in Section 3.1.1. These calibrations also assume that the piezoactuator (to move the cantilever base or the stage) is independently calibrated.

Calibration of the OLS, or conversion of photodiode voltage to actual cantilever free-end deflection, was noted previously and requires displacement of the cantilever against a rigid surface such as a glass slide. Upward deflection of the cantilever free-end will be equivalent to downward displacement of the cantilever base; as the base displacement is independently calibrated (*e.g.*, it is a piezoactuator), conversion between the deflection measured as photodiode voltage and the actual deflection measured in meters is readily obtained as the slope of the free-end deflection (V) *vs* base displacement (m) line. Note that, in practice, there will be a linear region of this relation near the initial contact point, so it is important to maximize the number of data points within that region and average over many replicate responses to obtain an average OLS. This method typically determines OLS with 3% standard error. This parameter is sufficiently sensitive that it must be repeated for each use of each cantilever, and can differ for the same cantilever in different imaging fluids. The cantilever spring constant k can be obtained subsequently, via collection of the frequencies and amplitudes of the cantilever oscillation at ambient temperature. Most commercial AFMs include semi-automated analyses to render the fast Fourier transform of this oscillation as a power spectral density, from which the spring constant k can be estimated by analogy to a simple harmonic oscillator:

$$2\pi f_o = \sqrt{k / m} \tag{3-1}$$

where f_o is the cantilever first mode of resonance and m is the effective cantilever mass. As the amplitude of oscillation is also directly related to k, the cantilever mass value can be obviated. Note that this is an estimate, in that AFM cantilevers are not simple harmonic oscillators, but in fact have several other modes of deformation including torsion. Signal drift rate is measured simply as the change in cantilever deflection with time, at least over times that are comparable to the experiment duration. The inherent sensitivity of AFM cantilevers requires thermal stability to minimize deflection fluctuations.

3.4.2. *Effects of cantilever stiffness on inferred mechanical properties*

The choice of cantilever has a significant effect on inferred mechanical properties of the biomaterial sample, due to the fact that cantilever deflection represents an additional mode of mechanical energy dissipation. Simply put, if the cantilever is much stiffer than the sample, it will penetrate the sample without deflecting at the free end, and thus no indentation force or depth will be registered. Conversely, if the cantilever is much more compliant than the sample, it will deflect through its full range without penetrating the sample, and thus no indentation depth will be attained as required for calculation of properties. The obvious recommendation is to choose a cantilever of "correct" stiffness, one that is neither too stiff nor too compliant. However, if the biomaterial sample is in fact a material of study that is not yet well characterized, this selection becomes iterative. At best, a range of cantilever stiffnesses may be employed to assay the range of mechanical properties. At worst, a single cantilever stiffness is chosen and the reported properties of the sample in fact reflect the stiffness of the sample and the cantilever.

When the AFM cantilever is in contact with a compliant surface (as compared to the rigid surface on which OLS was calibrated), the cantilever oscillation energy is dominated by thermal fluctuations on the order of $k_B T$, where T is temperature and k_B is the Boltzmann constant. It is this so-called "thermal noise" that limits the resolution of AFM cantilevers. Thus, the root-mean-square variation $<P>$ or load signal resolution depends on both temperature and cantilever spring constant k:

$$\langle P \rangle = \sqrt{kk_B T} \qquad (3\text{-}2)$$

and is approximately 5 pN for a cantilever of $k \sim 0.01$ N/m. However, in practice, the bandwidth of signal acquisition can further decrease load resolution of such compliant cantilevers to ~10 pN; this is because the cantilever resonant frequency is within the measurement bandwidth.

The standard equations of elastic contact analysis assume negligible adhesion, whereas JKR-type theories of elastic contact exploit probe-surface adhesion. In practice, with biological samples there is typically a measureable degree of adhesion when the charged, proteinaceous sample surface is displaced from the charged, synthetic cantilevered probe. For experimental analyses that assume negligible adhesion, probe interactions can be minimized by conducting experiments in ionic buffers such as 150 mN phosphate buffered saline.

As noted, the mechanical stiffness of AFM cantilevers can adversely affect the inferred mechanical properties of the indented material, including the *apparent* elastic modulus of the biomaterial surface E_a. For example, when the stiffness of a μm-scale bundle of actin protein filaments is analyzed with sharp ($R \sim 25$ nm) probe cantilevers of $k_1 = 0.01$ N/m and $k_2 = 0.5$ N/m to the same depth $h \sim 15$ nm, the inferred E_a differ by an order of magnitude ($E_{a1} = 2.4 \pm 1.0$ MPa; and $E_{a2} = 19.6 \pm 4.2$ MPa, reporting one standard deviation of the average value). Unless this filamentous bundle were also characterized by independent mechanical experiments, it would not be possible to rule out either data set. Instead, one could analyze other synthetic materials for which E is established by macroscale tests. These results are shown in Table 3-2, which indicates that E_a of polydimethylsiloxane elastomer ($E \sim 1.8$ MPa), amorphous polystyrene ($E \sim 2.7$ GPa), and glass ($E \sim 70$ GPa) also depend on AFM cantilever stiffness. As in the case of a biological material, E_a of these synthetic materials increased with increasinig k, and the compliant cantilevers typically employed for high-resolution imaging of biological samples failed to accurately measure E_a of materials of ~2 GPa and higher. This indicates that AFM cantilevers used to obtain spatial information and sample topography on hydrated biomaterials may in fact be too compliant to accurately infer mechanical properties of these imaged regions. Further, the reasonable supposition that one should aim to match cantilever stiffness [N/m] to sample stiffness [N/m^2] to measure mechanical properties accurately is not straightforward to obtain: E_a of polystyrene did not approximate the known stiffness of this material except for the stiffest cantilever, and E_a of polydimethylsiloxane fluctuated within an order of magnitude of known stiffness for all cantilevers considered.

Table 3-2. Apparent elastic moduli of three materials inferred from AFM-enabled indentation with cantilevers of varying stiffness k: glass (E ~ 70 GPa); polystyrene or PS (E ~ 2.7 GPa); and polydimethylsiloxane or PDMS (E ~ 1.8 MPa). Courtesy of D. Nikova and K. J. Van Vliet. All values in MPa.

***k*:**	**0.01 N/m**	**0.03 N/m**	**0.1 N/m**	**0.5 N/m**	**2 N/m**	**42 N/m**
Glass	19.20 ± 8.44	53.03 ± 10.13	840.71 ± 90.06	1850 ± 670	2220 ± 950	57880 ± 6200
PS	3.19 ± 2.60	18.95 ± 7.60	172 ± 4.06	244.6 ± 79	338.96 ± 83	2400 ± 370
PDMS	5.42 ± 4.06	2.40 ± 1.95	—	8.54 ± 2.42	11.13 ± 2.07	14.96 ± 1.66

3.4.3. *Objective identification of mechanical contact point*

Another key difference between instrumented indentation and AFM-enabled indentation is the inherent challenge of identifying the initiation of mechanical contact in the latter approach. As the force transducer (cantilever) is mechanically compliant, the probe deflects away from the surface prior to the establishment of firm contact that deforms the sample surface. Post-processing methods to identify this point include graphing log(P) *vs* log(h) to determine the point at which large load fluctuations cease and actual loading commences.[10] This method still requires subjective choice of the threshold fluctuations, and the calculated mechanical properties are sensitive to this choice.

An alternative approach, for the specific case of spherical elastic contact, expresses the Hertzian contact equation (see Chapter 5) to be linear not in the applied load P but in the resulting indentation depth h[11]:

$$P^{2/3} = a - \left[\frac{4E\sqrt{R}}{3\left(1-\nu^2\right)}\right]^{2/3} h \tag{3-3}$$

In this way, the slope of $P^{2/3}$ *vs* h provides an estimate for E that is independent of contact point determination a, though still requires accurate estimate of probe radius R and Poisson's ratio ν.

3.4.4. *Quasistatic load-depth hystereses*

The deformation via AFM cantilever penetration is neither load- nor indenter displacement-controlled, as the feedback loop only modulates the displacement of the cantilever base over a specified time and extent. Thus, load reversal at a specified cantilever base displacement will not deform all materials to the same depth; and load reversal at a specified cantilever free end deflection will deform all materials to the same maximum load. Further, this immediately indicates that the material is not deformed at a constant loading rate or a constant indenter displacement rate. It is technically possible to calculate these quantities on-the-fly and conduct signal feedback on either load or indenter probe displacement in any commercial AFM via software modification. However, owing to the historical application of AFM for imaging rather than indentation, these feedback loops are not currently standard features on commercial instrumentation. Thus, comparisons of biomaterial response between AFM-enabled indentation and instrumented indentation must bear in mind the differing load histories and control variables, even if applied strains are comparable. This is particularly important in consideration of nonlinear elastic and viscoelastoplastic materials that comprise the majority of biological materials.

In typical experiments, the extent of piezoactuator displacement and the time interval of that displacement are specified. The condition for actuator reversal (*i.e.*, unloading of the sample) may also specified, which could be the default displacement of the cantilever base or alternatively a limiting deflection (and thus force) attained on the sample surface. The *P-h* responses appear qualitatively similar to those obtained via instrumented indentation, but with important caveats. First, by default data is acquired as soon as the piezoactuator displaces, resulting in possible acquisition of ample data before and after the probe-sample contact. From such traces, it is clear that drift in the deflection signal [nm/s] can be comparable to displacement rates into the sample surface, and are typically subtracted from the data post-test. Second, the high force sensitivity of such cantilevered probes—sensitivity in terms of deflection that may be due to surface charge and other interactions besides mechanical contact—results in well-known difficulties in objective identification of the contact point $P = h = 0$. Various methods have been posed in the literature to unambiguously choose the point of actual probe-sample contact.[10,12] Automation of this step is considered important in terms of mechanical property calculation accuracy and in light of the large number of replicate experiments conducted for structurally heterogeneous samples.[13]

3.4.5. *Dynamic loading*

As in instrumented indentation, the AFM cantilever can be used to cyclically contact and load the biomaterial surface. This is commonly executed in AFM imaging of compliant material surfaces, and is then called tapping or intermittent contact or AC mode imaging. In such imaging, the probe scans the surface while maintaining a feedback loop between user-specified cantilever free-end amplitude and the cantilever base piezoactuation. A lock-in amplifier maintains a driving amplitude that is dissipated and shifted in phase via probe interaction with the surface. The error signal in such imaging is the actual free-end amplitude, which differs from the user-specified amplitude due to more rapid variations in mechanical energy dissipation across the surface than can be accommodated by the gains of the feedback loop. Additionally, the phase lag indicates energy dissipation at every pixel in the sample surface. This type of pixel-resolved, qualitative and yet visual representation of energy dissipation was the basis of the early work that preceded dynamic mechanical analysis via instrumented indentation.[14] Despite the widespread use of this dynamic contact imaging to visualize regions of disparate energy dissipation in structurally heterogeneous nanocomposites including tissues and cells, there is little quantitative interpretation of amplitude reduction and phase lag in terms of mechanical properties. This is due primarily to the multiple, complex interactions at the probe-biomaterial surface that dissipate contact energy, in addition to the mechanical dissipation of biopolymeric phases.

Imaging biomaterials in aqueous environments via dynamic contact is notororiously difficult, due to both the high mechanical compliance of fully hydrated biomaterials and the considerable mechanical coupling and noise associated with rapid oscillation of the microscale cantilever in fluid. The latter challenge is related to the standard means of forced cantilever oscillation, which is piezoactuation of the cantilever base; as a result, the entire cantilever length vibrates vertically to achieve the prescribed free-end amplitude and associated contact force. Such oscillation close to the sample surface can generate waves within the imaging fluid that distort the image and perturb the intended contact force. To minimize these deleterious effects for hydrated biomaterials, two unique hardware modifications have been developed. The first approach employs magnetic coating of the backside of AFM cantilevers, and actuation of the cantilever via an oscillating electromagnetic field localized at the free-end of the cantilever. This is called magnetic AC or MAC mode.[15] The second approach oscillates an uncoated cantilever by driving AC current directly through the cantilever legs in the presence of a small magnetic field; the interacting electromagnetic fields oscillate the cantilever free-end.[16] Both approaches significantly reduce mechanical coupling of the imaging fluid with cantilever oscillation, and enable increased resolution in height, amplitude error, and phase lag images from which mechanical properties can be inferred qualitatively[17] (Fig. 3-5).

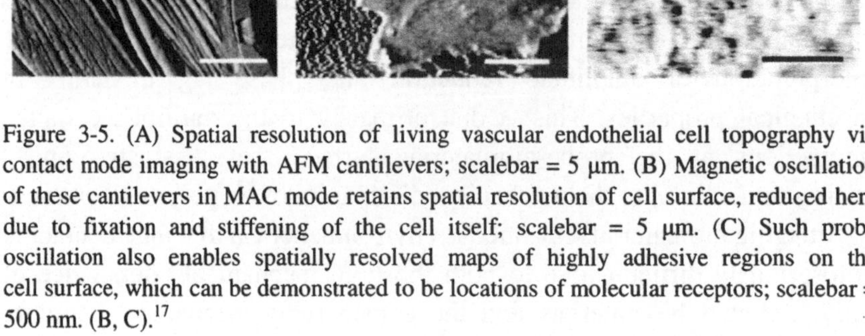

Figure 3-5. (A) Spatial resolution of living vascular endothelial cell topography via contact mode imaging with AFM cantilevers; scalebar = 5 μm. (B) Magnetic oscillation of these cantilevers in MAC mode retains spatial resolution of cell surface, reduced here due to fixation and stiffening of the cell itself; scalebar = 5 μm. (C) Such probe oscillation also enables spatially resolved maps of highly adhesive regions on the cell surface, which can be demonstrated to be locations of molecular receptors; scalebar = 500 nm. (B, C).[17]

3.4.6. *Adhesion*

Adhesion between the AFM cantilever probe surface and the sample surface can be measured for pN- to nN-scale forces, and has been leveraged to infer chemical specificity of biological sample surfaces.[18] As the focus of this chapter is the use of AFMs as indenters that apply primarily compressive contact, such applications of AFM to infer intra- and intermolecular rupture forces under tension will not be discussed in detail, but are summarized in reviews specific to biological cells and molecules.[18,19]

In the context of instrumented indentation analogues, the sensitivity of AFM cantilevered probes to adhesive interactions is inherently high, such that elastic properties of contact samples can be readily inferred from JKR and similar adhesion models.[20] In fact, both the force required to pull the probe onto the biomaterial (the so-called "snap to contact" upon approach of the surface) and the force required to pull the probe from the biomaterial (the so-called "rupture force" or "adhesion force" upon retraction from the surface) can be quantified directly from the *P-h* responses (see Chapter 4 for more detail). In order to maximize quantitative interpretation of such probe-surface interactions, it is preferable to employ colloidal (*i.e.*, spherical) probes of well-characterized materials, rather than to estimate contact areas of nominally sharp Si or Si_3N_4 fabricated cones. Further, it is important to note that the rate of loading and unloading strongly affects the measured force and work of adhesion, and this rate cannot be controlled directly via AFM cantilevers. Only the rate of piezoactuated cantilever base displacement is controlled directly, and the actual unloading rate must be calculated from the load-time slope just prior to rupture of the probe from the surface.

3.4.7. *Pixellated mapping of mechanical response*

In all of the above cases, one apparent advantage of AFM-based indentation over instrumented indentation is that data can be acquired for every pixel within an image area. For example, a square image of 512 × 512 pixels comprises 262,144 data points, which may represent the stiffness or adhesion of each pixel. If that image were of a 1 μm × 1 μm region of a biomaterial, each pixel would then represent a 2 nm × 2 nm sample region. One might then imagine that it is possible to measure the elastic modulus E of a complex surface at every pixel, and create a highly spatially resolved map of this important mechanical property to render its quantitative variation along a sample such as a hydrated tissue or an individual cell. Certainly, it is an attractive goal to correlate the spatial features of a structurally heterogeneous biomaterial with the mechanical properties of individual regions. However, although such maps of mechanical properties (sometimes called force-volume maps to indicate that they register a range of force-calculated property values along an x-y image plane) are straightforward to obtain via automated software, it is difficult for such data collection approaches to accurately reflect the mechanical properties of the biological material – uncoupled from mechanical history of adjacent pixels, sample topography, and ignorance of the indenter probe geometry at the indentation depths employed.

The data acquisition in such mapping is in contrast to instrumented indentation of heterogeneous samples, where indentations are small enough that the (plastically) strained indentation volume is less than the phase volume, and spaced widely enough that these strained volumes do not overlap but sample distinct phase regions. If the material were perfectly flat and the indentation responses perfectly elastic, such that no prior deformation was stored within the material adjacent to the current indentation volume, this pixellated mapping could render an approximately accurate estimate of elastic properties. If only the latter condition is met but the sample has surface roughness on the order of the nm-scale indentation depths, the assumptions of normal contact on the surface may not be met and thus the estimated elastic properties will be inaccurate due to errors in the probe-surface contact area. Additionally, the analysis of such indentations is never well defined at a phase interface, at which the strains and stresses are distributed nonuniformly, so image areas comprising many interfacial boundaries will be difficult to quantify accurately. Finally, viscoelasticity is typically neglected in such mapping, as the loading times sufficient to suppress viscous responses should differ from pixel to pixel but are maintained constant over the experiment. Thus, pixellated maps of mechanical properties inferred from AFM-enabled indentation should be considered qualitative estimates of mechanical variation that, unless carefully calibrated, to not reflect the mechanical properties of pixel-sized biomaterial volumes.

3.5. *Sample mounting and preparation*

As in the case of instrumented indentation (Section 2.5 of this chapter), biomaterial sample preparation is a critical part of accurate mechanical characterization. Only issues different from, or in addition to, those already presented for instrumented indentation will be considered here.

The change in fluid levels or salinity will affect the positional stability of the AFM cantilever. It is important to maintain the sample and the entire fluid cell under a constant volume of fluid, either by adding fluid periodically upon evaporation or, preferably, by adopting a closed cell configuration that minimizes evaporation.

In certain experiments, one may require exchange of fluids during mechanical analysis. To minimize mechanical agitation, injection via manually depressed syringe needles is preferable to mechanical pumps.

To secure biomaterial samples to rigid, flat surfaces required of AFM-enabled indentation, several options can be considered. In general, the goal is to maximize the adhesion strength of the sample to the rigid surface without denaturing or otherwise structurally altering the biomaterial. Proteinaceous samples can be secured with CellTak™ or similar biological adhesives, and can sometimes be physically adsorbed to surfaces via mild centrifugation or even drying and rehydration. As a rule, the less mechanical and physical stress exerted on the sample, the more likely the surface characterized via AFM-enabled indentation will accurately reflect the properties of the intact biomaterial sample.

4. Summary and Conclusions

This chapter has considered experimental issues relevant to contact mechanics testing techniques for biological materials. Separate sections have examined instrumented indentation (Section 2) and AFM-based testing (Section 3). Emphasis has been on issues of load and displacement resolution and range, signal stability and environmental control as well as other practical aspects of experimental testing such as sample preparation and mounting. No consideration has been made here for data analysis, as this will be considered for adhesion in Chapter 4 and for indentation in Chapter 5 of this volume.

References

1. K. Tai, H. Qi and C. Ortiz, *J. Mater. Sci. Mater. Med.* **16**, 947 (2005).
2. J. Thompson, *et al.*, *Nature* **414**, 773 (2001).
3. M. Oyen, *J. Biomech.* **39**, 2699 (2006).
4. W. Oliver and G. Pharr, *J. Mater. Res.* **7**, 1564 (1992).
5. K. J. Van Vliet, L. Prchlik and J. F. Smith, *J. Mater. Res.* **19**, 325 (2004).
6. M. Varenberg, *et al.*, Advanced testing of adhesion and friction with a microtribometer, *Review of Scientific Instruments* **77**, 066105 (2006).
7. A. M. Baro, *et al.*, Determination of surface-topography of biological specimens at high-resolution by scanning tunnelling microscopy, *Nature* **315**(6016), 253–254 (1985).
8. E. Silva and K. J. Van Vliet, Robust approach to maximize the range and accuracy of force application in atomic force microscopes with nonlinear position-sensitive detectors, *Nanotechnology* **17**, 5525–5530 (2006).
9. C. A. Tweedie, G. Constantinides, G. S. Blackman and K. J. Van Vliet, Controlled probe surface chemistry for nanoscale contact deformation, Symposium HH, Materials Research Society Spring Meeting, San Francisco. Paper 1021-HH06-01 (2007).
10. M. T. Thompson, *et al.*, Tuning compliance of nanoscale polyelectrolyte multilayers to modulate cell adhesion, *Biomaterials* **26**(34), 6836–6845 (2005).
11. S. Guo and B. Akhremitchev, Investigation of mechanical properties of insulin crystals by atomic force microscopy, *Langmuir* **24**, 880–887 (2008).
12. M. T. Thompson, *et al.*, Biochemical functionalization of polymeric cell substrata can alter mechanical compliance, *Biomacromolecules* **7**(6), 1990–1995 (2006).
13. G. Constantinides, *et al.*, Grid indentation analysis of composite microstructure and mechanics: Principles and validation, Materials Science and Engineering a-Structural Materials Properties Microstructure and Processing **430**(1–2), 189–202 (2006).
14. S. Asif, *et al.*, Quantitative imaging of nanoscale mechanical properties using hybrid nanoindentation and force modulation, *Journal of Applied Physics* **90**, 1192–1200 (2001).
15. H. Schindler, *et al.*, Optimal sensitivity for molecular recognition MAC-mode AFM, *Ultramicroscopy* **82**(1–4), 227–235 (2000).

16. Asylum Research, I., Small Talk, (2007). http://www.asylumresearch.com/Applications/SmallTalk/SmallTalk4.pdf (last accessed 7 December 2009).
17. S. Lee, J. Mandic and K. J. Van Vliet, Chemomechanical mapping of ligand-receptor binding kinetics on cells, *Proceedings of the National Academy of Sciences of the United States of America* **104**(23), 9609–9614 (2007).
18. K. J. Van Vliet, G. Bao and S. Suresh, The biomechanics toolbox: Experimental approaches for living cells and biomolecules, *Acta Materialia* **51**, 5881–5905 (2003).
19. K. J. Van Vliet and P. Hinterdorfer, Probing drug-cell interactions, *Nano Today* **1**(3), 18–25 (2006).
20. Y.-J. Sun, B. Akhremitchev and G. C. Walker, Using the adhesive interaction between atomic force microscopy tips and polymer surfaces to measure the elastic modulus of compliant samples, *Langmuir* **20**, 5837–5845 (2004).

CHAPTER 4

MECHANICS OF ADHESION

Robert F. Cook

Ceramics Division, National Institute of Standards and Technology
E-mail: robert.cook@nist.gov

A framework of increasing complexity is developed to describe the mechanics of adhesion and its reverse, de-adhesion or separation, during indentation contact cycles. The importance of the indentation probe stiffness in determination of system stability is emphasized and shown to control the details of the adhesive snap-on instability for a simple logarithmic probe-surface interaction potential and the snap-on and pull-off instabilities (and their existence) for a Mie interaction potential. The work of adhesion is shown to depend critically on the probe stiffness. Indentation contact-mechanics coordinates are introduced and are used to develop the Johnson-Kendall-Roberts (JKR) contact model. The similarities, differences, and conveniences of the logarithmic, Mie, and JKR adhesive indentation contact models are discussed.

1. Adhesion: The "Snap-On" Instability

In this section the process of *adhesion* is considered: Two bodies, initially in a state of separated, stable equilibrium, are brought together by an external influence such that the system passes through a state of unstable equilibrium to a new state in which the bodies are joined, or adhered, to each other in a new stable equilibrium configuration.

A very simple system that gives rise to an adhesion process will be considered: There is a long-range attraction only between the bodies and one of the bodies is rigid (the surface from Chapter 2) and the other linear-elastic (the probe from Chapter 2). Many of the required concepts and definitions pervasive in analyses of adhesion and fracture will be illustrated in consideration of this simple system: equilibrium—and the interplay of forces associated with surface interactions and solid deformations in determining characteristic system scales; and stability—and the central role played by stiffness fields in determining observed behavior.

The interaction between the bodies is described by a logarithmic potential, $\Phi_A(z)$:

$$\Phi_A = A\ln(z/z_A), \tag{4-1}$$

where z is the separation of the bodies, and A and z_A are amplitude and length-scale parameters of the potential, respectively. Such a potential describes the interaction between a point charge and an infinite line of charge.[1] For convenience, then, it is useful to view the first body—the "surface"—as flat with a long, narrow, charged strip on the surface and the second body—the "probe"—as containing a small charged area. (The two charges are of course of opposite sign, such that A is positive, and the small area and the strip are assumed to be separated by z measured normal to the surface.) The surface is taken to be rigid, given by $z = 0$, thus defining one boundary of the system. (As no deformation of the surface is assumed here, it is convenient to set the boundary of the system to focus exclusively on the deformation of the probe; we will see that in other systems, other choices for $z = 0$ are appropriate.) The probe is taken to be linear elastic with a harmonic deformation potential, $\Phi_S(z)$, given by

$$\Phi_S = \frac{k_S}{2}(z-s)^2. \tag{4-2}$$

It is convenient to view the probe (as in Chapter 2) as consisting of two parts: (i) a "tip," in this case containing a small charged area, attached to (ii) a "spring," the other end of which is (position-) controlled by the indentation system. k_S is then the stiffness of the spring and s is the position of the tip imposed by the indentation system in the absence of other influences (such as Φ_A above) and serves to define the other boundary of the system. A schematic diagram of the system and environment is shown in Fig. 4-1. The behavior of $\Phi_A(z)$ and $\Phi_S(z)$ are shown in Fig. 4-2. From a thermodynamic viewpoint, the "environment" surrounding the surface + probe system is the indentation apparatus, which, in this example, controls the size of the surface + probe system at its exterior boundaries, s, and the temperature, T. The simple form taken here for the interaction potential, allows closed-form analytical solutions to be obtained for the parameters describing the equilibrium and instability configurations of the system and thus the roles of the interaction and spring coefficients in defining these parameters to be seen explicitly.

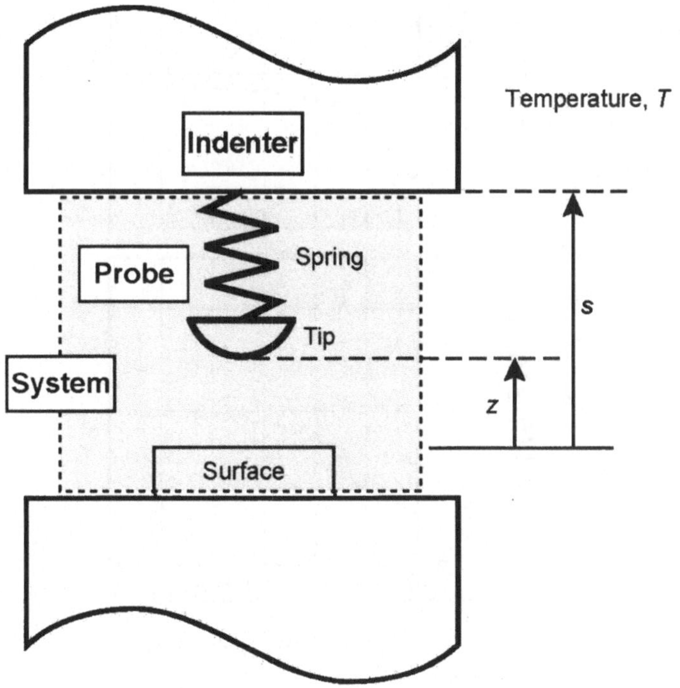

Figure 4-1. A schematic diagram of an indenter and contact system and the coordinate system used to describe adhesive indentation contact behavior between a probe and surface. The indenter imposes a probe position and system temperature and the tip and spring respond in ways that depend on the probe spring constant and the interaction between the tip and the surface.

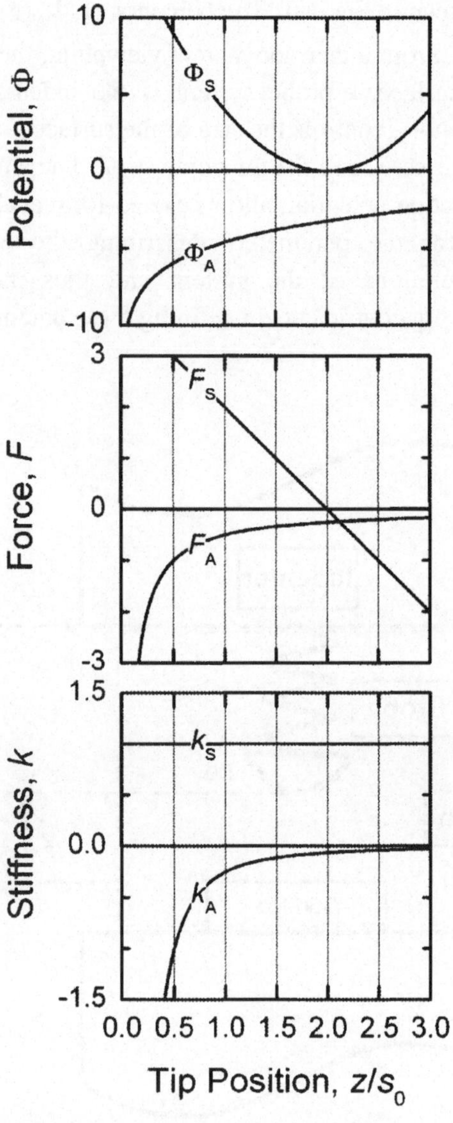

Figure 4-2. The variation in potential, force, and stiffness with tip position for a logarithmic potential (A, Eq. 4-1) between a tip and a surface and harmonic potential (S, Eq. 4-2) for a spring connected to the tip. The values are normalized using the characteristic values (Eq. 4-9).

The force, $F_A(z)$, imposed by the surface potential on the tip is given by

$$F_A = -\frac{d\Phi_A}{dz} = -\frac{A}{z}, \tag{4-3}$$

and the stiffness field, $k_A(z)$, associated with this force is given by

$$k_A = -\frac{dF_A}{dz} = -\frac{A}{z^2}, \tag{4-4}$$

where it is noted that for this example the positive z-direction is taken along the outward normal of the surface, towards the probe. The force, $F_S(z)$, imposed by the spring on the tip is

$$F_S = -\frac{d\Phi_S}{dz} = -k_S(z-s) \tag{4-5}$$

and of course the stiffness of this force is the invariant spring stiffness, k_S. The behavior of the force and stiffness fields is also shown in Fig. 4-2.

Consideration of the small- and large-z limits of the total force on the tip, $F = F_A + F_S$, shows that F must pass through a maximum for all values of s. The condition for this maximum is that the net stiffness experienced by the tip vanish. Using the stiffness relations above and setting the net stiffness to zero, $k = k_A + k_S = 0$, provides the tip-surface separation at which the tip experiences a maximum force:

$$z_{max} = \left(\frac{A}{k_S}\right)^{1/2}. \tag{4-6}$$

This separation is indeed invariant with respect to s. Using the force relations above gives the maximum total force on the tip at this separation:

$$F_{max} = k_S s - 2(Ak_S)^{1/2}. \tag{4-7}$$

Both the magnitude and sign of this maximum force depend on the indenter-imposed position s, which thus determines the state of the system relative to the adhesion sequence described above: $F_{max} > 0$, stable separated state (s large); $F_{max} = 0$, unstable transition state (s at a critical value); and $F_{max} < 0$, stable adhered state (s small). Setting $F_{max} = 0$ provides the critical indenter-imposed position, s_0, at which the system is in the transition state between the two stable states:

$$s_0 = 2\left(\frac{A}{k_S}\right)^{1/2} = 2z_{max}, \tag{4-8}$$

where the 0 subscript indicates zero force on the tip. The indenter-imposed position s_0 is a characteristic length scale of the system and is determined by a balance between the attraction of the surface force interaction and the resistance to solid (spring) deformation.

The condition $F = 0$ of course defines equilibrium and hence for the indenter-imposed condition of $s = s_0$ the spring is extended towards the surface by $s_0/2$ such that the tip is resting in unstable equilibrium at a separation from the surface of $z^* = s_0/2$, where the asterisk indicates an equilibrium state. The characteristic force scale for the system is thus the force exerted by the spring on the tip in this unstable equilibrium state, $F_S(s = s_0/2) = (k_S s_0/2) = (Ak_S)^{1/2}$. The characteristic energy scale for the system is the deformational energy contained in the spring at this state, $\Phi_S(s = s_0/2) = k_S(s_0/2)^2/2 = A/2$. The characteristic stiffness of the system, k_S, is evident.

Using the characteristic scales of the system allows the potential, force, and stiffness fields experienced by the tip to be specified in terms of the internal variable of the tip position:

$$\Phi = (A/2)\left[2\ln(z/z_A) + 4(z/s_0 - s/s_0)^2\right],$$

$$F = (Ak_S)^{1/2}\left[-s_0/2z - 2(z/s_0 - s/s_0)\right], \text{ and} \tag{4-9}$$

$$k = k_S\left[-s_0^2/4z^2 + 1\right],$$

in which the central role of the indenter-imposed probe position, s, relative to the critical position, s_0, in determining system behavior is obvious. Figures 4-3, 4-4, and 4-5 show the variations of $\Phi(z)$, $F(z)$, and $k(z)$, and their components for $s/s_0 = 1.5$, $s/s_0 = 1$, and $s/s_0 = 0.5$, respectively. The invariant position of the maximum in the force at $s_0/2$ and variation of the sign of the maximum force with s/s_0 is clear.

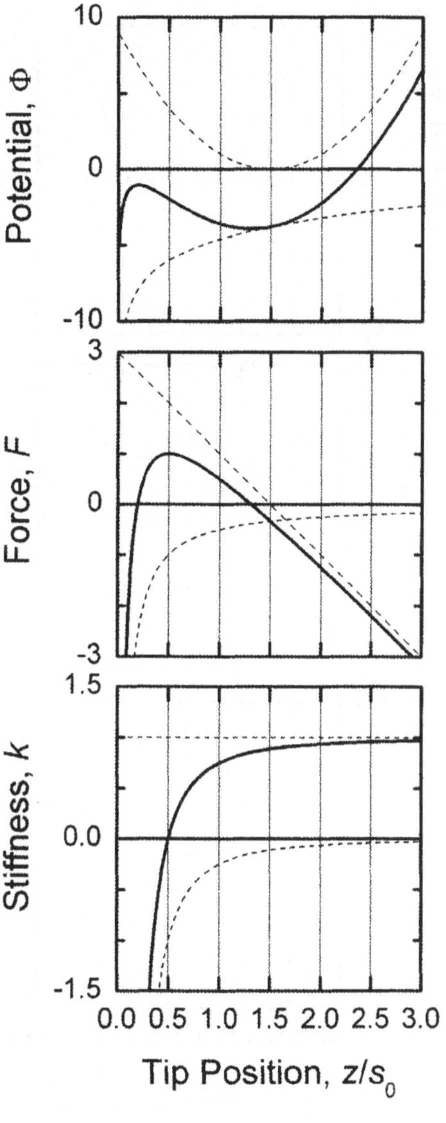

Figure 4-3. The variations in the full potential, force, and stiffness (solid lines) with tip position for combined logarithmic and harmonic potentials (dotted lines) for an imposed tip position larger than the critical tip-surface separation, such that there are two possible equilibrium tip positions. The values are normalized using the characteristic values (Eq. 4-9).

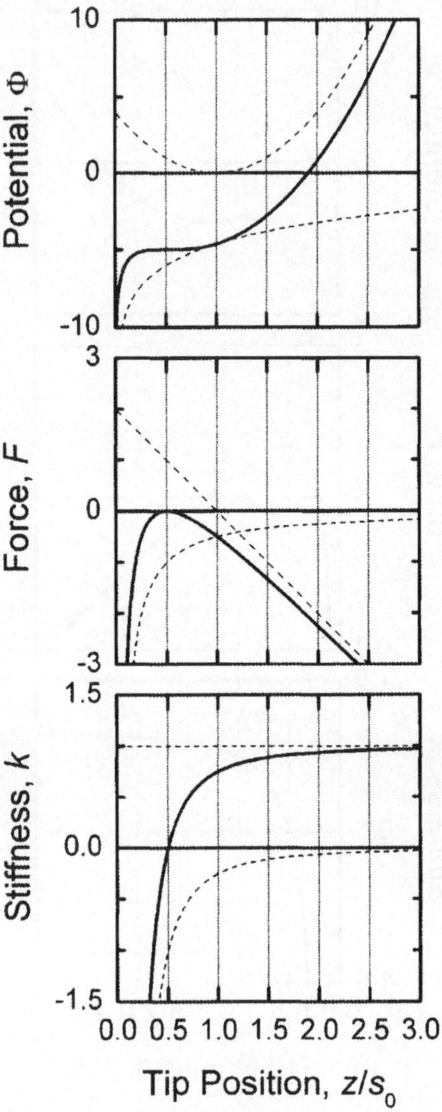

Figure 4-4. The variations in the full potential, force, and stiffness (solid lines) with tip position for combined logarithmic and harmonic potentials (dotted lines) for an imposed tip position at the critical tip-surface separation, such that there is one possible equilibrium tip position. The values are normalized using the characteristic values (Eq. 4-9).

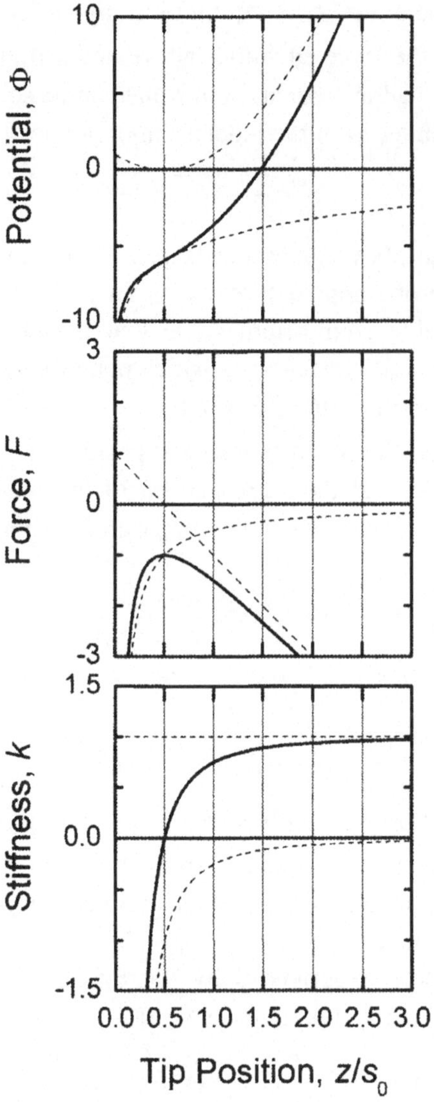

Figure 4-5. The variations in the full potential, force, and stiffness (solid lines) with tip position for combined logarithmic and harmonic potentials (dotted lines) for an imposed tip position smaller than the critical tip-surface separation, such that no equilibrium tip position is possible. The values are normalized using the characteristic values (Eq. 4-9).

For $s / s_0 > 1$, there are two equilibrium tip-surface separations, z^* (Fig. 4-3). Using the force equations above and setting the total force to zero generates a quadratic equation in z that can be solved to give the z^* values generally in terms of the indenter-imposed position:

$$z^* = \frac{s \pm (s^2 - s_0^2)^{1/2}}{2}. \tag{4-10}$$

Taking the + sign above gives the larger, stable equilibrium z^* value associated with probe approach to the surface. Taking the – sign gives the smaller unstable equilibrium value that is not accessed until the critical value of $s = s_0$ is reached, at which point the two solutions merge into a single metastable value $z^* = s_0 / 2$.

Experimentally, the indenter-imposed position, s, is the independent, controlled variable, and the *displacement* of the tip from the imposed position, $\Delta z = z^* - s$, is a dependent, measured-response variable (using the + sign):

$$\Delta z = (s_0 / 2)\left[(s^2 / s_0^2 - 1)^{1/2} - s / s_0\right]. \tag{4-11}$$

Often this displacement is represented as the force exerted by the spring, $k_S \Delta z = -F_S$:

$$-F_S = (k_S s_0 / 2)\left[(s^2 / s_0^2 - 1)^{1/2} - s / s_0\right]. \tag{4-12}$$

Noting that equilibrium requires $F_A = -F_S$ this equation enables the surface force to be determined from position-displacement, s-Δz, measurements, although as a function of imposed probe position, not tip-surface separation.

The net stiffness experienced by the tip, k, is another dependent variable, given by (again taking the + sign)

$$k = k_S \left\{1 - \left[s / s_0 + \left(s^2 / s_0^2 - 1\right)^{1/2}\right]^{-2}\right\}. \tag{4-13}$$

As noted in Chapter 2, the behavior of the stiffness field is usually not measured directly, but indirectly by changes in the resonant frequency and amplitude of vibration of the tip. In the presence of the surface force field, the net stiffness experienced by the tip changes with the equilibrium separation from the surface and hence with imposed probe position, such that (for small-amplitude oscillations) the motion about the stable equilibrium position is given by

$$z = z^* + a\sin(\omega t), \tag{4-14}$$

where a and ω are the amplitude of vibration and resonant frequency, respectively. The resonant frequency is related to the local stiffness by

$$\omega^2 = k / m^*, \tag{4-15}$$

where m^* is the effective mass of the probe. Hence, the variation in the stiffness field can be determined experimentally from the variation in the resonant frequency:

$$\omega = \omega_S \left\{1 - \left[s/s_0 + \left(s^2/s_0^2 - 1\right)^{1/2} \right]^{-2} \right\}^{1/2}. \tag{4-16}$$

Using the equipartition theorem gives the relationship between the resonant frequency and amplitude of oscillation of the tip as a function of temperature:

$$\left\langle m\omega^2 a^2 \right\rangle = \boldsymbol{kT}. \tag{4-17}$$

Hence, in thermal equilibrium the amplitude of tip oscillation should vary reciprocally with the resonant frequency:

$$a = a_S \left\{1 - \left[s/s_0 + \left(s^2/s_0^2 - 1\right)^{1/2} \right]^{-2} \right\}^{-1/2}, \tag{4-18}$$

where ω_S and a_S are the frequency and amplitude of free vibration, respectively.

The quantity $\boldsymbol{kT}$ is the natural energy scale for the environment, in this case the indenter. The natural energy scale for the system is the energy of deformation contained in the spring at the point of system instability. The ratio of these two energy scales, ~ $\boldsymbol{kT}/A$, provides a measure of whether thermal, $\boldsymbol{kT}/A \geq 1$, or mechanical, $\boldsymbol{kT}/A \leq 1$, effects are likely to dominate the behavior of the system. This becomes important as the system approaches the instability point. The behavior of the system in terms of the externally measured force, resonant frequency, and thermal amplitude of vibration are shown in Fig. 4-6; the instability point of the system at $s = s_0$ is clear in all three quantities.

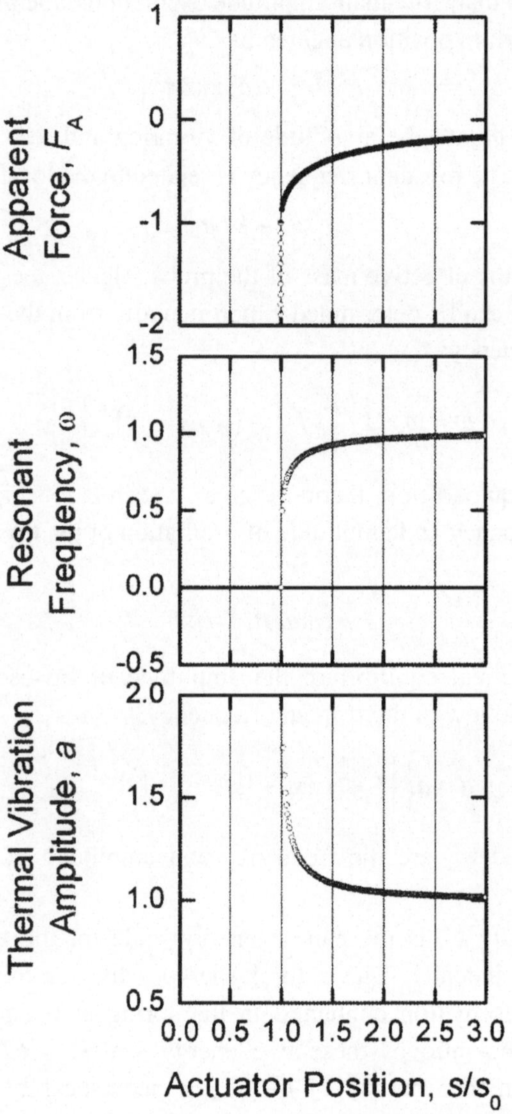

Figure 4-6. The variations in the apparent force exerted by a surface on a probe, the resonant frequency of the probe, and the thermally-induced amplitude of vibration of the tip of the probe as a function of the actuator-imposed position of the probe relative to the surface. The values are normalized using the characteristic values (Eqs. 4-12, 4-16, and 4-18).

The adhesion sequence may now be considered in terms of the two experimental variables: For indenter-imposed positions far from the surface, $s \gg s_0$, and Δz has two real solutions, one describing the stable equilibrium separated state and another describing an un-physical unstable equilibrium state. The stable state is characterized by $|\Delta z / s| \ll 1$; the tip is barely perturbed from its imposed position. The unstable state is characterized by $1 - |\Delta z / s| \ll 1$; the tip is very close to the surface (but for an adhesion sequence in which the indenter moves the probe to approach the surface, this state is not physically accessible). Far from the surface, the stiffness field of the tip is dominated by that of the restoring influence of the spring and for unforced thermal motion the tip vibrates with a frequency and amplitude determined by the spring stiffness and temperature.

For indenter-imposed positions closer to the surface, but still greater than the critical value, $s > s_0$, Δz again has two real solutions. However, although the stable state is characterized by $|\Delta z / s| < 1$, the tip is significantly perturbed from its imposed position. The stiffness field experienced by the tip is considerably reduced as the attractive influence of the surface competes with the restoring influence of the spring. As a consequence, the frequency of vibration of the tip decreases and the amplitude of vibration increases. The unstable state is now characterized by $1 - |\Delta z / s| < 1$ and although closer to the surface than the stable state, is now much closer to the stable state.

For indenter-imposed positions approaching the critical value, $s \to s_0$, Δz still has two real solutions but they are both converging on the critical value, $|\Delta z| \to s_0 / 2$. In this position, the tip experiences near to vanishing stiffness (as the surface field almost cancels the spring field), the frequency of vibration of the tip about the larger, stable equilibrium position is very small, and the amplitude of vibration is very large. As a consequence, the tip may move close enough to the surface so as to reach beyond the smaller, unstable equilibrium position. If it does so, the net force on the tip becomes negative and the tip is irreversibly attracted to the surface in a non-equilibrium manner with greater and greater (negative) force. This is the adhesive "snap-on" phenomenon. In the model here the tip comes to rest asymptotically on the surface at zero separation. In reality, there is a repulsive interaction that determines a non-zero separation (this will be considered in the next section). If the temperature is low enough the vibration of the tip about the stable equilibrium position may be insignificant and the critical equilibrium condition $s = s_0$ may be reached at which $F = k = 0$. Infinitesimal motion of the indenter such that $s < s_0$ is imposed leads to $F < 0$ for all z and the tip irreversibly snaps on to the surface as before.

In summary, adhesion between the two bodies under consideration here, the probe and the surface, occurred when the probe position imposed by the environment, in this case the indentation apparatus, reached a snap-on point. This point characterized a critical equilibrium configuration in which both the force and stiffness fields of the combined surface attraction and probe spring restoring influence were balanced. Mechanical perturbation of the system (by further imposed motion of the probe towards the surface) or thermal perturbation (by oscillation of the probe tip) led to a dominance of the surface attraction over the spring resistance and irreversible motion of the tip to the surface such that the two bodies were adhered. The imposed probe position at the onset of the snap-on process was determined by the *ratio* of the parameters characterizing the surface attraction and spring constant, $s = s_0 = 2(A / k_S)^{1/2}$, and the snap-on force was determined by the *product* of these parameters, $F_S(s = s_0) = (Ak_S)^{1/2}$. These simple expressions are a consequence of the simple system chosen here to illustrate the snap-on instability phenomenon. Greater analytic complication will result for systems with less analytically-tractable surface potentials and for which both the probe and the surface are deformable (and maybe non-linearly or irreversibly) but the physics will remain the same: A critical length scale in the system is determined by a vanishing stiffness that depends on a competition between surface attraction and elastic restoring forces. The next section will illustrate this principle for the de-adhesion or separation process in which the adhered bond is fractured.

2. Adhesion and Separation: The "Snap-On" and "Pull-Off" Instabilities

In this section, the complete process of adhesion and separation is considered: As before, the system consists of two bodies, initially in a state of separated stable equilibrium, that are brought together by an external influence such that the system passes through an unstable state—the "snap-on" instability considered in the previous section—to a new stable equilibrium adhered state. Here, however, we consider the external influence to continue to bring the bodies together, reducing their separation such that the system may even pass into a state of stable "indentation" compression, as opposed to the continuous tension considered in the previous section. After imposing a separation of closest approach, the external influence then reverses direction and begins to increase the separation of the bodies such that the system passes through a new unstable state—the "pull off" instability to be considered here—to the original stable equilibrium separated configuration. The approach and separation instabilities lead to non-equilibrium motion of the tip and do not occur under the same conditions; as a consequence the contact cycle is hysteretic and the energy lost is the contact "work of adhesion."

The concepts introduced in the previous section to describe and determine snap-on adhesion characteristics, particularly the importance of the net force and stiffness fields in determining equilibria and stability, will again be used to describe the complete adhesion and pull-off separation process. In the previous section a surface interaction potential was used that provided simple closed-form expressions for critical parameters, thus illustrating the snap-on process. Here, a more physically realistic interaction potential is used that, although somewhat less analytically tractable, provides aspects absent from the previous potential: a well-defined adhered state; the possibility of indentation; an intrinsic length scale; and the possibility of either stability over the complete interaction range or dual instability states corresponding to the snap-on and pull-off processes.

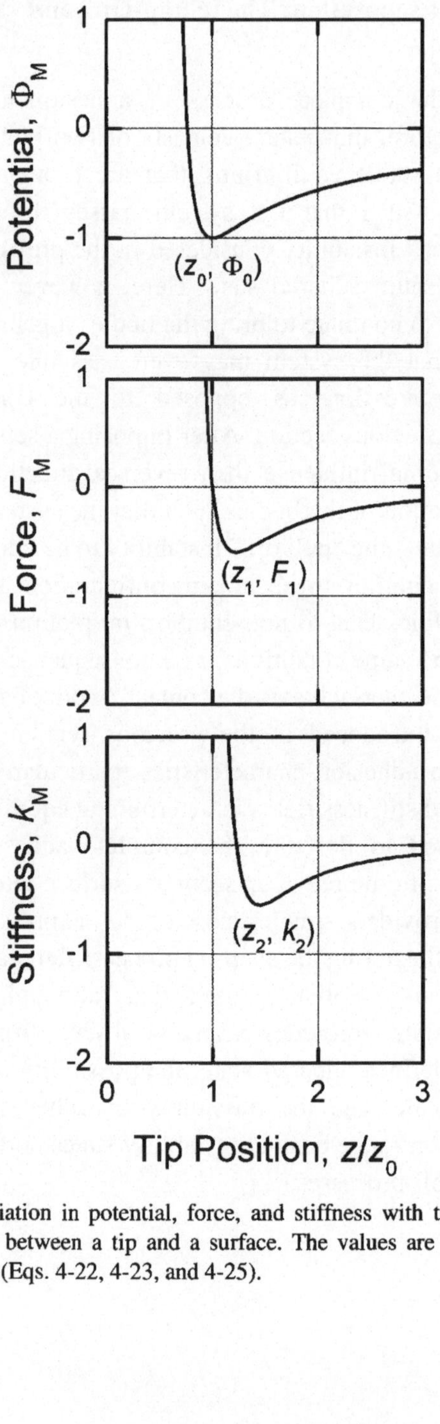

Figure 4-7. The variation in potential, force, and stiffness with tip position for a Mie potential (Eq. 4-19) between a tip and a surface. The values are normalized using the characteristic values (Eqs. 4-22, 4-23, and 4-25).

The interaction between the two bodies, once again envisaged as a surface and a probe, is described here by a Mie potential, $\Phi_{\mathrm{M}}(z)$:[2]

$$\Phi_{\mathrm{M}} = -\frac{A}{z^n} + \frac{B}{z^m}, \tag{4-19}$$

where A, B, n, m, $z > 0$. The simple, generalized form of Eq. 4-19 leads to frequent use of the Mie potential in considerations of the behavior of solids, in which both attractive and repulsive interactions exist between atoms, molecules, and surfaces. There is, as we shall see, analytic convenience in the fact that both terms are power-laws. The behavior of Φ_{M} is shown in Fig. 4-7. z characterizes the separation of the surface and the probe tip, although it will be shown that the term "separation" will require some refinement. As before z is taken as positive directed outward from the surface and $z = 0$ defines one boundary of the system. Both terms in the Mie potential go to zero as $z \to \infty$, that is, there is no potential for the interaction between the surface and the probe to perform work when they are widely separated. The first (A) term describes an attraction between the surface and the probe and may arise from Coulomb interactions (n = 1 to 3), fixed dipole interactions (n = 2 to 4), covalent bonding interactions (n = 1 to 3), or fluctuating dipole (van der Waals) interactions (n = 6 to 8). The values of the amplitude term, A, and the exponent, n, depend on the nature of the interaction and the geometry of the probe tip and surface. The second (B) term describes repulsion between the surface and the probe and may arise from Pauli exclusion or antibonding interactions (m = 9 to 11) or fluctuating dipole repulsion (m = 12 to 14). (The repulsive terms B and m are largely empirical. The Lennard-Jones potential for van der Waals interactions is a "6-12" Mie potential.) As $m > n$, at large separations the long-range attraction dominates and at small separations the short-range repulsion dominates; the full potential passes through a minimum characterizing a stable equilibrium at an intermediate separation. The force, $F_{\mathrm{M}}(z)$, and stiffness, $k_{\mathrm{M}}(z)$, fields associated with the Mie potential are

$$F_{\mathrm{M}} = -\frac{\mathrm{d}\Phi_{\mathrm{M}}}{\mathrm{d}z} = -\frac{nA}{z^{n+1}} + \frac{mB}{z^{m+1}} \tag{4-20}$$

and

$$k_{\mathrm{M}} = -\frac{\mathrm{d}F_{\mathrm{M}}}{\mathrm{d}z} = -\frac{n(n+1)A}{z^{n+2}} + \frac{m(m+1)B}{z^{m+2}} . \tag{4-21}$$

The behavior of these force and stiffness fields is also shown in Fig. 4-7.

Consideration of Φ_{M}, F_{M}, and k_{M} in Fig. 4-7 here and Φ_{A}, F_{A}, and k_{A} in Fig. 2-1 in Chapter 2 shows a similarity and a key difference. The similarity is that the potential, force, and stiffness curves in both cases are at first sight just similarly-shaped, shifted (to larger z) versions of each other; this similarity of shape is obvious here in the similarity of the algebraic representation of the three quantities (Eqs. 4-19 to 4-21). The key difference between the Φ_{M} quantities and the Φ_{A} quantities is that the Mie potential includes an intrinsic length, and thus energy, force, and stiffness scales. In the previous case, a length scale was only introduced into the system through interaction between the montonically attractive Φ_{A} and the spring potential, Φ_{S}. Here, interaction between the attractive (A) and repulsive (B) terms leads to a intrinsic length scale for the system in the absence of the spring. Setting $F_{\mathrm{M}} = 0$ provides the equilibrium surface-tip separation, z_0:

$$z_0 = \left(\frac{mB}{nA}\right)^{1/(m-n)} . \tag{4-22}$$

The potential minimum, Φ_0, at $z = z_0$ is given by

$$\Phi_0 = -\frac{A}{z_0^n}\left(\frac{m-n}{m}\right) . \tag{4-23}$$

Consideration of Fig. 4-7 provides insight into the physical meaning of z_0, "indentation," and "separation." At location $z = z_0$ the tip is essentially "on" the surface and $F_{\mathrm{M}} = 0$. For $z < z_0$, $F_{\mathrm{M}} > 0$, the surface repels the tip and a negative applied force would be required to maintain the tip in equilibrium in this state; the tip is indenting the surface. For $z > z_0$, $F_{\mathrm{M}} < 0$, the surface attracts the tip and a positive applied force would be required to maintain the tip in equilibrium in this state; the tip is separating from the surface. Hence physical "separation" of the tip and the surface with this form of potential strictly refers to the distance $z - z_0$.

This interpretation is illustrated in Fig. 4-8, which replots the force variation of Fig. 4-7 as a function of this distance.

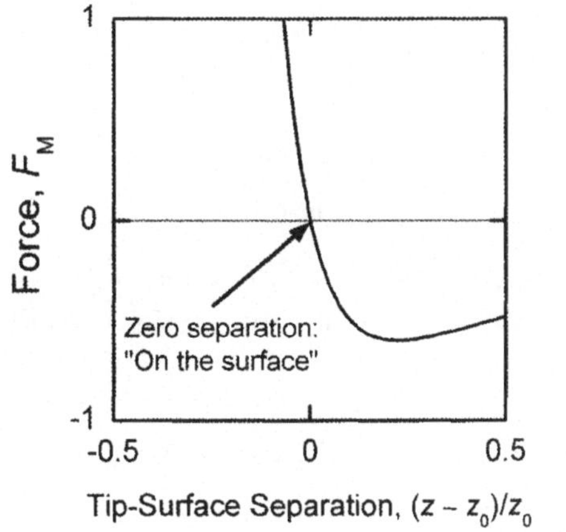

Figure 4.8. An expanded view of the force between a surface and a tip around the $F_M = 0$ point of Fig. 4-7; the distance between the tip and the surface is plotted as a function of separation from this equilibrium position.

Important as the equilibrium configuration (z_0, Φ_0), defined by the extremum in the potential Φ_M, is for setting the boundary between indentation and separation states, the previous section showed that it is extrema in the force and stiffness fields that are important for determining equilibrium and stability behavior. In this case, extrema in F_M and k_M place bounds on external influences, most importantly the probe spring, that can give rise to snap-on and pull-off behavior. Setting $k_M = 0$ provides the configuration (z_1, F_1) in which the surface exerts the greatest attraction (numerically minimum force, $F_1 < 0$) on the tip:

$$z_1 = z_0 \left(\frac{m+1}{n+1} \right)^{1/(m-n)} \tag{4-24}$$

$$F_1 = \frac{\Phi_0}{z_0} \frac{mn}{m+1} \left(\frac{z_0}{z_1} \right)^{n+1} . \tag{4-25}$$

Setting $\mathrm{d}k_M / \mathrm{d}z = 0$ provides the configuration (z_2, k_2) in which the surface stiffness field has the least value (numerically minimum stiffness, $k_2 < 0$):

$$z_2 = z_1 \left(\frac{m+2}{n+2} \right)^{1/(m-n)} \tag{4-26}$$

$$k_2 = \frac{F_1}{z_1} \frac{(m+1)(n+1)}{m+2} \left(\frac{z_1}{z_2} \right)^{n+2} . \tag{4-27}$$

It is interesting to compare this last value with the stiffness of the surface-tip interaction, k_0, at the $z = z_0$ equilibrium position, $k_0 = -mn\Phi_0 / z_0^2 > 0$;

$$k_2 = -k_0 \frac{n+1}{m+2} \left(\frac{z_0}{z_2} \right)^{n+2} . \tag{4-28}$$

The quantity $-k_2$ will be seen to be a critical value in determining the behavior of the system. (As noted above, the Mie potential is often used to represent interatomic interactions, in which case the characteristic interatomic bond quantities discussed in Chapter 2 are identified with the intrinsic length, stiffness, and force values above: d with z_0, k_{bond} with k_0, and $F_{\mathrm{max,\,bond}}$ with F_1.)

Once again, the probe will be taken to comprise the tip, interacting with the surface via Φ_M, attached to a spring, described by a harmonic potential, Φ_S, with indenter-imposed position s. In the previous section, equilibrium of the tip under these combined influences was shown to be determined by zero net force on the tip, in this case $F = F_M + F_S = 0$, and stability determined by positive net stiffness, in this case $k = k_M + k_S > 0$. Consideration of the negative minimum in k_M in Fig. 4-7 shows that two possibilities for stability behavior can occur. In the first, super-critical condition, $k_S \geq -k_2$ and the stability condition $k \geq 0$ is always fulfilled. That is, for all positions of the probe imposed by the indenter, s, there is an associated single, stable equilibrium position for the tip, z^*. This is shown in Fig. 4-9, in which the variation in $F(z)$ for $s/z_0 = 2.5$, $s/z_0 = 1.5$, $s/z_0 = 1.0$, $s/z_0 = 0.5$ for a super-critical spring stiffness of $k_S = 1.3k_2$.

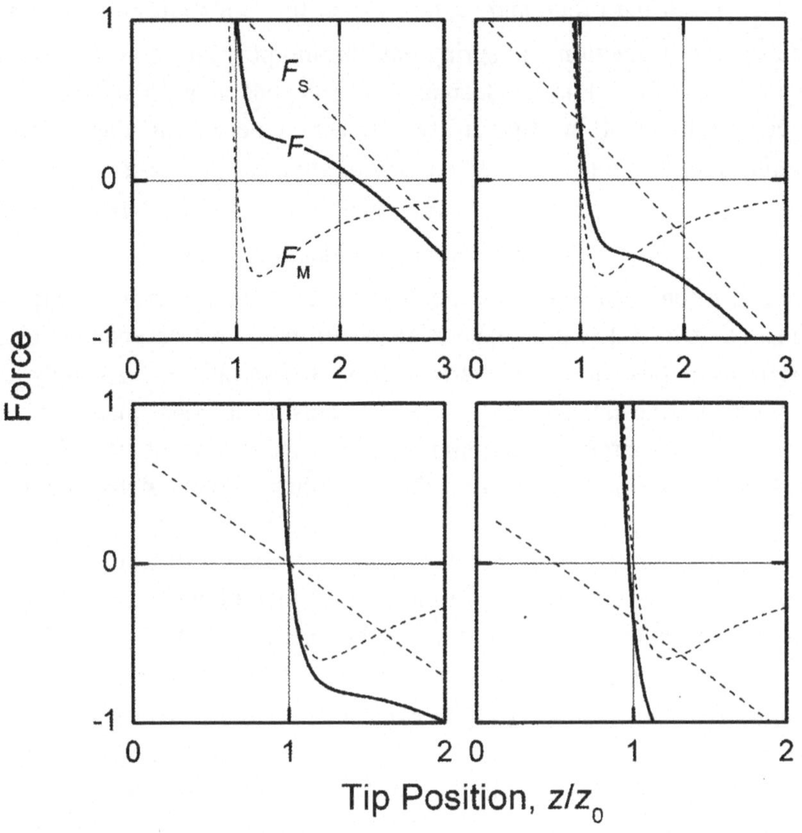

Figure 4-9. The variations in the net force (solid line) with tip position for combined Mie (M) and harmonic (S) interactions (dotted lines) between a probe tip and a surface as an actuator moves the probe towards the surface. The probe spring stiffness is super-critical such that a stable equilibrium position of the tip is possible for all imposed probe positions. The values are normalized using the characteristic values (Eqs. 4-22 and 4-25).

For s/z_0 = 2.5, the attractive force of the tip-surface interaction perturbs the tip to an equilibrium position $z^* < s$ (the $F = 0$ position in Fig. 4-9), but the perturbation is not large. As s is decreased to $s/z_0 = 1.5$, the surface attraction becomes greater and the separation between the imposed probe position and the tip position becomes greater; in fact $z^* / z_0 \approx 1$ under these conditions. As s is decreased even further, the tip position remains almost invariant as $z^* / z_0 \to 1$, until at s/z_0 = 1.0,

$z^* / z_0 = 1$ and the probe and surface are at the quiescent point at which the surface interaction and spring equilibrium positions coincide: the tip is on the surface. Further decreases in s beyond this point lead to the repulsive force of the tip-surface interaction perturbing the tip to an equilibrium position $z^* > s$ and the perturbations can be large as $z^* / z_0 \approx 1$. Hence for $s / z_0 > 1$, the tip is experiencing an adhesive interaction and for $s / z_0 < 1$, the tip is experiencing an indentation interaction. Note that the z-position of the inflection point of the total force F curve in Fig. 4-9 remains invariant with changes in s. In fact, the derivatives at all locations on this curve are invariant with s and variations in s simply lead to vertical translations of this curve. This is a consequence of the linearity of the probe spring; similar derivative invariance was observed in the earlier considerations of the logarithmic potential for the same reason.

The separation of the adhesive and indentation interactions at the quiescent point is made clear in Fig. 4-10, which plots the force-actuator position behavior as observed at the boundaries of the system for the $k_S = 1.3k_2$ system above—this is what would be observed in a typical atomic force microscopy (AFM) experiment. Note that in this case the perceived shape of the force interaction is distorted from that given in Figs. 4-7 and 4-9 by the addition of the displacement of the spring. As noted in Chapter 2, for most macroscopic indentations, nanoindentation of engineering materials with diamond probes, and force spectroscopy using stiff atomic force microscope cantilevers, $k_S \gg k_0$ and thus $k_S \gg |k_2|$ and the sequence above occurs during a contact. In fact, k_S is usually so large that the adhesive approach segment is not even observed as shown in Fig. 4-10 for the case $k_S = 100k_2$—this is what is observed in a typical nanoindentation experiment. Also shown in Fig. 4-10 is a magnification of the response around the quiescent point for the $k_S = 100k_2$ case; the maximum adhesive force exerted by the surface on the tip is identical to that exerted in the $k_S = 1.3k_2$ case, but is negligible relative to the force exerted during the subsequent indentation process. The very stiff probe, however, contributes negligibly to the displacement of the system and thus the perceived force interaction between the tip and

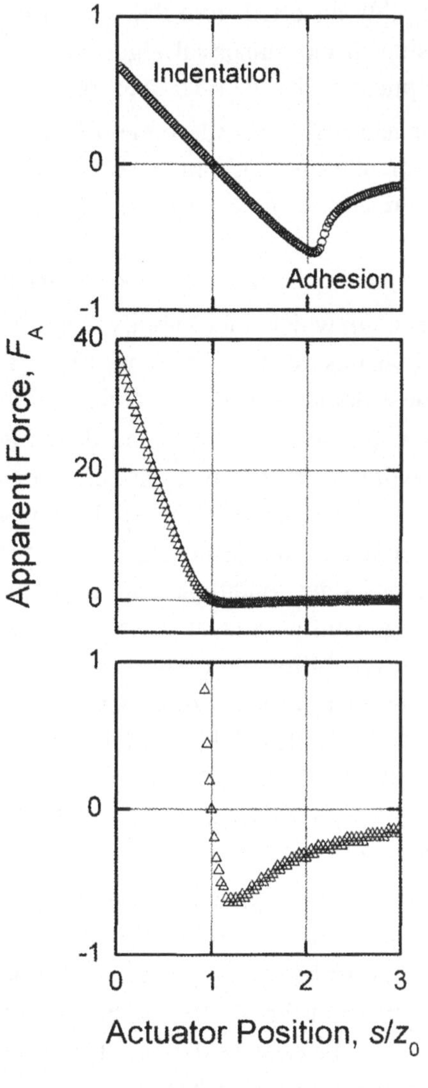

Figure 4-10. The variations in the apparent force exerted by a surface on a probe as a function of the actuator-imposed position of the probe relative to the surface. (Upper) The probe spring is barely super-critical leading to a large compliance range; (Center) The probe spring is extremely super-critical such that the indentation forces dominate the approach forces; (Lower) Magnification of the center graph, showing that the peak force is unaltered by the probe spring stiffness. Values are normalized using the characteristic values (Eqs. 4-22 and 4-25).

the surface is negligibly distorted from that shown in Figs. 4-7 and 4-9. The lack of precision in the numerical algorithm used to solve for the $F_A(s)$ responses replicates the thermal and other noise present in a contact experiment and highlights the trade-off in using stiff probes to investigate the form of tip-surface interaction forces: A stiffer probe usually provides greater accuracy for the force response but less precision.

A second, sub-critical condition, $k_S < -k_2$, is usually encountered in AFM measurements, in which the stability condition $k \geq 0$ is fulfilled only for limited domains of the probe position, s, imposed by the indenter. Within these domains there are stable equilibrium positions for the tip, z^*. However, the domains are bounded by instability conditions that lead to non-equilibrium tip motion between stable equilibrium configurations. As a consequence, dual equilibrium configurations are possible within the domains and the configuration of the system depends not just on the external state conditions but on the path imposed by the external constraint. A configurational dependence on path rather than state leads to irreversibility and thus indentation measurements performed with sub-critical probe springs give rise to adhesive contact hysteresis. This is shown in Figs. 4-11 and 4-12, in which the variation in $F(z)$ for $s/z_0 = 3.5$, $s/z_0 = 2.818$, $s/z_0 = 2.5$, and $s/z_0 = 1.0$ (Fig. 4-11) and $s/z_0 = 1.0$, $s/z_0 = 2.5$, $s/z_0 = 2.693$, and $s/z_0 = 3.5$ (Fig. 4-12) for a sub-critical spring stiffness of $k_S = 0.7k_2$.

Figure 4-11 displays the approach segment that leads to a snap-on instability and adhesion: For $s/z_0 = 3.5$, the attractive force of the tip-surface interaction perturbs the tip to a stable equilibrium position $z^* < s$ (the $F = 0$ position in Fig. 4-10), but the perturbation is not large, just as in the super-critical case considered in Fig. 4-9. Also, as before, as s is decreased the surface attraction becomes greater and the separation between the imposed probe position and the tip position becomes greater. However, in this case a critical condition is reached at $s^* / z_0 = 2.818$ for which the tip is at a condition of incipient instability. Thermal perturbation or any further incremental decrease in s lead to the tip moving via a non-equilibrium ($F \neq 0$) path to "snap on" to the surface at a new stable equilibrium position considerably displaced (towards the

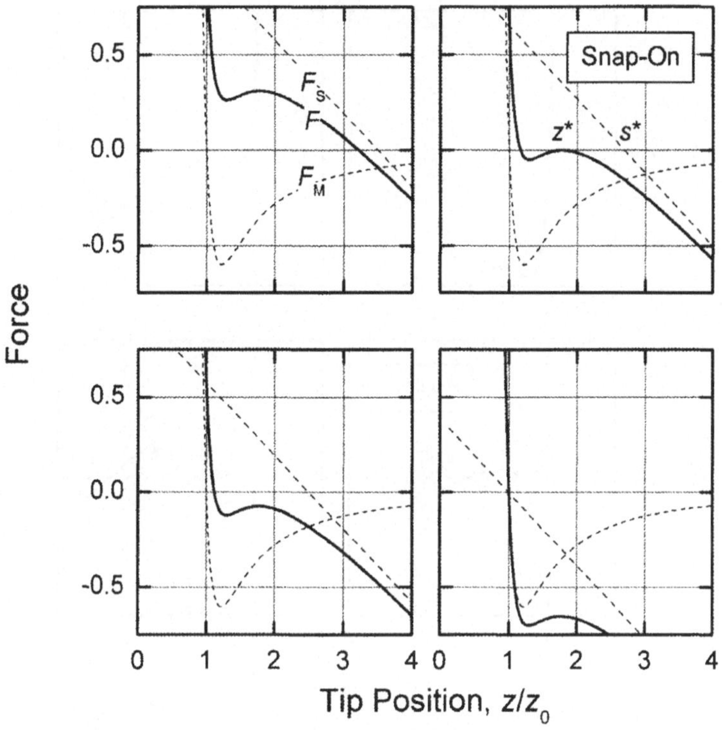

Figure 4-11. The variations in the net force (solid line) with tip position for combined Mie (M) and harmonic (S) interactions (dotted lines) between a probe tip and a surface as an actuator moves the probe towards the surface. The probe spring stiffness is sub-critical such that stable equilibrium positions of the tip are possible only for limited domains of imposed probe position. The domains are separated by a non-equilibrium "snap-on" configuration. The values are normalized using the characteristic values (Eqs. 4-22 and 4-25).

surface) from the instability point. As in the super-critical case, further decrease in s through s/z_0 = 2.5 leads to almost invariant tip position as $z^*/z_0 \rightarrow 1$, until at s/z_0 = 1.0, $z^*/z_0 = 1$ and the probe and surface are at the quiescent point. Further decreases in s beyond this point lead to an indentation interaction as before.

Figure 4-12 displays the retraction segment that leads to a pull-off instability and separation and begins with the system at the quiescent point of s/z_0 = $1.0, z^*/z_0 = 1$. As s is increased to s/z_0 = 2.5 the tip

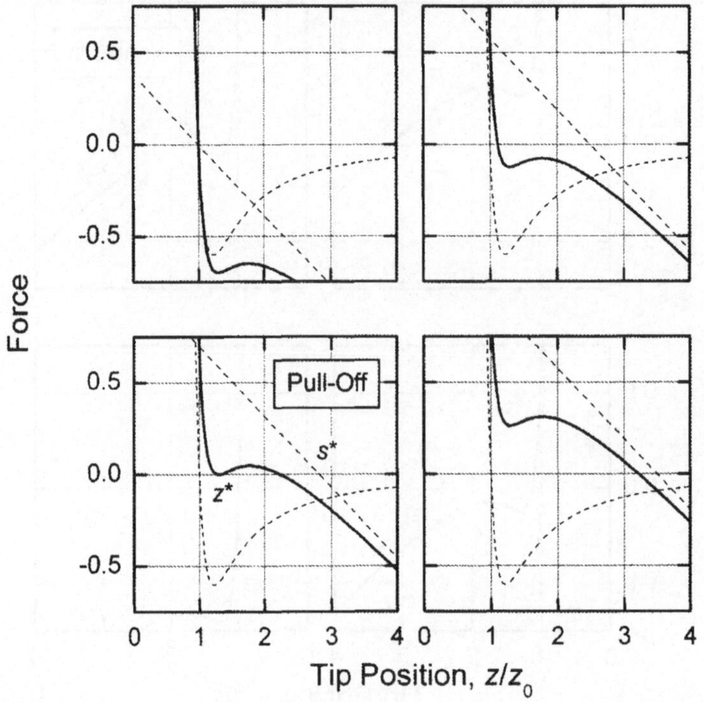

Figure 4-12. The variations in the net force (solid line) with tip position for combined Mie (M) and harmonic (S) interactions (dotted lines) between a probe tip and a surface as an actuator moves the probe away from the surface. The probe spring stiffness is sub-critical such that stable equilibrium positions of the tip are possible only for limited domains of imposed probe position. The domains are separated by a non-equilibrium "pull-off" configuration. The values are normalized using the characteristic values (Eqs. 4-22 and 4-25).

remains in stable equilibrium at almost invariant tip position of $z^* / z_0 \approx 1$. Further increases in s lead to a critical condition at $s^* / z_0 = 2.693$ for which the tip is once again in a condition of incipient instability. However, in this case, thermal perturbation or any further incremental increase in s lead to the tip moving via a non-equilibrium ($F \neq 0$) path to "pull off" the surface to a new stable equilibrium position considerably displaced (away from the surface) from the instability point. Further increases in s lead to stable incremental motion of the tip away from the surface until at s/z_0 = 3.5, the system has returned to its original state. We note once again that the z-positions of the extrema and all the derivatives of the total force F curve in Figs. 4-11 and 4-12 remain invariant with changes in s as a consequence of the linearity of the probe spring.

The irreversibility of the snap-on approach and pull-off retraction segments of the full contact cycle is made clear in Fig. 4-13, which plots the force-actuator position behavior for the $k_S = 0.7k_2$ system above, along with those for $k_S = 0.5k_2$ and $k_S = 0.4k_2$ systems The hysteresis loops are obvious in all three cases. The symbols represent conditions of stable equilibrium for the system and the (near) vertical lines bounding the hysteresis loops represent the non-equilibrium paths taken by the system during the snap-on (left) and pull-off (right) processes. A typical sub-critical AFM adhesive indentation contact cycle thus consists of (1) a reversible approach-retract segment, followed by (2) instability and an irreversible snap-on segment to an adhered, reversible "compliance" segment that leads through the quiescent point (3) to (4) an indentation segment experiment, which, on actuator reversal, passes back through the indentation, quiescent point (5), and compliance segments to (6) instability and an irreversible pull-off instability to (1) the reversible approach-retract segment. As the stiffness of the probe spring is decreased, the snap-on and pull-off segments occur further from the surface and the size of the hysteresis loop increases. These segments are labeled in the $k_S = 0.5k_2$ graph of Fig. 4-13 and in the schematic diagrams of Fig. 4-14.

From an engineering perspective, attention is usually focused on the *separation* segment of the contact cycle: the amount of force that must be applied or work that must be done to separate two adhered bodies. These two quantities are illustrated in Fig. 4-15. The pull-off force is the maximum force that the surface can exert on the tip and hence application of any force greater than this will lead to separation of the contact. The magnitude of the pull-off force depends only the nature of the tip-surface contact and is invariant with the stiffness of the probe spring, as may be seen by comparing the super-critical and sub-critical force-position curves in Figs. 4-9 and 4-13. Introducing a compliant element between the contacting elements (tip-surface or any two bodies)

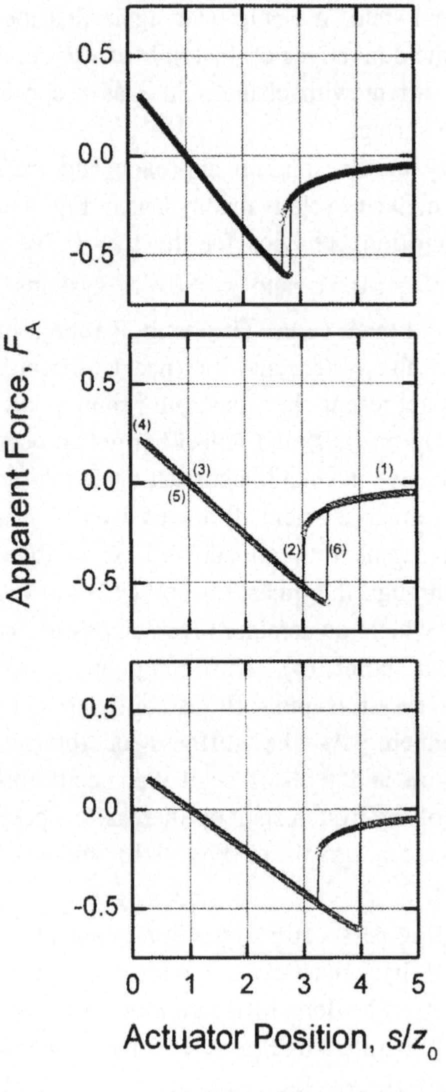

Figure 4-13. The variations in the apparent force exerted by a surface on a probe as a function of the actuator-imposed position of the probe relative to the surface. The probe springs are sub-critical, leading to hysteresis loops in the contact cycle that are bounded by "snap-on" and "pull-off" instabilities on approach and retraction, respectively. The hysteresis loops are larger and occur further from the surface as the probe spring stiffness is decreased (Upper to Lower). Values are normalized using the characteristic values (Eqs. 4-22 and 4-25).

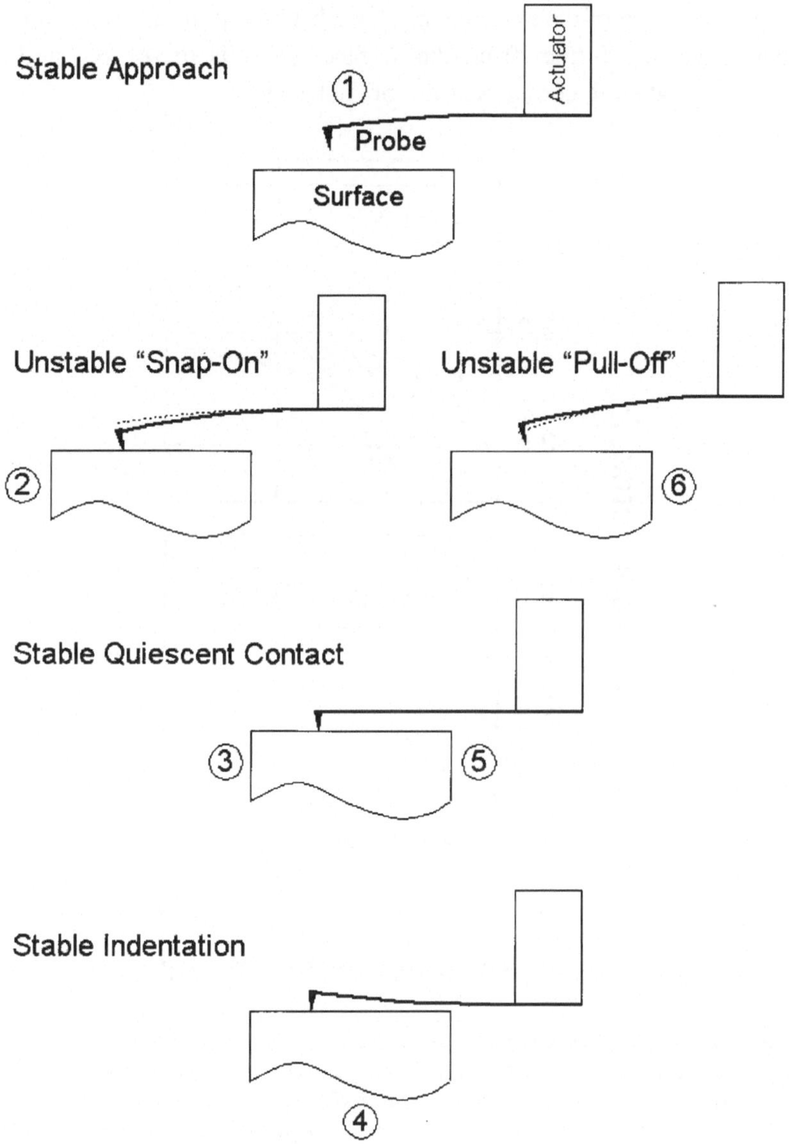

Figure 4-14. Schematic diagram of the approach, "snap-on," contact, indentation, and "pull-off" segments for an actuator displacing an AFM cantilever probe of sub-critical stiffness relative to an adhesive surface. The incipient instability configurations for the snap-on and pull-off events are shown as dotted lines in segments (2) and (6).

and the boundary of the system simply leads to the pull-off (and snap-on) events occurring further from the surface. For this reason the pull-off force is a good engineering measure of "adhesion."

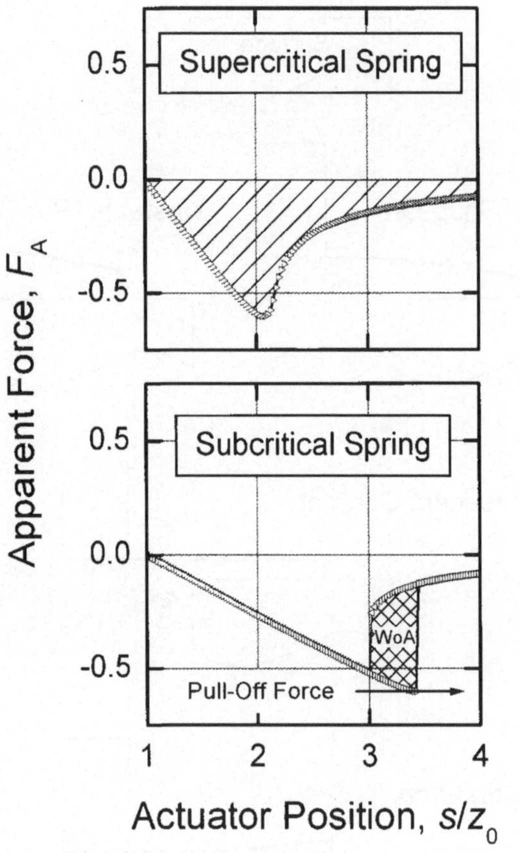

Figure 4-15. The variations in the apparent force exerted by a surface on a probe as a function of the actuator-imposed position of the probe relative to the surface. (Upper) The probe spring is super-critical leading to a reversible work of separation defined by the hatched area; (Lower) The probe spring is sub-critical leading to an irreversible work of adhesion defined by the hatched area. Values are normalized using the characteristic values (Eqs. 4-22 and 4-25).

The work that must be done to separate two bodies is *per se* easily specified: simply integrate the force-position response over a specified path from a selected adhered state to a selected separated state. For a supercritical contact the adhered state is obviously selected as the quiescent point and the separated state is infinite separation. The work of separation, represented by the hatched area in Fig. 4-15 for the $k_S = 1.3k_2$ case, is thus easily determined. However, such work obviously depends on the stiffness of the spring as can be observed by comparing the areas under the curves in Fig. 4-10: Less work is expended in extending a stiffer spring during separation, giving rise to a smaller value for the work of separation. The work of separation in this context is clearly a system parameter and not a quantity that characterizes the materials and their interactions that form the tip and surface. In the limit of a rigid spring, the interaction force-separation curve is recovered (approximated by the $k_S = 100k_2$ curve in Fig. 4-10, say) and the integral gives the work as Φ_0, the characteristic binding potential. This quantity, quoted per unit area of surfaces separated is often referred to (somewhat confusingly) as the "work of adhesion" (strictly a work per unit area) or the (also confusingly named) "surface energy."

For a sub-critical contact, the work of separation can obviously be determined in the same way by integration of the force-position response from the quiescent point to infinite separation. (As work is a path quantity, whether the path is reversible or irreversible is irrelevant, providing the path is specified.) Once again, the stiffness of the spring alters the force-position response and more compliant springs give rise to greater work of separation as indicated in Fig. 4-13. A measure of the full hysteretic adhesion-separation process for sub-critical systems is the area contained within the hysteresis loop, indicated by the cross-hatched area in Fig. 4-15. The work represented by this area is the energy lost by the actuator in moving the system through the sequence of reversible and irreversible segments of the contact cycle. Although still dependent on the stiffness of the spring in the system, it is a good measure of the "work of adhesion" as it represents both the adhesive and separation processes during the contact. In the next sections we will consider a very simple sub-critical system in which the pull-off force, work of adhesion, and surface energy are all related through an assumed contact geometry.

3. Adhesion and Indentation Coordinate Systems

In the previous two sections, attention has been focused on the adhesive and separation interactions between two bodies, a surface and a probe, with the recognition that the *indentation* interaction occurs between these two interactions in a contact sequence. In consideration of the adhesive and separation interactions it was convenient to use a coordinate system with force and displacement taken as positive when directed outward from the surface. In indentation processes this is not so convenient. In this section we consider the relationship between the coordinate system used so far in consideration of adhesion and separation and the coordinate system more conventionally used in consideration of indentation.

The force interaction between the surface and the probe, say described by the Mie potential, F_M, can be balanced in (neutral) equilibrium by an applied force, F_{app}, say exerted by an indenter:

$$F_M + F_{app} = 0 \tag{4-29}$$

or, more physically,

$$F_{app} = -F_M . \tag{4-30}$$

A diagram of these forces and this force balance is shown in Fig. 4-16. Provided the force applied by the indenter matches that exerted by the surface on the probe, the system can be maintained in stable equilibrium for any tip location, z. In particular, for $z < z_0$, $F_{app} < 0$ and this describes an indentation process. Obviously, when attention is focused on indentation processes, especially if surface interactions are weak or negligible, and the probe spring is very stiff such that the equilibrium tip position differs very little from the imposed position, s, this double negative coordinate system is not very convenient. Hence, contact mechanics systems are usually described by a coordinate system in which force and position are taken as positive when directed inwards from the surface, into the indented material.[3] The reference configuration, in which both the applied load and conjugate load-point displacement are zero, is taken as the quiescent point. In terms of the variables above, the load, P, and displacement, h, in this coordinate system are given by

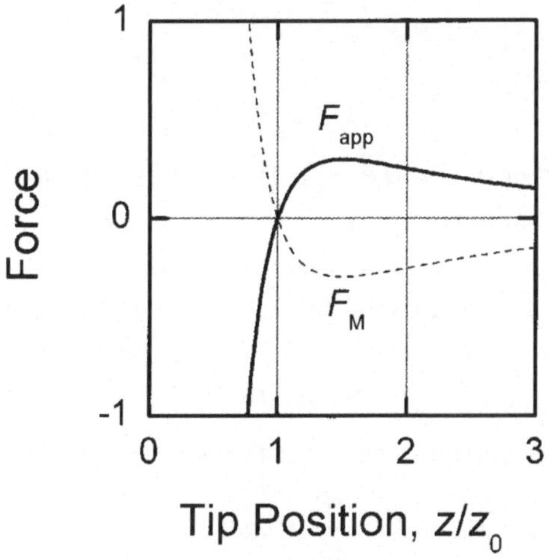

Figure 4-16. The variation in an applied force on a probe (solid line) that exactly balances that from a Mie force (dotted line) between a probe tip and a surface as a function of tip position. The net force is zero such that stable equilibrium positions of the tip are possible for all imposed probe positions. The values are normalized using the characteristic values (Eqs. 4-22 and 4-25).

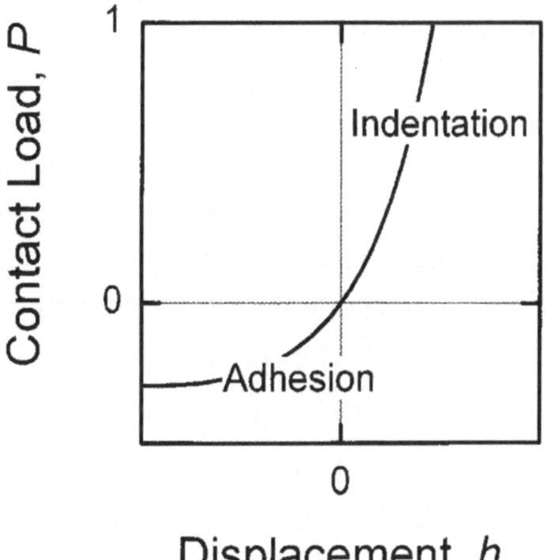

Figure 4-17. Indentation and adhesive interactions between a probe and a surface plotted in contact mechanics coordinates for which forces and displacements are taken as positive directed into the surface. The relationship between these coordinates and those of Fig. 4-16 is given by Eqs. 4-31 and 4-32.

$$P = -F_{app} \tag{4-31}$$

and

$$h = -s\,. \tag{4-32}$$

The indentation process for the Mie potential in (P, h) coordinates is shown in Fig. 4-17; adhesive processes are not excluded in this notation, they simply occupy the $P < 0$ and $h < 0$ quadrant.

Indentation of a flat surface by the flat punch considered in Chapter 2 is thus described by

$$P = 2aMh \tag{4-33}$$

in contact mechanics coordinates, and the contact stiffness, k, is given by

$$k = \frac{\mathrm{d}P}{\mathrm{d}h} = 2aM \tag{4-34}$$

where a is the punch radius and M is the indentation modulus. Flat punch indentation is a linearly-elastic system as the stiffness is invariant. Nonlinear systems, in which the stiffness varies with indentation load or displacement, include the Mie system considered above in which the non-linearity derives from the non-linearity of the interaction forces assumed to exist between the materials of the tip and surface; the materials are non-linearly elastic. Nonlinear indentation systems can also occur when the materials deform in a linearly elastic manner but the geometry of the contact causes stiffness to vary with indentation load and displacement. Common examples of this are indentation by spheres and cones. In spherical indentation of a flat surface, often referred to as Hertzian indentation, the load and displacement are described by

$$P = \frac{4MR^{1/2}}{3} h^{3/2}\,, \tag{4-35}$$

where R is the radius of the indenter. The crux of the non-linearity is that the contact radius varies with indentation displacement as

$$a^2 = Rh \tag{4-36}$$

such that the contact stiffness is still given by

$$k = \frac{\mathrm{d}P}{\mathrm{d}h} = 2aM \tag{4-37}$$

(as for all axisymmetric indentation processes), but in this case varies with indentation displacement as

$$k = \frac{\mathrm{d}P}{\mathrm{d}h} = 2MR^{1/2}h^{1/2} . \tag{4-38}$$

In indentation of a flat surface by an axisymmetric cone of included angle 2ψ, the load-displacement relation is

$$P = \frac{\pi M \tan\psi}{2} h^2 \tag{4-39}$$

and the relation between contact radius and indentation displacement is

$$a = \frac{2\tan\psi}{\pi} h \tag{4-40}$$

such that the contact stiffness varies with displacement as

$$k = \pi M \tan\psi h . \tag{4-41}$$

The load-displacement behavior for flat punch, spherical, and conical indentation are shown in Fig. 4-18. Schematic cross-sectional diagrams of these elastic indentation processes are shown in Fig. 4-19.

In real life, of course, adhesive interactions described by some sort of interaction potential (of the overall shape of the example Mie potential) occur between the indentation tip and the surface, both interior and exterior to the contact radius, which could then be defined by the locus of points at which the separation between the surfaces is the quiescent separation, $z = z_0$. Frequently, $z_0 \ll a$ and it is a good approximation that adhesive interactions only occur within the contact. This is certainly true of flat-punch loading. If the adhesive interactions are strong enough such that the surface has sub-critical stiffness (small elastic modulus) and the probe is very stiff, the contact can support a negative load. The load-displacement behavior of an adhesive flat punch under negative ($P < 0$) load is shown in Fig. 4-18; a schematic diagram of such a process is shown in Fig. 4-19. The surface is displaced outwards ($h < 0$).

In the next section, a simple model will be described that makes use of adhesive flat punch loading to describe an adhesive indentation contact. Before doing so, however, it is useful to consider the opposite

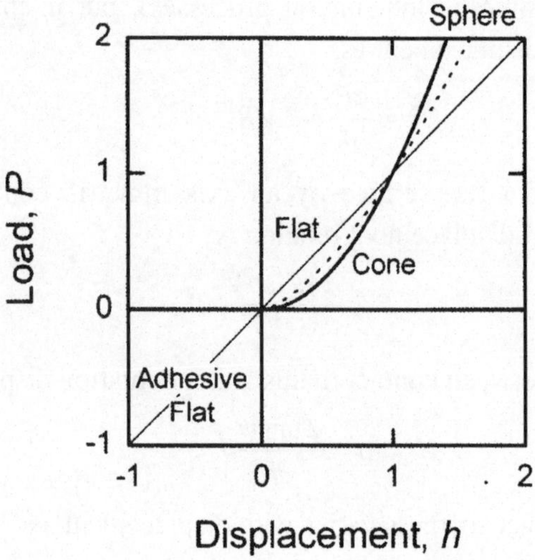

Figure 4-18. Load-displacement relations for flat-punch, spherical, and conical indentation of a linearly-elastic material with no adhesive interactions (positive P and h quadrant); the non-linear responses for the sphere ($P \sim h^{3/2}$) and cone ($P \sim h^2$) arise from geometrical non-linearity of the system. The flat-punch response for an adhesive interaction is also shown (negative P and h quadrant).

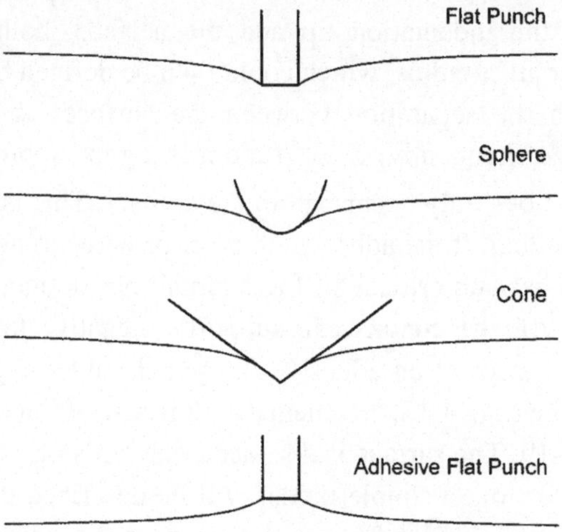

Figure 4-19. Schematic diagrams of flat-punch, spherical, and conical indentation indicating the deformation both within and exterior to the contact.

supercritical system of a surface and flat-punch probe that are rigid. Under this condition, separation of the tip from the surface leads to a load-displacement relation that is a reflection only of the adhesive interactions as considered in the last section. The work of separation per punch area, Γ, for the flat punch is then given by

$$\Gamma = \Phi_0 \left(\frac{a}{d}\right)^2 \left(\frac{1}{\pi a^2}\right) \sim \frac{\Phi_0}{d^2}, \tag{4-42}$$

using the notation and scaling arguments of Chapter 2 and the previous sections (the first term in parentheses is the number of bonds in the contact area and the second term is the punch area). It is clear that Γ reflects only the properties of the tip and surface materials, or at least their interface. Consideration of the area exterior to the contact of the flat punch prior to separation suggests that Γ can be considered as the excess energy of the system per area associated with uncontacted surface relative to the contacting interface. For this reason Γ is usually referred to as the "surface energy." Γ figures prominently in the next section, in which adhesive surface interactions and indentation deformation interactions are combined in a simple adhesive indentation model.

4. The JKR Model of Adhesive Indentation Contact

In this section we consider a simple model of adhesion due to Johnson, Kendall, and Roberts (JKR)[4] that, although not fully self-consistent, encapsulates all the essential elements of adhesive indentation, and introduces surface energy explicitly. As in previous sections, we shall see that the stiffness of the system is sub-critical, although here the compliant element is the elastic deformation of tip and surface; the probe spring is taken as rigid. The tip is taken as spherical with radius R and the tip displacement, h, is imposed at the sphere center (it is convenient to consider the load, P, measured here, too). The contact radius, a, between the tip and surface is taken to be very small such that $R >> a$ (the usual Hertzian contact assumption). At an even smaller length scale, the tip and flat surface interact via infinitesimally short-range attractive surface forces. The sphere and surface are taken to be elastic (with different, but isotropic elastic properties) and the free (non-contacting) surfaces of the sphere and flat surface are taken to introduce an inelastic internal energy component Γ, the excess surface energy density relative to the contacting interface.

The system is regarded as a Hertzian contact perturbed by an additional traction distribution, representing the surface-force effects, derived from that for rigid flat-punch contact. The problem is formally modeled as the sequential superimposition on the flat surface of two pressure distributions describable as purely elastic systems, one non-linear, one linear. In the first (Hertzian) element of the sequence, the load is increased to P_1, leading to a load-point displacement and contact radius, h_1 and a, respectively. In the second (flat punch) element of the sequence, the contact radius is held fixed at a and the load perturbed by ΔP, leading to a displacement perturbation of Δh. The final load and displacement are P and h, where

$$P = P_1 + \Delta P \tag{4-43}$$

and

$$h = h_1 + \Delta h\,. \tag{4-44}$$

The quantities associated with the initial Hertzian loading are all positive, whereas those associated with the following flat-punch loading are all negative. (The lack of internal self-consistency comes from the fact that the local surface displacement fields interior and exterior to the contact cannot simultaneously meet the assumed loading conditions.) Equations 4-43 and 4-44 may be re-written using Eqs. 4-35 and 4-36 as

$$h = \frac{a^2}{R} + \Delta h \tag{4-45}$$

and

$$P = \frac{4M}{3R}a^3 + \Delta P\,, \tag{4-46}$$

where the ΔP and Δh terms are related by Eq. 4-33 by

$$\Delta P = 2aM\,\Delta h\,, \tag{4-47}$$

The expression for the elastic internal energy of the system, U_E, (energy associated with deformation of the sphere and surface) is obtained from the work performed by the applied loading during this sequence (by integrating the load-displacement path) and is given by

$$U_E = \frac{2}{5} P_1 h_1 + P_1 \Delta h + \frac{1}{2} \Delta P \Delta h . \tag{4-48}$$

After some manipulation, using Eqs. 4-45 to 4-47, this becomes

$$U_E = M \left(\frac{a^5}{5R^2} - \frac{2ha^3}{3R} + h^2 a \right). \tag{4-49}$$

The inelastic internal energy of the system, U_I, (energy not associated with elastic deformation processes but with the surface interactions via the increase in the contact area πa^2) is given by

$$U_I(a) = -\Gamma \pi a^2 . \tag{4-50}$$

Equilibrium of the system at a given imposed displacement is determined by

$$\frac{\partial (U_E + U_I)}{\partial a} = 0 \tag{4-51}$$

Solving this equation gives the displacement and load increments associated with the second element of the contact sequence:

$$\Delta h = -\left(\frac{2\pi \Gamma a}{M} \right)^{1/2} \tag{4-52}$$

and

$$\Delta P = -\left(8\pi \Gamma M a^3 \right)^{1/2} , \tag{4-53}$$

and thus

$$h(a) = \frac{a^2}{R} - \left(\frac{2\pi \Gamma a}{M} \right)^{1/2} \tag{4-54}$$

and

$$P(a) = \frac{4Ma^3}{3R} - \left(8\pi \Gamma M a^3 \right)^{1/2} , \tag{4-55}$$

which form a parametric set for $P(h)$. Clearly, the Hertzian solutions are recovered for $\Gamma = 0$. In addition, it is clear that for a given load or displacement, the contact radius is greater than the Hertzian value for $\Gamma \neq 0$ —a consequence of the adhesive interaction. Figure 4-20 shows the variation of P and h with a for the JKR model. Figure 4-21 shows the variation in P with h for the JKR model and the Hertzian model; note that in the limit of large loads and displacements the two models converge.

As observable in Fig. 4-20, the system is capable of supporting a tensile load ($P < 0$) at a finite contact radius. The critical radius, a_c, at which the greatest tensile load occurs, is obtained by setting $\mathrm{d}P(a)/\mathrm{d}a = 0$ to gain

$$a_{cP} = \left(\frac{9\pi\Gamma R^2}{8M}\right)^{1/3} . \tag{4-56}$$

The conjugate load and displacement are thus

$$P_{cP} = -3\pi\Gamma R/2 \tag{4-57}$$

and

$$h_{cP} = -\frac{1}{3}\left(\frac{9\pi\Gamma R^{1/2}}{8M}\right)^{2/3} = -\frac{a_c^2}{3\mathrm{R}} . \tag{4-58}$$

These values define the system configuration at "pull-off" under load control, the $\mathrm{d}P(h)/\mathrm{d}h = 0$ condition in Fig. 4-21. We are interested in displacement control here, in which the indenter actuator imposes h. The contact radius, load, and displacement at pull off under displacement control, the $\mathrm{d}P(h)/\mathrm{d}h = -\infty$ condition in Fig. 4-21, are obtained by setting $\mathrm{d}h(a)/\mathrm{d}a = 0$ to gain

$$a_{ch} = (1/9)^{1/3} a_{cP} \approx 0.48 a_{cP} , \tag{4-59}$$

$$P_{ch} = (5/9) P_{cP} \approx 0.55 P_{cP} , \tag{4-60}$$

and

$$h_{ch} = 9^{1/3} h_{cP} \approx 2.08 h_{cP} . \tag{4-61}$$

Just as in the example Mie potential sub-critical example considered in a previous section, the JKR model has a quiescent condition of zero load and a snap-on condition intermediate to the quiescent condition and the pull-off condition. The parameters associated with these conditions may be obtained by setting $P(a) = 0$ and $h(a) = 0$, respectively and are given

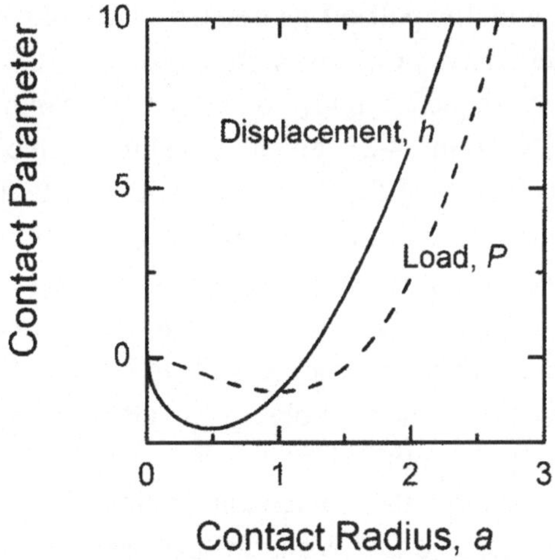

Figure 4-20. Variations in the contact load and displacement with contact radius for a Hertzian contact perturbed by an adhesive interaction over the contact—the JKR model. The adhesive interactions enable negative loads and displacements at small contact radii. The values are normalized using the characteristic values (Eqs. 4-56 to 4-58).

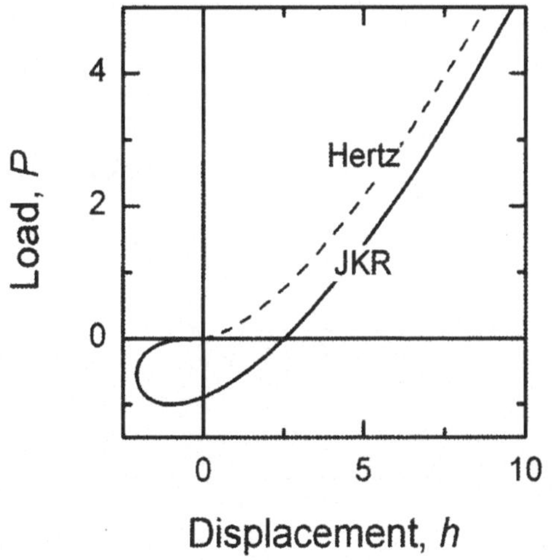

Figure 4-21. Variations in the contact load with displacement for a Hertzian contact (dotted line) and a JKR contact. The adhesive interactions in the JKR model enable a quiescent (zero-load) displacement separating positive-load indentation and negative-load adhesive separation. Load- and displacement-controlled pull-off instabilities occur in the negative load and displacement quadrant. The values are normalized using the characteristic values (Eqs. 4-56 to 4-58).

in Table 1, along with the pull-off parameters. The values are normalized using the characteristic values (Eqs. 4.56–4.58).

A JKR contact sequence under displacement control is then, with reference to Figs. 4-22 and 4-23, given by: (1) reversible approach of the sphere to the surface along the $P=0$ line with $a=0$ and $h<0$; (2) at $h=0$ an unstable equilibrium condition is reached at $P=0$ and subsequent positive infinitesimal perturbation of either a or h leads to an irreversible non-equilibrium increase in a and decrease in P until a stable equilibrium condition is reached with $a>0$ and $P<0$ (the sphere has snapped onto the surface). Subsequent infinitesimal increase of h leads to reversible quasi-equilibrium increase of a through a sequence of stable equilibria until the quiescent point (3) is reached at $P = 0$. A further increase in h leads to increases in a and a reversible adhesive indentation process ($P > 0$) until (4) maximum displacement and contact radius are reached.

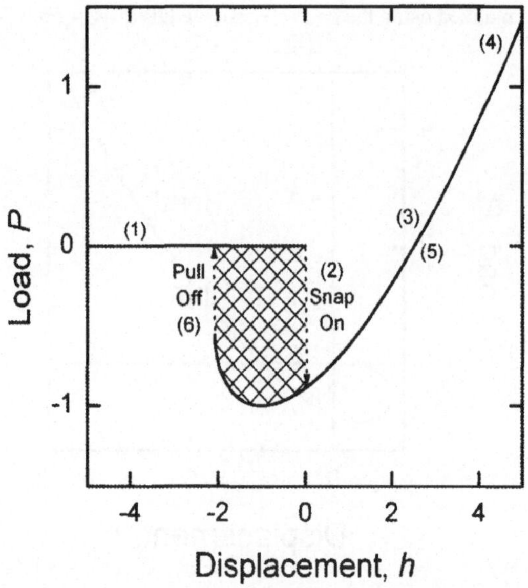

Figure 4-22. Variations in the contact load with displacement for a JKR contact cycle under displacement control. The system is sub-critical, leading to a hysteresis loop in the $h=0$ contact cycle that is bounded by "snap-on" and "pull-off" instabilities on approach and retraction, respectively. The values are normalized using the characteristic values (Eqs. 4-56 to 4-58).

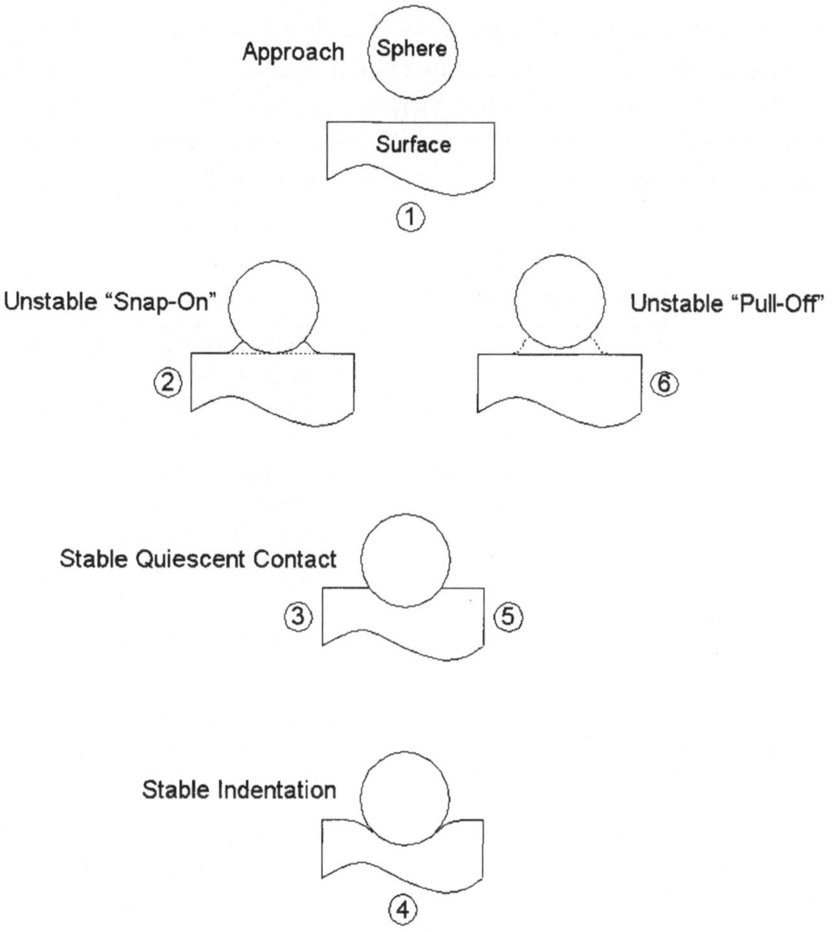

Figure 4-23. Schematic diagram of the approach, "snap-on," contact, indentation, and "pull-off" segments for a sphere contacting an adhesive surface. The incipient instability configurations for the snap-on and pull-off events are shown as dotted lines in segments (2) and (6). Such contacts are well described by the JKR contact adhesion model.

If the maximum displacement is large, the load-displacement behavior approaches that for a non-adhesive Hertzian contact. Infinitesimal reversal of h leads to quasi-equilibrium decrease of a through the same sequence of stable equilibria, passing through the quiescent point (5) and the maximum adhesive force ($P < 0$) condition (2), until an unstable equilibrium condition is met. Subsequent negative infinitesimal perturbation of either a or h leads to an irreversible non-equilibrium decrease (6) in a and increase in P until both $P = 0$ and $a = 0$ and the state (1) is regained (the sphere has "pulled-off" the surface). The work performed by the actuator during the full adhesive indentation sequence is shown as the cross-hatched area in Fig. 4-22.

Table 4-1. The salient points of the JKR response.

Load	Displacement	Contact Radius	Comments
0	$2^{4/3}$	$2^{2/3}$	The quiescent point; zero load adhered state
$-8/9$	0	$(4/3)^{2/3}$	The snap-on state under displacement control
-1	-1	1	The pull-off state under load control
$-5/9$	$-9^{1/3}$	$(1/3)^{2/3}$	The pull-off state under displacement control

A contact sequence under load control is similar and transitions in control are possible as well. For example, a sphere may be "placed" on the surface under displacement control as above, reaching the stable equilibrium condition (2), and then "let go" such that the system is now characterized by a load-controlled non-equilibrium $P = 0$ condition. The contact radius increases, following the non-equilibrium $P = 0$ line until the quiescent point is reached (the sphere has snapped on to the surface under load control).

The JKR model shares with the considerations of the logarithmic potential a simplicity that allows the roles of the various factors influencing the system to be calculated in closed-form analyses. In the case of the logarithmic potential the factors were the tip-surface interaction potential and the deformation potential of the probe spring, which allowed the snap-on phenomenon to be studied. In the case of the JKR model the factors were the surface energy of the tip-surface interface and the elastic deformation energy of the tip and surface, which allowed the snap-on, quiescent, indentation, and pull-off phenomenon to be studied. In both cases, the models are probably too simple to describe real adhesive indentation contacts exactly, although the JKR model does very well in cases of compliant spheres on stiff substrates (*e.g.*, polymers on glass) and the logarithmic potential is correct, if of limited use. The Mie potential is probably of the broadest applicability, if only because of the generalized empirical nature of its formulation, and incorporates the possibility of snap-on, quiescent, indentation, and pull-off phenomena along with an implicit surface energy, but has the disadvantage of lack of closed-form solutions. In all these cases, the central physics of adhesive indentation contact is the same; the net stiffness field in which the tip moves, set by surface interactions and elastic deformations, determined what is observed outside the system by the indenter. Indentation without apparent adhesion is possible if the probe spring is very stiff. Adhesion without apparent indentation is possible if the probe spring is very compliant. Both adhesion and indentation with no loss of energy are possible during a single contact event if the probe spring is stiff enough (a supercritical probe). Adhesion and indentation with both adhesive (snap-on) and separation (pull-off) instabilities are possible if the probe spring is compliant enough (a sub-critical probe). Indentation of biological materials with either compliant nanoindenters or atomic force microscopes will nearly always fall into this last category, in which the energies involved with deforming (indenting) the surface will be comparable to those involved with deforming the probe spring and with interactions between the tip and surface. The ideas laid out here in Chapters 2 and 4 provide a framework for interpreting adhesive indentation measurements of biological materials and optimizing instrument configurations for measurement of both surface interaction and material deformation properties.

References

1. B. I. Bleaney and B. Bleaney, *Electricity and Magnetism, Third Edition.* Oxford University Press (1976).
2. J. S. Rowlinson, *Cohesion*, Cambridge University Press (2002).
3. K. L. Johnson, *Contact Mechanics*, Cambridge University Press (1985).
4. K. L. Johnson, K. Kendall and A. D. Roberts, *Proc. Roy. Soc. A* **324**, 301 (1971).

CHAPTER 5

MECHANICS OF INDENTATION

Michelle L. Oyen

Department of Engineering, Cambridge University
E-mail: mlo29@cam.ac.uk

The fundamental contact mechanics principles underlying nano-indentation testing techniques are reviewed. A range of material constitutive responses are covered, including elastic, plastic, and viscous deformation, and incorporating indentation of linearly viscoelastic materials and poroelastic materials. Emphasis is on routine analysis of experimental nanoindentation data, including deconvolution techniques for material properties measurements during indentation. In most cases, an analytical approach for an isotropic half-space is considered. Special cases are briefly described, including anisotropic materials, inhomogeneous composite materials and layered film-substrate systems.

1. Introduction

Experimental contact mechanics measurement of material properties has become extremely popular in the last decade, due to both the commercialization of dedicated instrumented indentation testing devices and advances in the practical and experimental understanding of performing contact mechanics experiments. Many commercial indenters are set-up as "black-box" instruments for routine mechanical testing, in which data analysis is automated and "the mechanical properties" are directly exported from the instrument. However, it is extremely important to review the underlying contact mechanics that are used in indentation data analysis for several reasons. First, although an elastic-plastic analysis (typically based on the method of Oliver and Pharr[1]) can be used to obtain numerical values for the elastic modulus (E) and hardness (H) of just about any sample placed into the indenter, an elastic-plastic analysis may be inappropriate for the material (given its constitutive behavior, which may or may not be elastic-plastic) and the numbers thus obtained are "garbage in, garbage out." As most biological materials are not well-described by an elastic-plastic response—with perhaps the exceptions of dried plant materials and biomineral composites with large mineral fractions—other analyses are often more appropriate than an elastic-plastic analysis.[2]

This chapter will provide a background in contact mechanics—including elastic, plastic and time-dependent deformation—with an emphasis on experimental applications of contact mechanics. As such, there will be no effort to derive key equations, and the reader is referred to other works on contact mechanics[3,4] for derivations and methods of deriving the equations presented. The chapter starts with the practical, experimental considerations of tip geometry and loading function selection, and how these selections influence indentation data analysis. Next, elastic contact mechanics is considered, including solutions for indenting elastic half-spaces with probes of different geometries as well as for more complicated, inhomogeneous geometries. Next, more complicated constitutive responses are considered, including elastic-plastic, linear and nonlinear viscoelastic, and poroelastic contact mechanics. The chapter concludes with considerations of time-dependent contact deformation, including viscoelastic and poroelastic approaches that have been used for the study of hydrated biological tissues.[5,6]

2. Practical Considerations

In this section, two practical considerations are explored: probe tip geometry and time-history of the loading function during contact. Indentation data analysis is based in part on the geometry of the contact, and as such the choice of indenter tip geometry is a crucial consideration in designing any indentation experiment. For an elastic solid, the time-history of the loading function is irrelevant and the choice can be made based on convenience; however, for time-dependent materials the loading history is integrated into the data analysis and must be considered in further detail.

2.1. *Indenter probe geometry*

Indentation testing considers the contact between a probe and a sample. The probe geometry thus determines the deformation profiles obtained during the test, and as such is an extremely important experimental controllable. Flat punch tips are convenient in that the cross-sectional area of the contacted region is constant, simplifying analysis. However, due to the sharp corners, there can be stress singularities at the edges. Particularly in extremely small-scale nanoindentation testing, it can also be nearly impossible to align the tip's flat surface parallel to the sample surface. Therefore two tip shapes dominate small-scale contact testing: spherical and conical or pyramidal (Fig. 5-1).

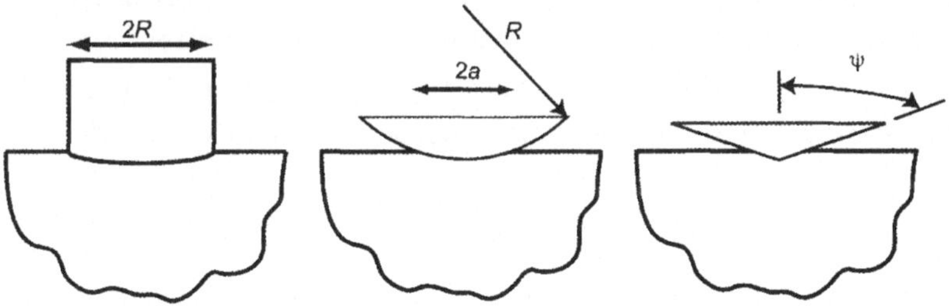

Figure 5-1. Schematic illustrations of cylindrical, spherical and conical or pyramidal indenter geometries. The critical parameters are the indenter radius, R, for a cylindrical or spherical indenter, the contact radius $a < R$ for a spherical indenter, and the included indenter half-angle ψ for a conical indenter (or effective indenter half-angle for a pyramidal indenter).

Spherical tips are advantageous in that there is a delayed onset of plastic deformation, while conical or pyramidal tips are exactly the opposite since there is usually an immediate onset of plasticity and thus elastic-plastic deformation even at small displacements. In indentation testing with conical or pyramidal tips there is also the advantage of geometric similarity—there is no intrinsic length scale associated with the size of the tip, and indentation tests at large loads result in deformation patterns that are scaled up from indentation tests at small loads. A disadvantage in the use of spherical and conical tips is that load is no longer a linear function of displacement, as is the case for a homogeneous tensile test or a flat-punch indentation test, due to the changes in tip-sample contact area as the tip is pressed further into the surface. These load-displacement functional forms for different indenter tip geometries—and thus the data analysis options for each tip—will be explored further in Section 3 of this chapter.

2.2. *Load functions*

The second critical experimental controllable in indentation testing, as in any sort of mechanical testing, is the input loading function. Commercial nanoindentation instruments are essentially load-controlled, in stark contrast to the displacement-control used in servohydraulic universal test-frames. Because the probe is pressed into the sample and then retracted, the indentation cycle is most commonly considered as a load-unload cycle. Standard test profiles (Fig. 5-2) are for ramping at constant loading and unloading rates, ramping to a peak load and then holding for a fixed time before unloading, and loading exponentially:

$$P(t) = P_0 \exp(rt) \tag{5-1}$$

where P_0 is an extremely small tare load at the which the test commences and r is the effective strain rate. This loading condition (Eq. 5-1, Fig. 5-2c) approximates a constant indentation strain-rate under geometrically similar indentation with constant material hardness.[7] Some nanoindentation systems are equipped with a feedback-control mechanism, which allows for displacement-controlled indentation testing. Again the displacement can be ramped up and down at fixed rate, and can potentially include a holding period at peak displacement, resembling Fig. 5-2 (a and b) but with displacement replacing load as the controlling variable along the y-axis.

In addition to the quasistatic indentation conditions presented in Fig. 5-2, a number of indentation loading schemes include dynamic loading, typically as a sinusoidal variation in the load. The sinusoidal loading can be of quite small amplitude, and is then superposed on top of an existing quasistatic load function to increase the amount of information obtained from a single indentation test. For example, small-amplitude sinusoidal loading superimposed on the holding segment of a trapezoidal load function (Fig. 5-2b) can be used to measure the dynamic response of a viscoelastic material, giving the real and imaginary parts of the complex modulus ($E^*(\omega)$ where ω is the sinusoidal loading frequency; see Section 6.1.4 below). This mode of dynamic testing will be discussed later in this chapter. First, we examine the elastic contact mechanics expressions as a function of indenter tip shape.

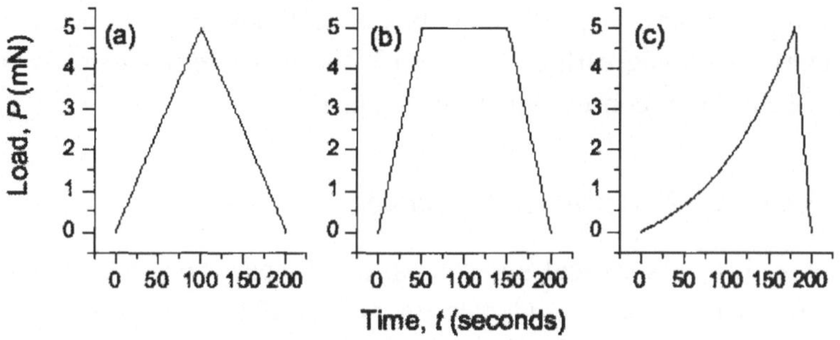

Figure 5-2. Load-time profiles for (a) a simple load-unload cycle under triangular loading, (b) a trapezoidal load-hold-unload profile in which creep can be measured during the holding period, and (c) exponential loading.

3. Elastic Contact

Much of the work on elastic contact originated with the work of Hertz in the late 19th century.[8] An excellent overview is presented in the text of Johnson.[4] The work of Hertz concerned the contact of two bodies with effectively spherical profiles, in which an effective radius R is defined in the context of the radii of the two bodies (R_1, R_2) as:

$$\frac{1}{R} = \frac{1}{R_1} + \frac{1}{R_2} \tag{5-2}$$

Similarly, the elastic modulus of the contact is defined by the series combination of the plane strain moduli of the two bodies:

$$\frac{1}{E_R} = \frac{1-\nu_1^2}{E_1} + \frac{1-\nu_2^2}{E_2}. \tag{5-3}$$

where ν_i is the Poisson's ratio and E_i is the elastic modulus for each body, and for which isotropic elasticity has been assumed. (Note the direct analogy to Eq. 2-9 in Chapter 2 of this volume.) In the following sections we will see how these ideas have influenced modern nanoindentation testing. To continue with the Hertzian contact theme, we will start with spherical indentation before moving on to consider flat punch and conical/pyramidal indentation. As noted in the introduction to this chapter, the mechanisms for solving elastic contact problems will not be considered here, only the solutions at textbook-level. In addition to the texts mentioned earlier,[3,4] the reader is also referred to the classic work of Sneddon[9] concerning elastic contact.

3.1. *Spherical (Hertzian) elastic contact*

In the limit of a spherical indenter on a half-space, Eq. 5-2 reduces to $R = R_i$ where R_i is the indenter tip radius. The indenter is typically penetrated only a small fraction of its radius into the material, such that the projected contact radius (a) is much smaller than the indenter radius. The contact radius can be related to the radius (R) and total indentation depth (h) as:

$$a = \sqrt{Rh} \tag{5-4}$$

The load-displacement (P-h) relationship for spherical (Hertzian contact) indentation is:

$$P = \frac{4\sqrt{R}}{3} \frac{E}{(1-\nu^2)} h^{3/2} \tag{5-5}$$

In this case, the indenter has been assumed to be much stiffer than the sample ($E_s \ll E_i$) such that the reduced modulus E_R is simply the plane strain modulus for the sample, $E_R = E_s/(1 - \nu_s^2)$.

For a perfectly elastic spherical indentation experiment, the load-displacement response can be fit to Eq. 5-5 directly (Fig. 5-3). One mechanism to determine if the contact is perfectly elastic is to estimate the indentation strain ε_i as:

$$\varepsilon_{\mathrm{i}} = 0.2(a / R) = 0.2(h / R)^{1/2} \tag{5-6}$$

At small strains, the yield stress has not been reached and the contact is perfectly elastic. When the yield stress has been reached, the functional form of the load-displacement response shifts due to dominant plastic deformation, as will be discussed further in Section 4.1.

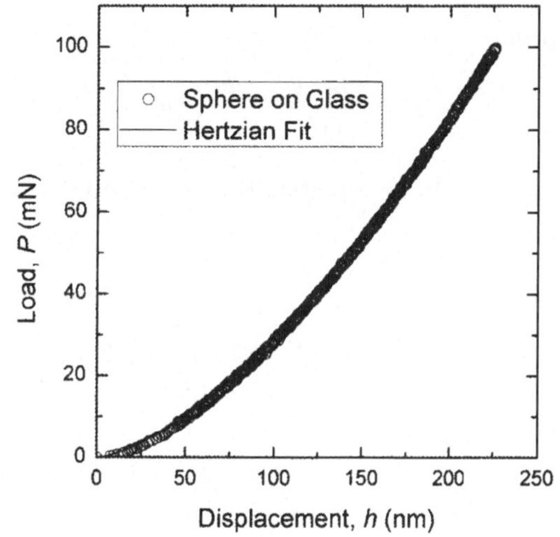

Figure 5-3. Perfectly elastic (Hertzian) contact on glass, in load-displacement (P-h) coordinates, showing the fit to Eq. 5-5.

3.2. *Flat punch elastic contact*

Although not often used on stiff materials at the nano-scale, flat punch indentation is occasionally utilized for extremely compliant materials, including hydrogels and hydrated biological tissues. The load-displacement response for a cylindrical indenter with radius R is:

$$P = 2R\frac{E}{(1-\nu^2)}h \tag{5-7}$$

In contrast to the Hertzian case, in which the relationship between load and displacement is not linear, the constant contact area means that in the flat punch case, the load-displacement response is linear.

3.3. *Sharp elastic contact*

Most commonly a sharp indenter tip—either a cone or a three-sided pyramid—is used in nanoindentation. The three-sided pyramid, called a "Berkovich" tip, was designed to mimic the relationship between indentation depth and projected contact area as is found for a Vickers pyramid—the four-sided pyramid most commonly used in classic microindentation hardness experiments. The full elastic solution for conical indentation is[9]:

$$P = \frac{\pi \tan \psi}{2\gamma^2} \frac{E}{(1-\nu^2)} h^2 \qquad (5\text{-}8)$$

for a conical indenter with included half-angle ψ and where is $\gamma = \pi/2$.

4. Elastic-Plastic Contact

The standard metric for evaluating the extent of plastic deformation in an indentation test is by the "hardness" which is representative of the resistance to plastic deformation, and can be related to a metal's yield stress. In microindentation, the residual indentation impression is measured with an optical microscope, and the hardness—a representative contact stress—is computed as the ratio of indentation load to projected area[10]:

$$H = \frac{P_{max}}{2b^2} = \frac{P_{max}}{\pi a^2} \qquad (5\text{-}9)$$

where b is half the length of the impression diagonal for a Vickers four-sided pyramid, or a is the effective contact radius for a circular impression. In soft metals, the bulk of indentation deformation is plastic, not elastic. In ceramics there is a more even balance of elastic and plastic deformations[11] while in polymers and elastomers there may be little or no plastic deformation. In the case of an absence of plastic deformation, the "hardness" can still be computed as the mean contact stress, although obviously there is no residual impression to measure and the contact radius at maximum load must be inferred from a calibration function.

In elastic-plastic indentation testing, as in metals, ceramics and to some extent glassy polymers, it is desirable to separate the elastic and plastic deformation components to allow for calculations of the elastic modulus as well as the hardness. Key developments in this field were made by Doerner and Nix[12] prior to the landmark work of Oliver and Pharr.[1] The Oliver-Pharr procedure is most commonly employed for conical or pyramidal indentation, although it is occasionally used for spherical indentation. A second elastic-plastic deconvolution scheme for spherical indentation is that of Field and Swain,[13] which will be mentioned briefly. In this section, the mechanisms for measuring elastic modulus in elastic-plastic materials will be considered, along with procedures for identifying plastic deformation and for quantifying the plasticity in terms of hardness.

4.1. *Elastic-plastic spherical contact*

In spherical indentation, there is a change in the functional form of the load-displacement relationship from $P \sim h^{3/2}$ to $P \sim h$ when deformation shifts from elastic to plastic. This change is clearly evident in soft, single crystal aluminum; nanoindenter data from fused silica glass and aluminum have been plotted on a log scale along with lines indicating the two critical functional forms of the load-displacement relationship for spherical indentation. The aluminum data clearly demonstrates the transition to $P \sim h$ behavior (Fig. 5-4) while the glass data (the same response as in Fig. 5-3) retains the elastic functional form.

In spherical indentation of elastic-plastic materials, the separation of elastic and plastic deformation responses is performed utilizing a procedure that utilizes a modification of the Hertzian elastic expression (Eq. 5-5) based on an assumption of elastic unloading.[13]

$$E_R = \frac{3}{4}\frac{P}{h_e^{3/2}}\left(\frac{1}{R}-\frac{1}{R'}\right)^{1/2} \tag{5-10}$$

where h_e is the elastic penetration depth and

$$\left(\frac{1}{R}-\frac{1}{R'}\right) \tag{5-11}$$

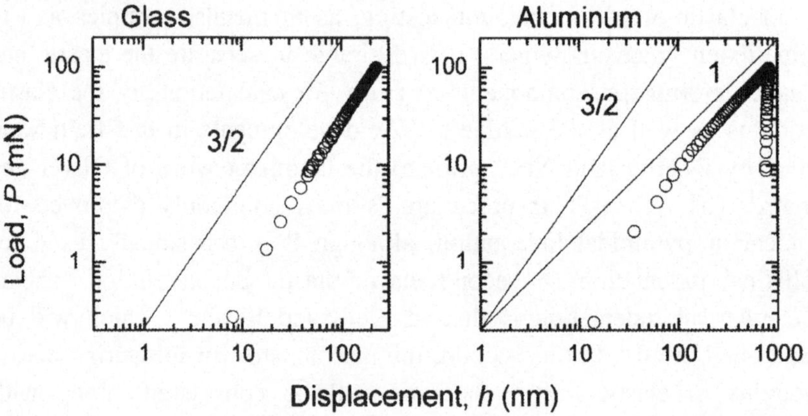

Figure 5-4. Elastic-plastic indentation responses for a sphere on fused silica glass and aluminum. The fused silica remains remains elastic while the aluminum demonstrates the onset of plastic deformation and a transition from $P \sim h^{3/2}$ to $P \sim h$ behavior at displacements larger than 100 nm.

is the relative curvature between the indenter and the residual impression on the sample surface. The challenge lies in identifying the elastic depth h_e as a function of the total depth, and separating out this relative curvature term. In this regard a "partial unloading" protocol has been developed for using multiple loading and (partial) unloading cycles—a stepwise indentation protocol—to assist with identification of the plastic depth and allow for elastic modulus measurements. The interested reader is referred to Ref. 13 for further details; in many cases the advantage of using a spherical indentation probe is the avoidance of the onset of plastic deformation. For this reason spherical probes are often used for elastic, viscoelastic and poroelastic analysis and less commonly for elastic-plastic problems.

4.2. *E and H deconvolution for sharp indentation*

Elastic-plastic indentation testing typically uses a sharp "geometrically similar" pyramidal (or, less commonly, conical) probe with a fixed cone angle (Fig. 5-1). For conical pyramidal indentation with very sharp probe tips (indenter radius R_i = 50–100 nm), the large strains in the material below the point results in a near-immediate onset of plasticity and a dominantly plastic overall response. The functional form for elastic and plastic deformation are both $P \sim h^2$ under conical or pyramidal loading, further complicating the distinction (Fig. 5-5) and making it impossible to separate these deformation components by functional form in the same manner as for spherical indentation. As mentioned briefly above (Eq. 5-9), the parameter generically called "hardness" (H) is frequently used to characterize dominantly plastic materials.[10]

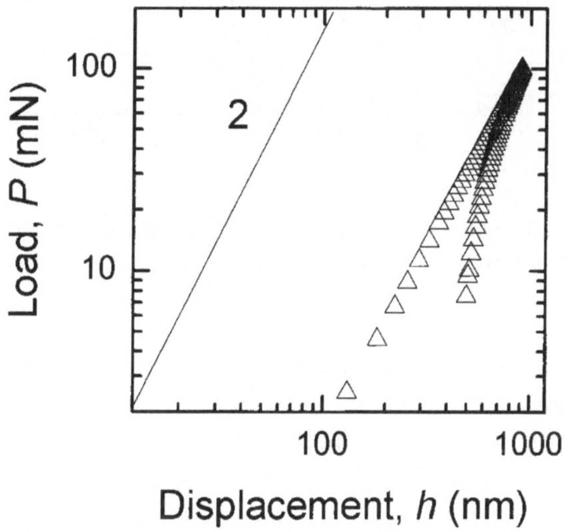

Figure 5-5. Elastic-plastic indentation responses for a Berkovich on glass demonstrating a consistent $P \sim h^2$ response across the entire loading trace. The unloading trace has a different, and changing, functional form.

The overall scheme for separating elastic and plastic deformations in the Oliver-Pharr method is based on the assumption that the unloading response is purely elastic, such that the reduced modulus E_R (Eq. 5-3) of the contact can be related to the unloading stiffness S (Fig. 5-6) via:

$$E_R = \frac{S\sqrt{\pi}}{2\sqrt{A_c}} \tag{5-12}$$

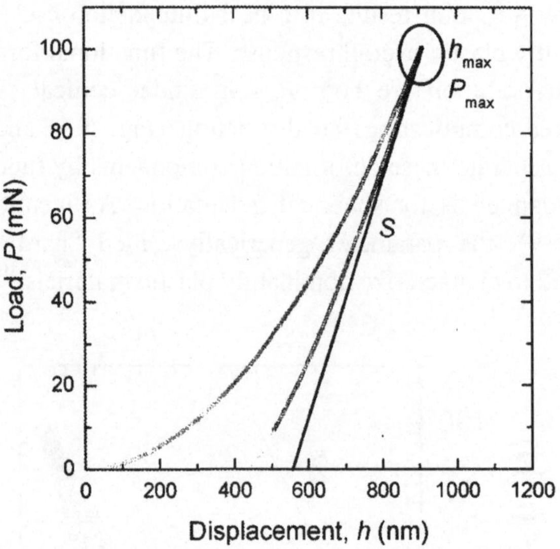

Figure 5-6. Elastic-plastic indentation responses for a Berkovich on fused silica glass, illustrating the three parameters taken directly from the load-displacement (P-h) response for Oliver-Pharr[1] analysis: h_{max}, P_{max}, S.

In Eq. 5-12 the contact area (A_c) appears, where the contact area is the projected area calculated from the contact displacement (h_c) via a calibration function of the form $A_c = A_c\,(h_c)$. In Eq. 5-9, it was assumed for both a four-sided and a three-sided pyramid that the indenter was infinitely sharp. However, in reality the tip is blunt at small indenter depths as a pyramid cannot be made perfectly sharp. The probe therefore approximates a sphere at some small contact depth h_c. The hardness is re-defined as $H_c = P_{max}/A_c(h_c)$ where the relationship between contact depth and contact area is often taken as a summed polynomial of the form

$$A_c(h_c) = C_0 h_c^2 + \sum_{k=1} C_k h_c^{1/2^{(k-1)}} \tag{5-13}$$

in which the coefficients C_i are determined by calibration. Other functional forms are possible, such as the physically-meaningful[14]:

$$A_c(h_c) = (\pi \tan^2 \psi) h_c^2 + 4R\pi h_c + 4R^2 \pi \cot^2 \psi \tag{5-14}$$

from the harmonic average of two tip geometries (conical and spherical) and in which the only two parameters are the effective tip radius R and the included cone angle ψ. Of course, this expression is less useful at extremely small indentation depths since it gives $A_c(h_c) \neq 0$ at $h_c = 0$, but it is very effective in estimating a tip effective radius. Any area function $A_c(h_c)$ can be determined by solving the E_R-S relationship (Eq. 5-12) for A_c at each contact depth h_c:

$$h_c = h_{max} - 0.75\frac{P_{max}}{S} \tag{5-15}$$

The calibration is performed by assuming the fixed value of E_R for a material with known (and constant with depth) elastic modulus (and thus reduced modulus). This calibration standard is typically the glass fused silica (a.k.a. fused quartz) with an elastic modulus E = 72 GPa and Poisson's ratio $\nu = 0.17$ such that, with diamond (E = 1.02 TPa and $\nu = 0.07$) the reduced modulus E_R = 69.6 GPa.

In addition to the area function, there is an additional calibration-derived quantity, the machine (frame) compliance. The measured stiffness, S, is the series combination of the sample stiffness, S_s, and the frame compliance, S_f, as:

$$\frac{1}{S} = \frac{1}{S_s} + \frac{1}{S_f} \tag{5-16}$$

The contact area (A_c) and the frame compliance ($C_f = S_f^{-1}$) are related, and the calibration protocol used to obtain these values has been discussed at length in other sources[1] and in the manuals for commercial nanoindentation instruments.

Once the full calibration (frame compliance and tip area function) is known, the analysis of elastic-plastic DSI data using the Oliver-Pharr protocol is straightforward. Three parameters are obtained directly from raw P-h data, the peak load (P_{max}), the peak displacement (h_{max}) and the unloading stiffness ($S = dP/dh$) as shown in Fig. 5-6. The measured stiffness S is corrected by Eq. 5-16 to obtain the sample stiffness and the contact depth (h_c) is computed from Eq. 5-15. The corresponding contact area $A_c(h_c)$ is calculated from the area calibration (Eq. 5-13 or 5-14) and used to calculate properties of reduced modulus (Eq. 5-12) and the contact hardness ($H_c = P_{max}/A_c$). Typically, in commercial indentation instruments the full procedure is carried out in the background, such that the output to the instrument operator is simply the properties (E_R and H_c) as a function of depth (h_c) assuming the frame compliance and area functions were programmed into the instrument.

5. Elastic Contact of Complex Materials and Systems

The elastic solutions presented in Section 3 were for indentation of a homogeneous isotropic elastic half-space. Before moving on to examine more complicated material constitutive behavior, such as viscoelastic materials, we will first briefly mention elastic indentation in three cases that frequently arise in the context of biological tissues: anisotropic materials, inhomogeneous (composite) materials and layered systems (i.e. thin films on substrates). Unsupported membrane systems will be considered in Chapter 10 of this volume, and will therefore be excluded from the current discussion.

5.1. *Anisotropic materials*

Indentation of an anisotropic material has been considered in explicit analytical formulations[15,16] and experimentally for bone.[17] The premise is that the anisotropic indentation modulus, M, is defined where M is specific to the orientation of the anisotropic material being indented, with the limiting case that $M = E/(1 - \nu^2)$ for an isotropic material. Analogous to the relationship between reduced (or plane strain) modulus in Eq. 5-12,

$$M = \frac{S\sqrt{\pi}}{2\sqrt{A_c}} \tag{5-17}$$

The indentation modulus M is defined in the context of the ratio of radii of the projected contact, a_1/a_2 (Fig. 5-7). The solution for the indentation modulus is a complicated integral expressions over components B_{ij} derived from the stiffness matrix (C_{ij}):

$$M = \frac{4\pi}{\int_0^{2\pi} \frac{a_{3i} B_{ij}^{-1}(\gamma) a_{3j}}{\left[(a_1/a_2)\cos^2\gamma + (a_2/a_1)\sin^2\gamma\right]} d\gamma} \tag{5-18}$$

where a_{3i} and a_{3j} are the direction cosines of the angles between the indentation direction and the elastic stiffness matrix. The expression must be solved numerically and therefore is not easily implemented in routine data analysis. Commonly reported is simply the effective indentation modulus for different orientations, in which the results are not converted back to components of the stiffness tensor (C_{ij}). Overall, the indentation modulus varies less with orientation than does the effective elastic modulus, where the maximum indentation modulus is less than the maximum apparent modulus in tension and the minimum indentation modulus is greater than the apparent modulus in tension.[17]

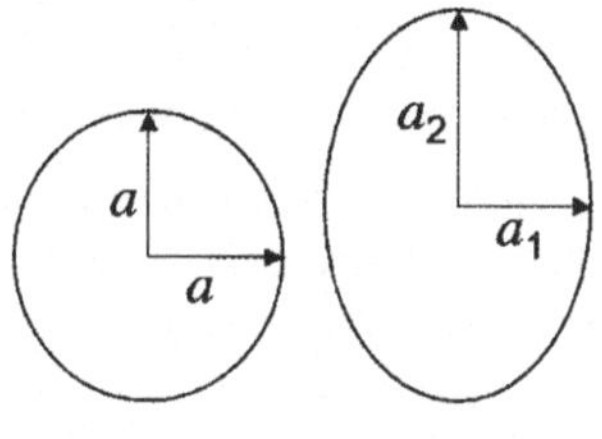

Figure 5-7. Projected contact area for an isotropic and anisotropic material following spherical or pyramidal indentation. For the isotropic case, the contact radii in perpendicular directions are equal ($a_1 = a_2 = a$) whereas for the anisotropic case $a_1 \neq a_2$.

5.2. *Inhomogeneous materials*

Two features differentiate the indentation responses of composite materials (*i.e.* materials with two or more distinct phases) compared with homogeneous materials. First, a length-scale is introduced, in the form of the size of the features of the multiphase material (Fig. 5-8 where the feature size is the particle radius r). Second, there is a critical dependence on indenter tip placement in terms of the observed indentation response, especially at small indentation depths where the indentation contact radius a or contact depth h_c is comparable in scale to the composite feature size r. Several recent studies have explicitly examined indentation of materials with substantial heterogeneity.[18-20] The general approach involves the sampling of a large number of individual indentation responses across the material surface and using the obtained range of responses to bound the inferred microstructure in terms of phase length-scale[18] and mechanical properties.[19] Substantially more effort is required to develop a robust method for the mechanical characterization of composite materials by indentation without the need for finite element analysis.[20]

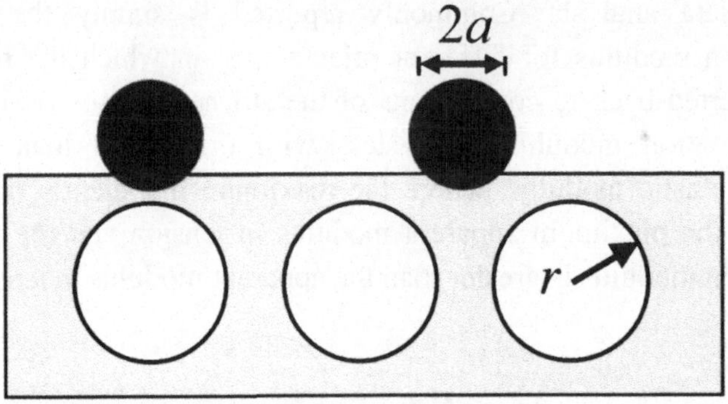

Figure 5-8. Schematic illustration of indentation of a composite material, in which the observed response depends on the placement of the indenter relative to microstructural features and the relative indentation and microstructural length-scales (a vs r).

5.3. *Layered systems*

The analyses presented thus far have been oriented towards bulk materials. However, in the majority of soft tissue indentation tests, the soft tissue layer of interest (*e.g.* cartilage) is supported by a much stiffer substrate (frequently bone). Just as in the case of composite materials, the indentation of a layered structure introduces a length-scale into the indentation problem, where the indentation contact depth or contact radius (h_c or a, respectively) are considered relative to the layer thickness (t_f) as shown in Fig. 5-9. The most commonly employed analysis for soft tissue indentation testing relies on the finite thickness of the tissue layer, considering the underlying bone as a rigid substrate.[21] The elastic expression employed for analysis is thus a modification of the bulk flat-punch elastic solution (Eq. 5-7) and is written:

$$P = \kappa 2R\frac{E}{(1-\nu^2)}h \tag{5-19}$$

where $\kappa = \kappa(\nu, R/t_f)$. In this instance, the contact dimension is fixed and $R = a$. The original work[21] required a look-up table to determine values of κ based on the numerical solution, but of course in the intervening years the analysis for the film-substrate problem has become near routine, at least in the instance in which the modulus mismatch between the film and substrate (E_f/E_s) is not more than an order of magnitude. The problem of larger modulus mismatch has also been examined within a framework of analytical elasticity.[22] In practice, it is common to use relatively simple empirical functions to extrapolate apparent elastic modulus-depth data to estimate the modulus of a thin layer[23] or to keep the indentation depth well below 10% of the total layer thickness.[1]

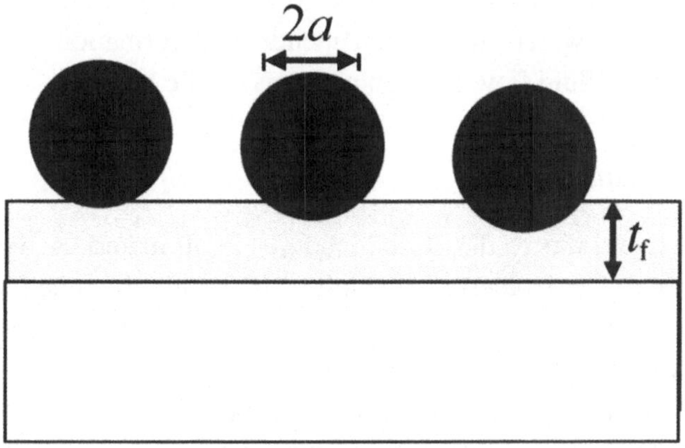

Figure 5-9. Schematic illustration of indentation of a layered structure, in which the observed response depends on the ratio of the indentation length-scale (as given by either the indentation depth h or contact radius a) relative to the layer thickness (t_f).

6. Time-Dependent Contact

A complication arises in the application of nanoindentation techniques for examination of biological materials, in that substantial time-dependent deformation can occur during the experimental time-frame. As such, the assumption of elastic unloading in the Oliver-Pharr procedure is violated, and different analyses must be employed. Time-dependent deformation appears most readily in nanoindentation data as increasing creep displacement at fixed load; in some cases, further forward displacement can be seen during unloading, a feature that has been termed a "nose" in the curve.[24] When experiments are conducted in displacement control instead of load control, the time-dependence is seen as load relaxation at fixed depth, and no "nose" is present in the data. Related to the existence of time-dependent deformation observed at fixed load or displacement, there are clear differences observed in the load-displacement response of time-dependent materials when probed at different load- or displacement- rates.

In this section, two mechanisms of time-dependent deformation will be explored, viscoelasticity and poroelasticity. Empirical viscoelastic approaches will be presented first, addressing both linear and nonlinear viscoelasticity. Mechanism-driven poroelastic approaches will then be considered, in which the time-dependent deformation is explicitly associated with fluid flow through a porous elastic network.

6.1. *Viscoelastic contact*

There are two limits to the elastic modulus in all materials, the adiabatic modulus and the isothermal modulus. In relatively time-independent materials, the two limits are comparable and are typically the difference is not easily noted in routine mechanical testing experiments. In time-dependent materials, the two limits can be widely divergent and the region of transition between the limits can be complicated. As it is experimentally impossible to measure mechanical properties either infinitely quickly or infinitely slowly, the apparent elastic modulus of a viscoelastic material changes with time over reasonable experimental time-frames. A full physical and mechanistic model for viscoelastic behavior remains elusive; for engineering practice, it is common to fit the elastic modulus-time responses to empirical functions based on simple mathematical constructs such as an exponential decay or a power-law decay in time.

6.1.1. *Linearly viscoelastic materials*

The time-dependent mechanical behavior of polymeric and many biological materials is often represented with a mechanical model formed of spring elements that are linear elastic (Hookean):

$$\sigma = G\varepsilon \qquad (5\text{-}20)$$

and linearly viscous (Newtonian) "dashpots":

$$\sigma = \eta \frac{\mathrm{d}\varepsilon}{\mathrm{d}t} \qquad (5\text{-}21)$$

Combinations of these simple relationships are made in arranging springs and dashpots in series and parallel configurations (Fig. 5-10) and then solving for the stress-strain-time (σ-ε-t) relationship, a differential equation in time.

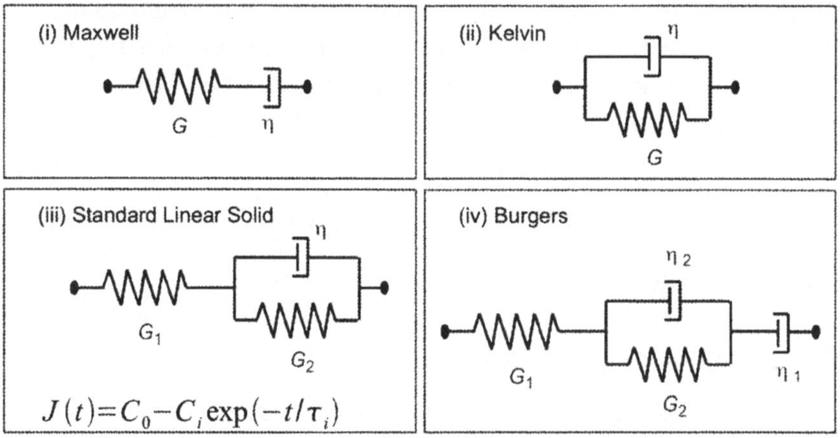

Figure 5-10. Examples of linear viscoelastic models formed from combinations of springs and dashpots with their independent constitutive responses as in Eqs. 5-20 and 5-21.

The order of the differential equation depends on the number of individual dashpots in the full system: the standard linear solid model has a first order differential equation constitutive response while the Burgers model is second order. From this constitutive relationship the homogeneous (single-axis, uniform loading) creep and relaxation functions can be obtained by solving the stress-strain-time (σ-ε-t) relationship for a condition of step stress

$$\sigma(t) = \sigma_0 \mathrm{H}(t) \tag{5-22}$$

or step strain

$$\varepsilon(t) = \varepsilon_0 \mathrm{H}(t) \tag{5-23}$$

where the H(t) is the Heaviside step function. From this time-dependent response to a step input, the relaxation function $G(t)$ and creep function $J(t)$ are defined as

$$\frac{\sigma(t)}{\varepsilon_0} = G(t) \tag{5-24}$$

$$\frac{\varepsilon(t)}{\sigma_0} = J(t) \tag{5-25}$$

These functions typically have a quite simple summed exponential functional form for linearly viscoelastic solids (*i.e.* those defined by a combination of linear springs and dashpots) even when many elements are present, such that they can be generalized as

$$G(t) = C_0 + \sum C_i \exp\left(-t / \tau_i\right) \tag{5-26}$$

$$J(t) = C_0 - \sum C_i \exp\left(-t / \tau_i\right) \tag{5-27}$$

where τ are material time constants.

The creep and relaxation functions are often taken as a simple summed exponential functional form:

$$G(0) = J(0)^{-1},\ G(\infty) = J(\infty)^{-1}, G(t) \neq J(t) \tag{5-28}$$

In many cases, it is of limited value to try and relate the coefficients C_k back to individual spring and dashpot terms in a 1-D mechanical model representation (Fig. 5-10), with the exception of a free spring defining either an instantaneous or infinite time modulus value [*i.e.* $G(0)$ or $G(\infty)$].

6.1.2. *Linear viscoelastic indentation correspondence*

The development of viscoelastic functions here has been for relationships between single scalar stress and strain components. In adapting this viscoelastic approach for three-dimensional stress and strain analysis, these basic functional forms are assumed to hold between individual tensor components of the overall stress and strain tensors. In practice, indentation problems are most commonly written in the single scalar form relating axial load to axial displacement, and in this context the conversion from viscoelastic stress-strain space to axial load-displacement space is trivial.

Approaches to the analytical viscoelastic indentation problem utilize the correspondence between elastic and viscoelastic elasticity problems[25,26]; the only difference between an elastic and a viscoelastic problem is in the choice of a time-dependent constitutive law for the material.[26] Elastic-viscoelastic correspondence analysis was examined within the context of the Hertzian contact problem by Lee and Radok[27] and is summarized in Johnson's text.[4] For analysis of indentation loading conditions in which the contact area is non-decreasing, the problem is fairly straightforward; for decreasing contact areas an adapted analysis[28] is required. Since the two most common time-dependent experiments (creep or relaxation at fixed load or displacement, or loading at various load- or displacement-rates) are for non-decreasing contact areas, the straight-forward analysis of Lee and Radok[27] is the sole analysis considered here in the context of elastic-viscoelastic correspondence.

The elastic solutions for indentation—written in terms of an axial load and axial displacement—are rewritten as viscoelastic expressions by replacing the elastic constants by viscoelastic operators, and integrating over the applied load- or displacement-perturbation. For the viscoelastic indentation analysis, an elastic indentation solution (Eqs. 5-5,7,8) is rewritten in terms of the shear modulus, where $G = E/[2(1+\nu)]$ for the general case (requiring fixed Poisson's ratio) and $G = E/3$ for the incompressible case (and the plane strain modulus thus equates to $4G$). From this point we will emphasize the incompressible case; data can always be back-corrected in the instance where the Poisson's ratio is known (and known to be less than 0.5).[29]

In cases where step-loading can be assumed, there is a straightforward replacement of $P/2G$ with the product of the creep load, P_0, and the creep function, $J(t)$:

$$h = \frac{1}{4R}\left[\frac{P}{2G}\right] \rightarrow h(t) = \frac{1}{4R} P_0 J(t) \tag{5-29}$$

$$h^{3/2} = \frac{3}{8\sqrt{R}}\left[\frac{P}{2G}\right] \rightarrow h^{3/2}(t) = \frac{3}{8\sqrt{R}} P_0 J(t) \tag{5-30}$$

$$h^2 = \frac{\gamma^2}{\pi \tan\psi}\left[\frac{P}{2G}\right] \rightarrow h^2(t) = \frac{\gamma^2}{\pi \tan\psi} P_0 J(t) \tag{5-31}$$

For "real" loading conditions, there is a finite ramping time or ramping rate up to the peak load. In this case, a hereditary integral formulation is used, where integration takes place over the creep function and the applied loading history $P(t)$. For the general indentation, indentation and load-control, this amounts to replacing $P/2G$ with an integral:

$$h = \frac{1}{4R}\left[\frac{P}{2G}\right] \rightarrow h(t) = \frac{1}{4R}\int_0^t J(t-u)\frac{dP(u)}{du}du \tag{5-32}$$

$$h^{3/2} = \frac{3}{8\sqrt{R}}\left[\frac{P}{2G}\right] \rightarrow h(t) = \left[\frac{3}{8\sqrt{R}}\int_0^t J(t-u)\frac{dP(u)}{du}du\right]^{2/3} \tag{5-33}$$

$$h^2 = \frac{\gamma^2}{\pi \tan\psi}\left[\frac{P}{2G}\right] \rightarrow h(t) = \left[\frac{\gamma^2}{\pi \tan\psi}\int_0^t J(t-u)\frac{dP(u)}{du}du\right]^{1/2} \tag{5-34}$$

As can be seen from the forms of Eqs. 5-32 to 5-34 is that the integral term $\int_0^t J(t-u)\frac{dP(u)}{du}du$ is identical, and as such a single set of solutions for commonly used experiments, such as ramping at constant loading rate, or a creep test following a ramp, can be used in each geometry. For a ramp test where $P(t) = kt$ the integral is simply over the creep function $k\int_0^t J(t-u)du$ and the exponential representation of the creep function (Eq. 5-27) makes the integration trivial. Solutions for multiple-stage indentation tests, such as a stair-step creep test at several (increasing) peak loads, involves a substantial book-keeping effort but relatively simple calculus[30]: solutions maintain the exponential functional form of the creep function and are thus relatively easy to implement experimentally. For example,[30,31] for spherical indentation creep following a ramp to peak load P_{max} in time t_R (such that the loading rate is constant and equal to $k = P_{max}/t_R$) the solution for displacement-time can be written in the compact form:

$$h^{3/2}(t) = \frac{3P_{max}}{8\sqrt{R}}\left[C_0 - \sum C_i \exp(-t/\tau_i)\mathrm{RCF}_i\right] \tag{5-35}$$

$$\mathrm{RCF}_i = \frac{\tau_i}{t_R}\left[\exp(t_R/\tau_i) - 1\right] \tag{5-36}$$

where RCF_i is a dimensionless "ramp correction factor"[30]; comparison of Eqs. 5-35 and 5-30 illustrates that this factor fully accounts for a finite ramping time instead of an assumption of step loading.

Recall that the restriction on this simple mathematical approach[27] is that the contact area is non-decreasing such that unloading is not a focus. A different approach to the same problem[28] a few years later yielded a complementary solution that does not carry the restriction and allows for full unloading analysis in a linearly viscoelastic material. A further discussion on these two approaches in the context of indentation viscoelasticity is available.[32]

For displacement controlled testing, under a step displacement $h(t) = h_0\,\mathrm{H}(t)$ the load-time relationship is again trivial by the substitutions paralleling those used in Eqs. 5-29 to 5-31:

$$P = 4Rh_0G(t) \tag{5-37}$$

$$P(t) = \frac{8\sqrt{R}}{3}h_0^{3/2}G(t) \tag{5-38}$$

$$P(t) = \frac{\pi\tan\psi}{\gamma^2}h_0^2G(t) \tag{5-39}$$

However, the hereditary integral approach is slightly more complicated. The form of the integral depends on the linearity or nonlinearity of the load-displacement relationship—integration is conducted over $2Gh^m$ where m is the power law exponent for the indentation load-displacement relationship:

$$P = 4R2Gh \rightarrow P(t) = 4R\int_0^t G(t-u)\frac{dh(u)}{du}du \quad (5\text{-}40)$$

$$P = \frac{8\sqrt{R}}{3}2Gh^{3/2} \rightarrow P(t) = \frac{8\sqrt{R}}{3}\int_0^t G(t-u)\frac{dh^{3/2}(u)}{du}du \quad (5\text{-}41)$$

$$P = \frac{\pi\tan\psi}{\gamma^2}2Gh^2 \rightarrow P(t) = \frac{\pi\tan\psi}{\gamma^2}\int_0^t G(t-u)\frac{dh^2(u)}{du}du \quad (5\text{-}42)$$

This nonlinearity can result in integral expressions with no closed form solution, particularly in a spherical indentation configuration.[33]

The expressions here considered the case of bulk material indentation in that the elastic solutions used were applicable to an elastic half-space. However, it is equally possible to use this correspondence approach in the context of other elastic solutions, such as a layered system. Such an approach has been extremely common in flat-punch indentation for a soft layer on a stiff underlayer, such as articular cartilage on bone[21] or a polymer thin film on a silicon substrate.[34]

6.1.3. *Viscous-elastic-plastic materials*

In sharp indentation problems, especially when the materials considered are glassy polymers, plastic deformation may be present along with viscoelastic deformation. Two approaches have been used to address this plastic deformation in the viscoelastic data analysis. The first mechanism is to remove the plastic deformation component from consideration using a load-unload-reload protocol,[35] in which the reloading segment is assumed to be purely viscoelastic. Alternately, the plastic deformation can be modeled explicitly and thus included as an additional parameter in the analysis. A viscous-elastic-plastic series model with independent elements for each deformation mode has been developed.[36] In its simplest incarnation, the model incorporates three independent deformation components such that the total displacement during loading is the sum of the component displacements:

$$h = h_V + h_E + h_P \quad (5\text{-}43)$$

There are three defining material properties, one associated with each deformation component (a plane strain elastic modulus E', a resistance to plastic deformation H—not the contact hardness, H_c—and a viscous time constant τ_Q). For constant loading-rate indentation testing where $P(t) = kt$, k = constant,

$$h_{\text{load}}(t) = \left[\frac{1}{(\alpha_2 \bar{E})^{1/2}} + \frac{1}{(\alpha_1 H)^{1/2}} + \frac{2t}{3\tau_Q (\alpha_2 \bar{E})^{1/2}} \right] (kt)^{1/2} \quad (5\text{-}44)$$

where the α_i values are dimensionless constants related to the indenter included angle.[36] For a load reversal at $t = t_R$ (*i.e.* the second phase of a triangle-wave test, Fig. 5-2b), the load-time response is:

$$\begin{aligned} h_{\text{unload}} = & \left[\frac{1}{(\alpha_2 \bar{E})^{1/2}} + \frac{1}{(\alpha_1 H)^{1/2}} + \frac{2t_R}{3\tau_Q (\alpha_2 \bar{E})^{1/2}} \right] (kt_R)^{1/2} \\ & + \frac{1}{(\alpha_2 \bar{E})^{1/2}} \left[(2kt_R - kt)^{1/2} - (kt_R)^{1/2} \right] \quad (5\text{-}45) \\ & - \frac{2}{3k\tau_Q (\alpha_2 \bar{E})^{1/2}} \left[(2kt_R - kt)^{3/2} - (kt_R)^{3/2} \right] \end{aligned}$$

This approach has also been expanded to incorporate the effects of a polymer film on a stiff and time-independent elastic substrate.[37]

6.1.4. *Dynamic contact measurements*

In the case of dynamic sinusoidal loading for frequency-based measurements, the input is an oscillation at force amplitude P_0, where the value P_0 is a small perturbation on the primary peak force $P_{\max}$ such that $P_0 << P_{\max}$ and $P(t) = P_{\max} + P_0 \sin(\omega t)$ for frequency ω. The response is a displacement of amplitude h_0 and a phase shift ϕ such that the response to the load perturbation has an $h(t) = h_0 \sin(\omega t - \phi)$ dependence. The real, or storage (E_S) and imaginary, or loss (E_L) parts of the elastic modulus are then calculated where the complex modulus is $E^* = E_S + i\, E_L$ and

$$\frac{E_S}{(1-\nu^2)} = \frac{S\sqrt{\pi}}{2\sqrt{A_C}} \quad (5\text{-}46)$$

analogous to Eq. 5-12 and where D is a damping coefficient. The real and imaginary moduli can be calculated directly from

$$E_S = (1-\nu^2)\frac{\sqrt{\pi}}{2\sqrt{A_C}}\left|\frac{P_0}{h_0}\right|\cos\phi \tag{5-48}$$

$$E_L = (1-\nu^2)\frac{\sqrt{\pi}}{2\sqrt{A_C}}\left|\frac{P_0}{h_0}\right|\sin\phi \tag{5-49}$$

The derivations of these expressions and caveats concerning their use can be found in other sources.[38-40] In practice, the typical experiment would include a number of different measurements for different frequencies ("frequency sweep") and the evaluation of the real and imaginary parts of the complex modulus as a function of frequency and potentially also as a function of static (P_{max}) and dynamic (P_0) load levels.

6.2. *Poroelastic contact*

Poroelasticity concerns the flow of fluid through a saturated porous elastic solid.[41] Although the components (fluid, solid) themselves are taken to be time-independent, the flow problem creates time-dependence in the response. The basic elastic problem is augmented in poroelasticity with two additional scalar variables, the pore pressure, p, and an increment of fluid content, ζ, the change in fluid volume relative to a control volume. The poroelastic framework considered here is for an isotropic elastic porous material with elastic material properties shear modulus, G, and Poisson's ratio, ν. There are two other parameters to characterize the fluid and the fluid-solid interactions, α and B, which are bounded between zero and unity. With the additional variables (p, ζ) and material constants (α, B), the stress-strain ($\sigma_{ij} - \varepsilon_{ij}$) relationship for elasticity is modified by the addition of a term including the pore pressure ($\alpha p\delta_{ij}$) to become:

$$\sigma_{ij} + \alpha p\delta_{ij} = 2G\varepsilon_{ij} + \left(\frac{2G\nu}{1-2\nu}\right)\varepsilon_{kk}\delta_{ij} \tag{5-50}$$

where δ_{ij} is the Kronecker delta.[26] A single additional scalar expression is added to the six elastic equations to total seven constitutive equations:

$$2G\zeta = \alpha\left(\frac{1-2\nu}{1+\nu}\right)\left(\sigma_{kk} + \frac{3p}{B}\right) \tag{5-51}$$

The four parameters (G, ν, α and B) make a complete set of material properties for a poroelastic solid; other sets of four parameters are possible and it is particularly useful to use a parameter set that includes ν_u, the undrained Poisson's ratio. Finally, the permeability (κ) is the parameter characterizing flow through the porous elastic skeleton in Darcy's law relating the flux (q_i) to the pressure gradient ($p_{,i}$):

$$q_i = -\kappa p_{,i} \tag{5-52}$$

The equations of poroelasticity are coupled: a change in applied stress leads to a change in fluid pressure or mass, and a change in fluid pressure or mass leads to a change in the volume of the solid. Because of this coupling, the poroelastic problem can be solved analytically in only a few cases, such as homogeneous uniaxial creep. Numerical solutions[42] as well as computational approaches[43] have been particularly useful in combining the poroelastic equations with the indentation problem. Spherical indentation has been considered for characterizing hydrated bone[6] and acrylamide gels,[44] in both cases resulting in estimated hydraulic permeability values consistent with values measured using traditional (*i.e.* non-indentation) methods. The application of poroelastic material models to the routine analysis of indentation data is an area of active current research.

7. Conclusions and Outlook

This chapter has focused on the analysis of indentation data, with an emphasis on material behavior that deviates from the elastic-plastic responses typical of engineering ceramics and metals. In cases such as linear viscoelasticity, analysis of nanoindentation data has become routine, while new approaches are just emerging for poroelasticity. In general, when handling nanoindentation data for biological materials, it is critical to identify the modes of deformation active during the indentation test, and to analyze the data appropriately for those deformation modes.

References

1. W. C. Oliver and G. M. Pharr, *J. Mater. Res.* **7**, 1564 (1992).
2. M. L. Oyen and R. F. Cook, *J. Mech. Behavior Biomed. Mater*, **2**, 396 (2009).
3. R. O. Davis and A. P. S. Selvadurai, *Elasticity and Geomechanics*, Cambridge University Press (1985).
4. K. L. Johnson, *Contact Mechanics*, Cambridge University Press (1985).
5. A. K. Bembey, A. J. Bushby, A. Boyde, V. L. Ferguson and M. L. Oyen, *J. Mater. Res.* **21**, 1962 (2006).
6. M. L. Oyen, Poroelastic nanoindentation responses of hydrated bone, *J. Mater. Res.* **23**, 1307 (2008).
7. B. N. Lucas, W. C. Oliver, G. M. Pharr and L.-L. Loubet, *Mater. Res. Soc. Symp. Proc.* **436**, 233 (1997).
8. H. Hertz, Miscellaneous Papers, Translated by Jones, D. E. and Schott, G. A., Macmillan and Co. Ltd. (1896).
9. I. N. Sneddon, *Int. J. Engng. Sci.* **3**, 47 (1965).
10. D. Tabor, *The Hardness of Metals*, Oxford University Press at the Clarendon Press (1951).
11. M. L. Oyen, *J. Biomech.* **39**, 2699 (2006).
12. M. F. Doerner and W. D. Nix, *J. Mater. Res.* **1**, 601 (1986).
13. J. S. Field and M. V. Swain, *J. Mater. Res.* **8**, 297 (1993).
14. J. Thurn and R. F. Cook, *J. Mater. Res.* **17**, 1143 (2002).
15. J. J. Vlassak, M. Ciavarella, J. R. Barber and X. Wang, *J. Mech. Phys. Solids* **51**, 1701 (2003).
16. J. G. Swadener and G. M. Pharr, *Phil. Mag.* **81**, 447 (2001).
17. J. G. Swadener, J.-Y. Rho and G. M. Pharr, *J. Biomed. Mater. Res.* **57**, 108 (2001).
18. M. L. Oyen and C. C. Ko, *J. Mater. Res.* **23**, 760 (2008).
19. G. Constantinides, K. S. Ravi Chandran, F.-J. Ulm and K. J. Van Vliet, *Mater. Sci. Eng. A* **430**, 189 (2006).
20. K. Durst, M. Goeken and H. Vehoff, *J. Mater. Res.* **19**, 85 (2004).
21. W. Hayes, L. Keer, G. Herrmann and L. Mockros, *J. Biomech.* **5**, 541 (1972).
22. A. Rar, H. Song and G. M. Pharr, *Mater. Res. Soc. Symp. Proc.* **695**, 431 (2002).
23. J. Mencik, D. Munz, E. Quandt, E. R. Weppelmann and M. V. Swain, *J. Mater. Res.* **12**, 2475, 1997.
24. B. J. Briscoe, L. Fiori and E. Pelillo, *J. Phys. D: Appl. Phys.* **31**, 2395 (1998).
25. W. N. Findley, J. Lai and K. Onaran, *Creep and Relaxation of Nonlinear Viscoelastic Materials*, Dover, New York (1989).
26. G. T. Mase and G. E. Mase, *Continuum Mechanics for Engineers, 2nd Ed.* CRC, Boca Raton, FL (1999).
27. E. H. Lee and J. R. M. Radok, *J. Appl. Mech.* **27**, 438 (1960).
28. T. C. T. Ting, *J. Appl. Mech.* **88**, 845 (1966).

26. G. T. Mase and G. E. Mase, *Continuum Mechanics for Engineers, 2nd Ed.* CRC, Boca Raton, FL (1999).
27. E. H. Lee and J. R. M. Radok, *J. Appl. Mech.* **27**, 438 (1960).
28. T. C. T. Ting, *J. Appl. Mech.* **88**, 845 (1966).
29. M. L. Oyen, *Phil. Mag.* **86**, 5625 (2006).
30. M. L. Oyen, *J. Mater. Res.* **20**, 2094 (2005).
31. M. L. Oyen, *Acta Mater.* **55**, 3633 (2007).
32. C. Tweedie and K. J. Van Vliet, *J. Mater. Res.* **21**, 1576 (2006).
33. J. M. Mattice, A. G. Lau, M. L. Oyen and R. W. Kent, *J. Mater. Res.* **21**, 2003 (2006).
34. C. Y. Zhang, Y. W. Zhang and K. Y. Zeng, *J. Mater. Res.* **19**, 3053 (2004).
35. C. Y. Zhang, Y. W. Zhang, K. Y. Zeng and L. Shen, *J. Mater. Res.* **20**, 1597 (2005).
36. M. L. Oyen and R. F. Cook, *J. Mater. Res.* **18**, 139 (2003).
37. M. L. Oyen, R. F. Cook, N. R. Moody and J. A. Emerson, *J. Mater. Res.* **19**, 2487 (2004).
38. G. Huang, B. Wang and H. Lu, *Mech. Time-Dep. Mater.* **8**, 345 (2004).
39. Y.-T. Cheng, W. Ni, C.-M. Cheng, *Phys. Rev. Lett.* **97**, 075506 (2006).
40. E. G. Herbert, W. C. Oliver and G. M. Pharr, *J. Phys. D. Appl. Phys.* **41**, 074021 (2008).
41. H. F. Wang, *Theory of Linear Poroelasticity with Applications to Geomechanics and Hydrogeology*, Princeton University Press (2000).
42. L. K. Agbezuge and H. Deresiewicz, *Israel J. Technol.* **12**, 322 (1974).
43. A. P. S. Selvadurai, *Int. J. Solids Structures* **41**, 2043 (2004).
44. M. Galli, K. S. C. Comley, T. A. V. Shean and M. L. Oyen, *J. Mater. Res.* **24**, 973 (2009).

CHAPTER 6

COMPUTATIONAL MODELING OF INDENTATION

Xi Chen

Department of Civil Engineering and Engineering Mechanics,
Columbia University
Email: xichen@civil.columbia.edu

Indentation is a popular method for evaluating mechanical properties of materials and structures, including elastic modulus, yield strength, hardening coefficient, residual stress, fracture toughness, and viscoelastic behavior, among others. It can also be applied to measure the properties of individual phases as well as global properties of composite materials, coatings and multilayers. Since it requires much less effort on sample preparation than other techniques, it is in particular useful for small material structures and biological materials (including living tissues). The primary challenge remains in the explanation of indentation data, since the material properties are implicitly related with indentation response; due to the finite local deformation involved and nonlinear contact conditions, computation modeling of indentation is an indispensable tool to correlate the indentation characteristics with material parameters, from which a reverse analysis can be carried out to identify material properties from indentation tests.

1. Introduction

Numerous advances in solid mechanics and materials science have been possible through instrumented indentation, a technique now widely used to probe mechanical properties of materials.[1] The depth and force of an indenter penetrating a material are measured continuously over the complete load cycle, resulting in a force-displacement diagram. Its popularity arises from that instrumented indentation is easy to conduct, and that only minor preparation (if any) of the tested material is needed.

The technique has been used to study the constitutive relationship of bulk material,[1-8] mechanical behavior of thin films and coatings[9-17] including heterogeneous microstructure,[18-20] interface adhesion,[21,22] residual stresses,[23] phase transformation,[24,25] dislocation behavior in metals,[26-28] fracture behavior in ceramics,[3,29-31] strength of bone,[32] time-dependent behavior in soft metals[33-35] and polymers,[36-40] and small material structures including the colloidosome particles[41] and carbon nanotubes.[42,43]

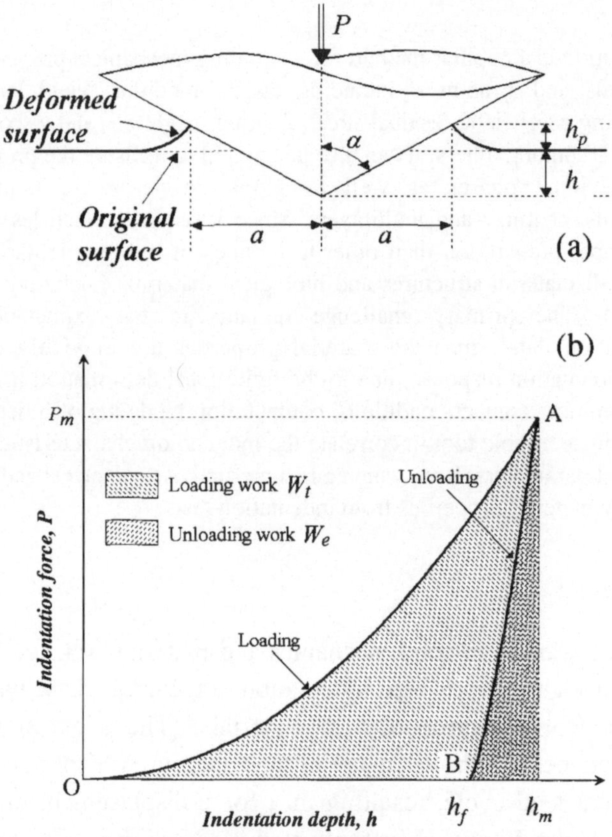

Figure 6-1. (a) A schematic of the sharp indentation test with a conical indenter whose half-apex angle is α. (b) A typical indentation load-displacement (P-h) curve measured by sharp indentation.

When the small compliance of the measuring system and the indenter tip are neglected, rigid sharp indenters (*e.g.* Berkovich, Vickers, conical) and spherical indenters (with radius *R*) are the most widely employed in practice. The contact between the indenter tip and specimen is usually taken to be frictionless. As the indenter penetrates normally into a solid, the indentation load, *P*, and displacement, *h*, are continuously recorded during the loading and unloading cycle. An example is given in Fig. 6-1. With a state-of-the-art nanoindentation, both *P* and *h* can be measured very accurately with high resolution, the key question is therefore: what can we learn from the *P-h* curve and how is that related with intrinsic material properties?

With respect to Fig. 6-1, it is convenient to define several shape parameters related with a typical $P-h$ curve. P_m, h_m and h_f are the maximum indentation load, the maximum indentation depth, and the residual penetration after complete unloading, respectively. During loading, the $P-h$ curve is quadratic for a continuum bulk and the loading curvature is $C = P/h^2$.[1] The contact stiffness $S = dP/dh\big|_{h=h_m}^{unloading}$ is obtained from the slope of initial portion of the elastic unloading curve. The work done by the indenter during loading, $W_t = \int_0^{h_m} Pdh$, is the area under the loading curve OA. Similarly, the area underneath the unloading curve AB is the total unloading work occurred during elastic recovery, $W_e = -\int_{h_m}^{h_f} Pdh$. Note that not all of the shape factors characterizing the indentation response are independent of each other; moreover, they need to be uniquely correlated with the material constitutive or fracture properties, which is the key for a successful indentation analysis or technique.

We use the simplest case of a homogeneous and isotropic bulk material as an example. Denote the material Young's modulus by *E* and yield stress by σ_y, without losing generality, the uniaxial true stress-strain ($\sigma - \varepsilon$) curve of a stress-free solid can be expressed in a phenomenological power-law form (Fig. 6-2)[1]:

$$\sigma = E\varepsilon \text{ for } \varepsilon \leq \sigma_y / E \text{ ; } \sigma = R\varepsilon^n \text{ for } \varepsilon \geq \sigma_y / E \text{ ,} \quad (6\text{-}1)$$

where n is the hardening exponent and $R \equiv \sigma_y (E/\sigma_y)^n$ is the work-hardening rate. When n is zero, Eq. 6-1 reduces to an elastic-perfectly plastic material. For most materials (including biological materials during loading) n is between 0.0 ~ 0.5, E is between 0.01 ~ 600 GPa, σ_y is between 1 MPa ~ 2 GPa, and E/σ_y is between 1 ~ 5000.[44] Note that the material deformation is not uniaxial during indentation, therefore, the influence of the Poisson's ratio, ν, must be taken into account.

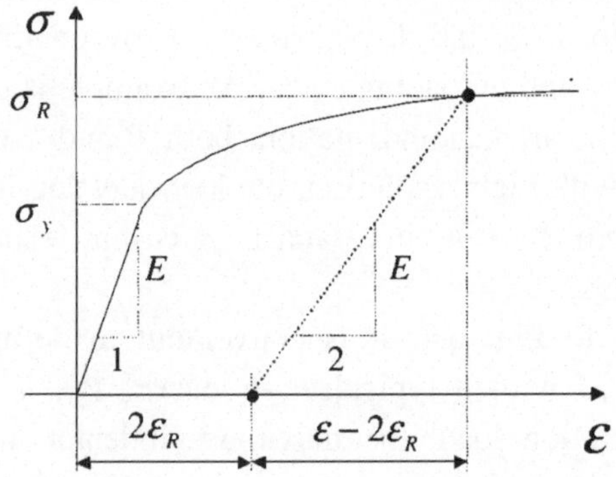

Figure 6-2. A schematic of the uniaxial true stress-strain curve ($\sigma - \varepsilon$) of a representative power-law material, and the definitions of the representative strain (ε_R) and representative stress (σ_R) used in this study.

Four independent parameters (E, ν, σ_y, n) are needed to completely characterize the constitutive properties of a power law stress-free material. Probing these material parameters by indentation has in recent years become a focal point of interest in the indentation literature and various techniques were proposed, see the review by Cheng and Cheng.[1] In most cases, the relationship between characteristic shape factors of the *P-h* curve and (E, ν, σ_y, n) needs to be established from extensive numerical analyses. Although indentation is a very broad and versatile technique, in this review we only focus on the measurement of constitutive properties.

2. Computational Method

The finite element method (FEM) is the primary tool of computational indentation analysis. Commercial software such as ANSYS[45] or ABAQUS[46] could effectively account for the nonlinear finite deformation due to contact, and to establish relationships between indentation responses, material properties, and indenter geometries. The strain gradient effect[47] is usually ignored by assuming that the indentation depth is sufficiently deep so that the continuum mechanics still applies to the bulk specimen. The rigid contact surface option can be used to simulate the rigid indenter, and the Coulomb's friction law can be easily embedded. In fact, friction is a minor factor for indentation[1,48] as long as this value is relatively small – when the Coulomb's friction coefficient is less than about 0.15 the contact solution is close to that of frictionless indent. This has been verified by the studies by Bucaille *et al.*[49] and Wang *et al.*[50,51]

In order to reduce computational cost, most researchers reduce the indentation problem to axisymmetric. Previous numerical work has shown that the loading and unloading *P-h* curves are almost same for pyramidal and conical indentations, as long as the ratios of cross-sectional area to depth are the same for these two types of indenters.[1,52] To simplify the analysis, the sharp pyramidal indenter is usually modeled as a rigid cone with half apex angle $\alpha = 70.3°$, so that the ratio of cross-sectional area to depth is the same as for a widely used Berkovich or Vickers indenter.[1,3] Other variants of α are also commonly used in the indentation analyses,[5,49,53-56] such as $\alpha = 60°$ (Rockwell tip), $\alpha = 63.14°$ (the cross-sectional area is half of that of the Berkovich indenter), $\alpha = 75.79°$ (the cross-section area is twice of that of the Berkovich tip), and $\alpha = 80°$, etc. For other common shapes such as spherical or conical, the indentation analyses are naturally axisymmetric. The axisymmetric assumption can also be employed when an indenter penetrates normally to thin films and multilayers.

However, in many other cases axisymmetry does not hold, such as in particle reinforced composites, anisotropic and inhomogeneous materials, and special indentation cases where indenter is close to specimen boundary, indentation induced surface radial fracture,[57] or when uniaxial in-plane residual stress present,[58] among others. In those cases three-dimensional mesh must be used.

Typical meshes for the axisymmetric indentation model and three-dimensional indentation model are shown in Fig. 6-3. Several key issues should be noted during computation: First, the effect of boundary

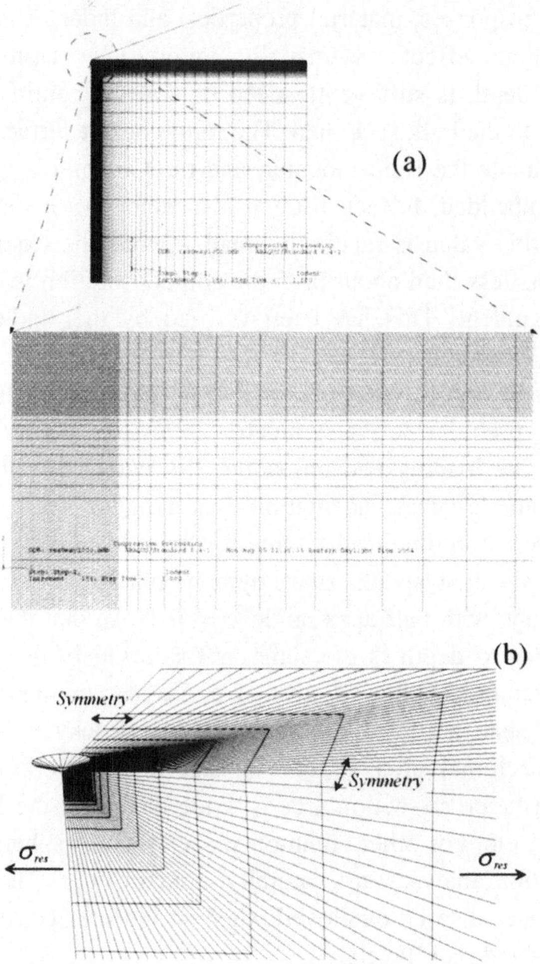

Figure 6-3. Examples of mesh used in FEM: (a) Axisymmetric mesh for conical indentation (b) Three-dimensional mesh for material with uniaxial stress.

(specimen size) has a profound effect on the resulting *P-h* curve; a rule of thumb is that the specimen size should be at least 50 times larger than the maximum indentation depth (in all directions) so as to eliminate the boundary effect. Otherwise, the boundary effect is more prominent on indentation stiffness compared with indentation hardness. Second, the mesh must be sufficiently refined below the indenter. A typical plastic zone has a radius of about two times the contact radius, and the mesh needs to be sufficiently refined within this region. Since the indenter would contact the surface nodes one by one, the spacing between the nodes dictates the resolution of key indentation measurements such as stiffness and hardness. Away from the indenter, the mesh can become more coarse.

Once the mesh is established, the material properties can be varied in a large range, and the indenter geometry can also be changed if necessary. Extensive FEM simulations can be carried out such that a relationship between the shape factors of indentation curves and material properties can be established. Appropriate functions can be chosen to fit the numerical results, and such functions can be mechanism-based to make it simpler. This is referred to as the forward analysis. Next, effective algorithms can be established to reversely solve the established functions, such that desired material properties can be deduced from indentation measurements. During such reverse analysis, there are two key issues, one is the uniqueness of solution (*i.e.* whether there is one-to-one correspondence between material properties and indentation behavior), and the other is error sensitivity (*i.e.* how sensitive is the determined material parameters with respect to the error of indentation data). Often during the numerical minimization process, special care needs to be paid to distinguish local minimum and global minimum. The working range of reverse analysis should be kept smaller than that used in the forward analysis, such that the numerical problems at the boundary can be avoided.

In the following text, we focus on measurement of constitutive properties of bulk materials to illustrate computational modeling of indentation. A brief introduction will be given for indentation on coatings, multilayers, composites, and other applications.

3. Sharp Indentation Analysis

For sharp indentation tests, the classic theory[2,3,8] relates the measured indentation hardness H and contact stiffness S with the constitutive properties of an elastic-perfectly plastic and stress-free bulk material (with $n=0$):

$$H = P / A = c_b \sigma_y , \tag{6-2}$$

and

$$S = 2\gamma\beta\bar{E}\sqrt{A/\pi} . \tag{6-3}$$

Here, $\bar{E} \equiv E/(1-\nu^2)$ is the plane-strain modulus and the projected contact area is A. The constant c_b in Eq. 6-2 is a constraint factor that depends on indenter angle and material properties: c_b increases with E/σ_y and approaches a constant (≈ 3) when $\bar{E}\tan\alpha/\sigma_y > 20$.[59,60] γ is a correction factor and it depends on the indenter angle (for a Berkovich indenter $\gamma \approx 1.08$),[61] and for axisymmetric indenters $\beta = 1$.[62] From dimensional analysis,[1] both hardness and stiffness are independent of the indentation depth if the strain gradient effect[47] is ignored for the continuum bulk.

The classic hardness-based indentation technique outlined above has several disadvantages. First, as the indenter penetrates the specimen, the material either produces plastic pile-up at the crater rim (when the yield strain, σ_y / E, is small), or exhibits the elastic sink-in effect (when σ_y / E is large)[60] – both strongly affect the projected contact area A and are difficult to measure in practice (especially at the maximum penetration when it needs to be measured), which makes it one of the primary error sources. Moreover, this approach cannot account for hardening and residual stress.

Since only the indentation load P and displacement h may be accurately measured during experiments, the key issue of a successful indentation analysis (theory) is to uniquely determine the material elastoplastic properties (E, ν, σ_y, n) from the shape factors of the $P-h$ curves. The self-similarity of $P-h$ curves of sharp indentation on bulk materials makes the conical indenters more widely used in indentation analysis: the resulting loading $P-h$ curve is quadratic and the unloading $P-h$ curve can be fitted into a power-law relationship.[63] Note that an analytical solution for indentation on an elastoplastic work-hardening solid (which involves finite plastic deformation) does not exist, and most studies in this area are based on numerical analyses. From dimensional analysis, Cheng and Cheng proposed the following relationship to correlate the indentation load, displacement, elastoplastic property, and indenter angle during loading[1,64]:

$$P = Eh^2\Psi\left(\sigma_y / E, \nu, n, \alpha\right). \tag{6-4}$$

Many researchers argue that the Poisson's ratio ν is a minor factor in sharp indentation[1,64] and in spherical indentation[48]; and thus ν can taken to be a constant (typically about 0.3). Although a recent study by Chen *et al.*[65] showed that the effect of ν is not negligible, by temporarily neglecting its influence the analysis could become simpler. The dimensionless function Ψ in Eq. 6-4, which is the normalized loading curvature and primarily depends on the material parameter set $\left(\sigma_y / E, n\right)$ for a given α, can then be calculated by using forward analysis. When the material properties are varied in a large range, a smooth functional form of Ψ may be obtained by fitting of the numerical forward analyses. A similar dimensionless function for the unloading curve was also proposed.[1,64] Finally, in order to obtain the constitutive behavior $\left(\sigma_y / E, n\right)$ of the bulk specimen from the indentation load-displacement curve, an effective reverse analysis algorithm must be developed. For example, Wang *et al.*[50,51] have developed numerical methods to solve for $\left(\sigma_y / E, n\right)$ from several dimensionless functions analogous to Eq. 6-4.

A mathematical trick known as the representative strain may be employed to take advantage of the self-similarity of the loading $P-h$ curves and reduce the number of apparent unknown variables in Ψ down to one. The concept of the representative strain was introduced by Atkins and Tabor,[66] which was later extended by Dao *et al.*,[4] and generalized by Ogasawara *et al.*[5,7] For a given indenter angle α, a representative strain ε_R and the corresponding representative stress σ_R (which depends on n) can be identified such that the new dimensionless loading function

$$\frac{C}{\sigma_R} = \Pi\left(\frac{\bar{E}}{\sigma_R}\right) \tag{6-5}$$

depends essentially only on $\bar{E}/\sigma_R$. There are various formulations of the representative strain-stress pair (ε_R, σ_R). Ogasawara *et al.*[5] proposed that the representative strain is related with the plastic strain of equi-biaxial deformation, and on the uniaxial stress-strain curve, the corresponding representative stress is (Fig. 6-2):

$$\sigma_R = R\left(2\frac{\sigma_R}{E} + 2\varepsilon_R\right)^n. \tag{6-6}$$

From extensive finite element analyses, the representative strain was found to be a function of α (from 60° to 80°):

$$\varepsilon_R = 0.0319 \cdot \cot\alpha . \qquad (6\text{-}7)$$

Specified for a Berkovich indenter ($\alpha = 70.30^\circ$), $\varepsilon_R = 0.0115$. Comparing with other definitions,[4,49,53,67] the formulation of the representative stress-strain given in Eq. 6-6, which was derived from an optimization process, not only has a better physical meaning, but also leads to the best apparent independence of n in Eq. 6-5 (and thus the best accuracy).[68] For $\alpha = 70.30^\circ$, the dimensionless loading curvature can be fitted as[5,68]:

$$\Pi\left(\frac{\bar{E}}{\sigma_R}\right) = -0.66\left[\ln\left(\frac{\bar{E}}{\sigma_R}\right)\right]^3 + 8.40\left[\ln\left(\frac{\bar{E}}{\sigma_R}\right)\right]^2 - 12.31\left[\ln\left(\frac{\bar{E}}{\sigma_R}\right)\right] + 9.21 . \qquad (6\text{-}8)$$

Similar polynomial fittings were widely adopted in the literature, *e.g.* Refs. 4, 49, 53, 55 and 69 and its variations were used for spherical indentations.[67,70] Such high-order fitting function often makes it very difficult to explicitly solve for material properties during the reverse analysis. Moreover, the fitted coefficients of the polynomial function (such as those in Eq. 6-8) do not have clear physical meaning, and the fitting function may behave incorrectly for extremely elastic or plastic materials used outside the material space of forward analysis. A simple relationship between C and $\bar{E}/\sigma_R$ needs to be established on a more physical basis.

Note that the specimen is essentially elastic when the variable $\bar{E}/\sigma_R \to 0$, whereas the material approaches to rigid plastic when $\bar{E}/\sigma_R \to \infty$ – the functional form $\Pi(\bar{E}/\sigma_R)$ should incorporate the limits of both mechanisms, such that it remains valid for all materials regardless of the range of data used for fitting. In the representative case of $\alpha = 70.3^\circ$ (Fig. 6-3), both of the elastic and rigid plastic limits can be well defined:

When $\bar{E}/\sigma_R \to 0$ (elastic), $C/\sigma_R = \Pi(\bar{E}/\sigma_R)$ varies linearly with $\bar{E}/\sigma_R$:

$$\frac{C}{\sigma_R} = m_e \cdot \frac{\bar{E}}{\sigma_R} \qquad \text{as} \qquad \bar{E}/\sigma_R \to 0 , \qquad (6\text{-}9)$$

where m_e can be derived from the classic solution of indentation on elastic materials,[71] and it agrees well with FEM calculations[7]:

$$m_e = \frac{2\gamma \tan\alpha}{\pi}. \tag{6-10}$$

When $\bar{E}/\sigma_R \to \infty$ (rigid plastic), C/σ_R approaches a constant:

$$\frac{C}{\sigma_R} = m_p \quad \text{as} \quad \bar{E}/\sigma_R \to \infty, \tag{6-11}$$

where m_p is the rigid plastic limit of conical indentation into a material that obeys the Mises yield criterion. Theoretical solutions of the plastic limits are available for either conical indentation on a specimen that obeys the Tresca yield criterion, or for wedge indentation with either Tresca or Mises yield criterion.[7] Although the errors of the existing solutions of the rigid plastic limits are not very large, in order to ensure maximum accuracy, the variation of m_p with respect to the indenter angle can be fitted from FEM simulations of conical indentations with the Mises yield criterion[72]:

$$m_p = 13.2\tan^2\alpha + 6.18\tan\alpha - 8.54 \tag{6-12}$$

for $50° \le \alpha \le 80°$. m_p is equal to 112.1 for the Berkovich indenter. In view of the importance of these two limits, a very simple form of $\Pi(\bar{E}/\sigma_R)$ can be proposed to incorporate both limits[7,72]:

$$\Pi = \frac{C}{\sigma_R} = \left(\frac{1}{m_e \cdot \frac{\bar{E}}{\sigma_R}} + \frac{1}{m_p} \right)^{-1}, \tag{6-13}$$

from which the representative stress can be obtained as $\sigma_R = \frac{m_e C\bar{E}}{m_p\left(m_e\bar{E} - C\right)}$ without iteration. Equation 6-13 not only incorporates the elastic and plastic limits (thus having physical meaning and wider range of application), but it is also involves no fitting parameter if m_p could be solved analytically. The example for Berkovich indenter (with half apex angle 70.3°) is given in Fig. 6-4.

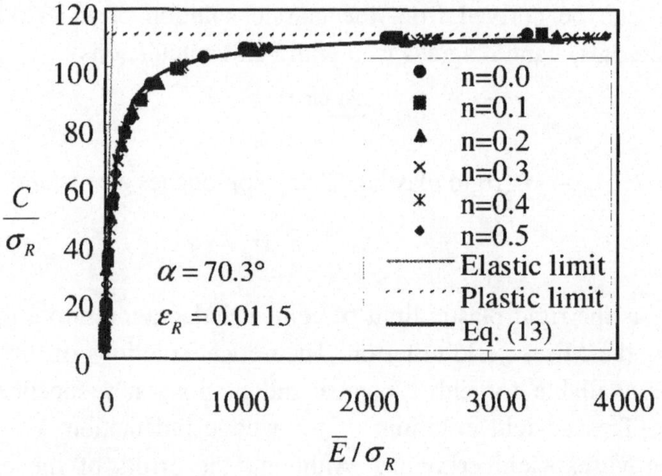

Figure 6-4. The relationship between C/σ_R and $\bar{E}/\sigma_R$ as n is varied, for $\alpha = 70.3°$. The numerical results (symbols) are shown with both elastic and rigid plastic limits, and the empirical fitting function incorporating these limits is given in Eq. 6-13 (which can be extended to other indenter angles).

For a given indenter angle α, once the dimensionless loading curvature Π is determined, if both E and ν known *a priori*, the only apparent unknown variable σ_R can be solved numerically from the measured $P-h$ data along with Eq. 6-5. Next, from Eqs. 6-6 and 6-7 a pair of flow stress-total strain (σ,ε) on the uniaxial stress-strain curve can be determined from the reverse analysis as

$$\sigma = \sigma_R \qquad \text{and} \qquad \varepsilon = 2\varepsilon_R + 2\sigma_R / E\,. \tag{6-14}$$

Finally, by using the dual (or plural) sharp indenter method, more pairs of true stress-strain can be determined from the indentation analysis when different α are used. It is commonly believed that the dimensionless loading function Π and representative strain ε_R are sufficiently different when another α is used.[1,53,56,73] For power-law materials, two independent conical indentation tests with distinct apex angle α are needed to determine the plastic parameters σ_y and n (assuming both $\bar{E}$ and ν are known); more indenter angles with larger separation may be used and the averaged result improves the accuracy of the reverse analysis.[5,74]

An example of the reverse analysis result is given in Fig. 6-5, which is taken from Ref. 7. In Fig. 6-5a, several ideal power-law materials are assumed, whose properties correspond to that of brass, gold, and aluminum, respectively. The uniaxial stress-strain curves of the input (true constitutive) properties are given as thick solid lines in Fig. 6-5a. The knowledge of their (E, ν , σ_y , n) is employed in a numerical indentation simulation using FEM, and the numerical $P-h$ curves are obtained. The simulations are carried out for three indenter angles, with half apex angle equals to 63.14, 70.3, and 75.79 degrees, respectively. Finally, the procedure outlined above, Eqs. 6-9 to 6-14 are used to identify three flow stress-total strain points (symbols) for each material in Fig. 6-5a, from which the plastic properties can be determined as the thin solid lines.

In Fig. 6-5b, two copper materials are studied and their uniaxial stress-strain tests (lab experiments) are shown in thick solid curves. These materials do not obey the power-law model exactly. Nevertheless, such stress-strain curve may still be employed in a numerical simulation using FEM, and the corresponding $P-h$ data can lead to the results of reverse analysis. The results of the identified material from the reverse analysis (symbols) are reasonably close to the input values, and the discrepancy mainly arises from the fact that these two materials do not conform to the power-law assumption exactly.

Other researchers have also proposed to measure the elastic modulus of the material from the unloading curves with a fixed ν, with approaches similar to Eq. 6-3.[49,54,55,73,75] When combined, the elastic and plastic properties of a specimen could be determined from the reverse analysis of the dual (or plural) indenter method, however, uniqueness of reverse analysis emerges as the main issue.

4. Uniqueness of Indentation Analysis

It is well known that single sharp indentation cannot yield unique solution. Based on extensive FEM simulations over a large range of materials, Alkorta *et al.*[76] and Tho *et al.*[77] found that only one shape factor of the $P-h$ curve can be claimed independent from unloading (such as the unloading work) and another one from loading (such as the

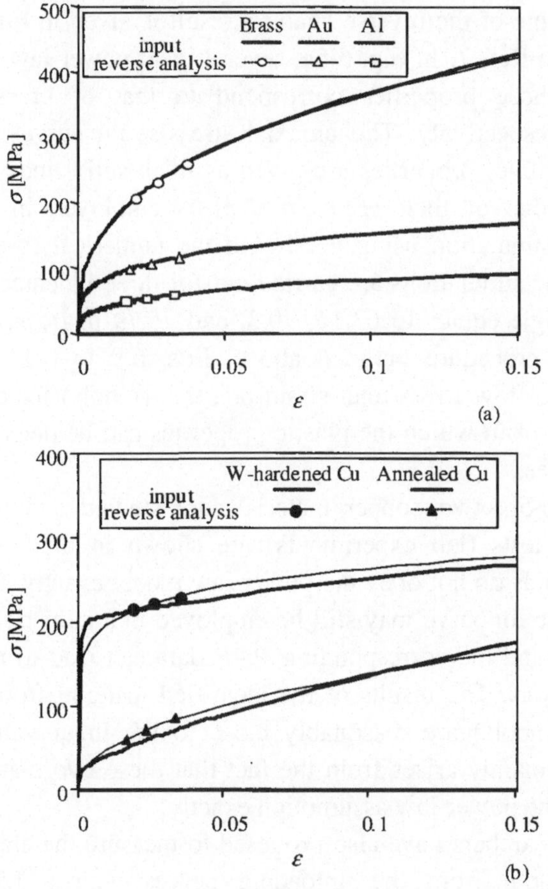

Figure 6-5. Illustration of using plural indenter method to determine material plastic properties: Good agreement between original input material property and constitutive properties derived from reverse analysis of numerical experiment. (a) brass, gold and aluminum which obey the power-law hardening rule; (b) work-hardened copper and annealed copper with stress-strain curves measured from uniaxial tensile test, who do not obey the power-law exactly.

curvature of loading curve or loading work). In other words, from one sharp indentation test on a bulk material, it seems that only two material parameters (*e.g.* σ_y and *n*) can be measured if the others (*e.g.* *E* and ν) are known *a priori*.[78] Thus, there is no one-to-one relationship between the material properties and the shapes of indentation load-depth curves.

For Berkovich indenter ($\alpha = 70.3°$), Alkorta *et al.*[76] have searched and found a set of distinct materials with almost identical $P-h$ curves.

Consequently, more information must be sought to obtain the constitutive relationship from the sharp indentation $P-h$ curves. It is widely assumed that the dual (or plural) indenter method provides such needed independent information.[1,49,50,53-56,69,73,76,77,79] In principle (based on the dimensional analysis, Eq. 6-4), unique loading and unloading curves can be obtained for a given set of material properties (E, ν, σ_y, n) and two (or more) sharp indenters. That is, although special sets of materials (with distinct elastoplastic properties) have been found to yield indistinguishable $P-h$ curves when a particular α is used, these materials are unlikely to again yield nearly identical $P-h$ curves when another α is chosen. It is believed by some researchers that with the additional independent equations established from other αs, there are sufficient equations needed to uniquely solve the material elastoplastic properties from the reverse analysis.[53,73]

Such a "well-established" theorem was recently challenged by Chen *et al.*[65] who have established step-by-step algorithms to explicitly derive sets of *mystical materials* exist, which, despite of their different elastoplastic properties, yield indistinguishable indentation load-depth curves for not only one particular indenter angle, but also indistinguishable $P-h$ curves other selections of α; in some cases, mystical materials cannot be distinguished effectively when α is varied from 60° to 80° (covering most indenter angles used in practice), and the resulting shape factors of the $P-h$ curves are within several percent of difference. When any power-law material is given, its mystical siblings can be derived explicitly (for a fixed Poisson's ratio) and when the Poisson's ratio is allowed to vary, more mystical groups can be derived with larger material property differences. It is further shown[65] that due to their similarities to plural indentation, shallow spherical indentation also cannot effectively distinguish these mystical materials.

The uniqueness problem (or finding of the mystical materials) is analogous to a multi-variable problem (Fig. 6-6), where the indentation shape factors (the general "z-axis") are related with the material elastoplastic properties (the general "x-axis") and indenter angle ranges (the general "y-axis") through a multi-dimensional "surface". On such

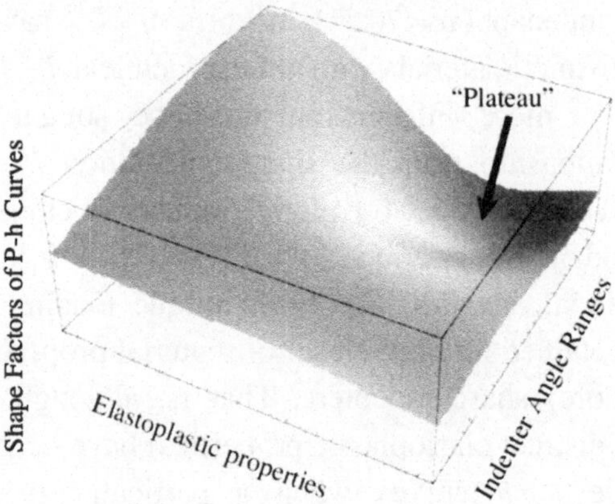

Figure 6-6. The schematic of the possible existence of the mystical materials and relevant indenter angle ranges – they correspond to the "plateaus" on the multi-dimensional "surface" of the shape factors of the indentation curves.

"surface", the "slope" may not be large and distinct everywhere, and small regions that are relatively "flat" may exist, as sketched in Fig. 6-6. The mystical materials and the relevant indenter angle ranges would correspond to such "plateaus" (with zero or almost zero "slope") of the multi-dimensional "surface" – although these plateaus are small, compared with the entire parameter space, they still exist and thus have important theoretical value for probing the uniqueness of indentation analysis.

The mystical materials have many interesting properties.[65] When the difference between the dual indenter angle ($\Delta\alpha$) is small, more mystical pairs can be found with larger differences in their elastoplastic properties. A gold mine of mystical materials can be identified when n is about 0 to 0.2, and when E/σ_y is about 0.01. These mystical materials are difficult to be identified by a wide range of indenter angle, and they correspond to technically important high-strength metals and alloys.

Figure 6-7a gives an example: this pair of mystical material leads to almost the same indentation loading-unloading behaviors when the indenter angle changes from 60° to 80°, which has reached the limit of the indenter angles used in most studies. Consequently, most existing dual (or plural) indentation methods in the literature would fail to distinguish them and more extreme indenters are needed. However this is often not practical because without advanced knowledge, one could not

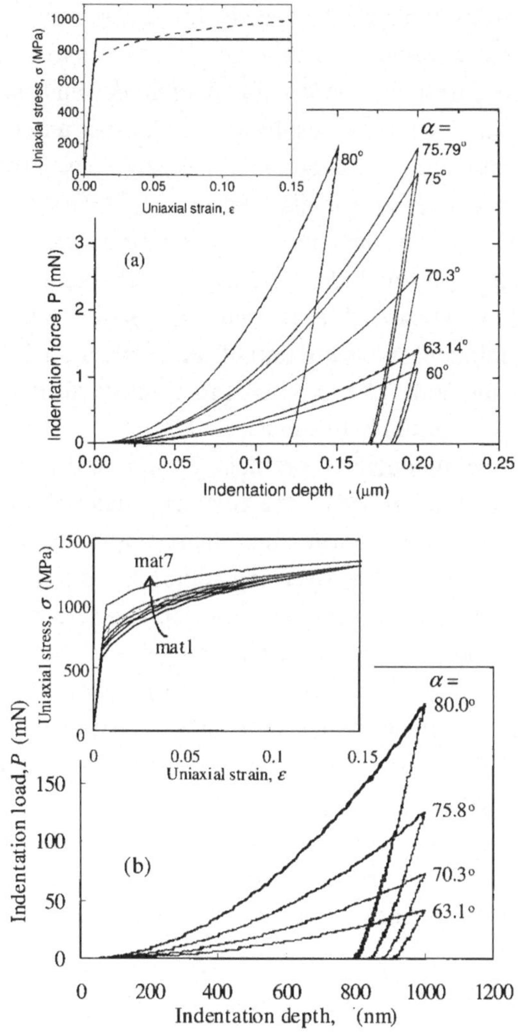

Figure 6-7. Examples of mystical materials; the $P - h$ curves of mystical materials and the uniaxial $\sigma - \varepsilon$ curves of mystical materials are given in insert on the top-left corner. (a) The Poisson's ratio is fixed at 0.3. A pair of mystical materials with $E_1 = 100\ GPa$, $\sigma_{y1} = 872.47\ MPa$, $n_1 = 0.0$ (solid curve) and $E_2 = 103.75 GPa$, $\sigma_{y2} = 715.61\ MPa$, $n_2 = 0.10663$ (dash curve). They cannot be distinguished by indenter angles from 60° to 80°. (b) The Poisson's ratio is allowed to vary (from 0.499 to 0.0 for mat1 to mat7). The material properties ($E(GPa)$, ν, $\sigma_y(MPa)$, n) are (104.23, 0.499, 574.4, 0.2434), (113.4, 0.40, 627.6, 0.2173), (116.9, 0.35, 659.4, 0.2038), (120.0, 0.30, 691.8, 0.1913), (122.2, 0.25, 725.5, 0.1784), (123.9, 0.20, 762.8, 0.1653), and (127.5, 0.0, 947.2, 0.1047), respectively, for mat1, mat2, mat3, mat4, mat5, mat6, and mat7.

pre-determine what kind of $\Delta\alpha$ needs to be used in an experiment. Moreover, when extreme indenter angles are used, the measured hardness would differ by orders of magnitude and new numerical problems associated with the resolution of the instrument emerge, and very sharp indenters may cause cracking and alignment problems.

When the Poisson's ratio is allowed to vary, an example is given in Fig. 6-7b, where the mystical materials can again exist within a wide range of indenter angles include those used in the literature, and for most materials their mystical siblings can be readily identified, which invalidates the uniqueness assumption of the conventional dual (or plural) sharp indenter method. The most distinct mystical materials can be obtained when ν is varied between 0 and 0.5.

In essence, the uniqueness problem is similar to error sensitivity. From a physical point of view, the mystical materials show a feature related to power-law strain hardening materials and the self-similarity of cone indenters. It means that by using plural sharp indenter techniques with experimental uncertainties, it is difficult to accurately obtain the elastoplastic properties for a wide range of materials. From a mathematical point of view, mystical materials exhibit low deviations between their corresponding indentation parameters. Since all experiments (lab or numerical) are subject to error, a few percent of such data fluctuation will lead to large perturbation of the determined material elastoplastic properties, and the mystical materials are especially sensitive to errors or data uncertainties. Alternative indentation techniques must be sought to more effectively distinguish these mystical materials.

5. Spherical Indentation Analysis

There have been several attempts to measure the elastoplastic properties with spherical indentation[23,80,81]; these methods, however, require an accurate determination of the contact radius which is very difficult in experiments. Once again, the key is to determine the unique elastoplastic properties solely from the trustworthy $P-h$ curves. Based on the contact stiffness, maximum load and penetration, Lee *et al.*[82] proposed a numerical technique which involves quite a few fitting parameters and not all of the varying rich information during loading were taken. Cao and Lu[67] have applied the idea of representative strain to normalize the loading $P-h$ curves of spherical indentation, and took measurements at two very shallow indentation depths. However, for spherical indentation, the loading curvature $C = P/h^2$ varies continuously with h during both loading and unloading processes. The self-similarity of the $P-h$ curves is not well-preserved and the apparent dependence of n cannot be completely removed with the incorporation of the representative strain.[70] Moreover, the conventional shallow spherical indentation cannot distinguish the mystical materials, as remarked earlier.

Let the radius of spherical indenter to be R. Chen *et al.*[83] have speculated that in terms of indentation work and contact area, there exists an analogy between the geometrical factor h/R of spherical indentation and α in conical impression. In essence, during the penetration of a spherical indenter, C decreases continuously as if the sharp indenter angle α is keep decreasing. Thus, a deep spherical indentation test can yield the similar effect of using very distinct sharp indenters, mimic the effect of a very sharp indenter (without the disadvantage of actually using a very sharp indenter), which may also circumvent the mystical material problem and lead to unique solution of the reverse analysis.

Zhao *et al.* proposed that by utilizing the contact stiffness, along with the normalized indentation loads obtained at two moderate indentation depths during loading, it is possible to measure the material elastoplastic properties (E, σ_y, n) (with $\nu = 0.3$) from one spherical indentation.[70] During penetration, the normalized loading curvatures are taken from $h/r = 0.13$ and 0.3, (the first one roughly corresponds to the sharp Berkovich indenter[83]); the contact stiffness is measured at unloading where $h_m/R = 0.3$. Within such framework, the normalized loading curvature and contact stiffness become

$$\frac{C_1}{\sigma_R^{(1)}} \equiv \frac{P_1}{h_1^2 \sigma_R^{(1)}} = f_1\left(\frac{\bar{E}}{\sigma_R^{(1)}}, n\right), \tag{6-15}$$

$$\frac{C_2}{\sigma_R^{(2)}} \equiv \frac{P_2}{h_2^2 \sigma_R^{(2)}} = f_2\left(\frac{\bar{E}}{\sigma_R^{(2)}}, n\right), \tag{6-16}$$

and
$$\frac{S}{h_2 \overline{E}} = g\left(\frac{\overline{E}}{\sigma_R^{(2)}}, n\right), \tag{6-17}$$

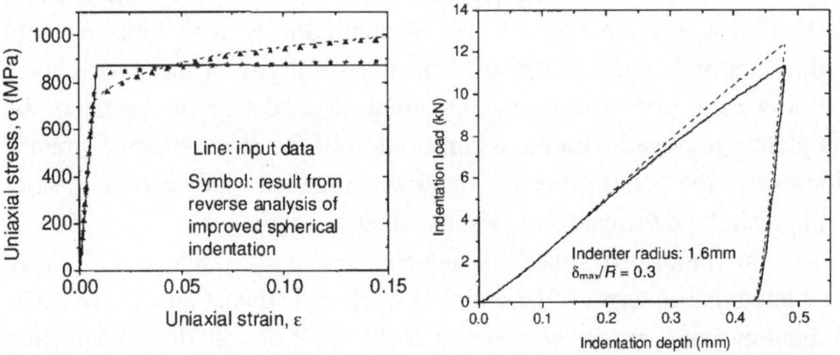

Figure 6-8. The $P-h$ curves of spherical indentation on the mystical materials found in Fig. 6-7a (solid curve for material #1 and dash curve for material #2). The deep spherical indentation method[70] can distinguish the mystical materials – in the figure on the left, the uniaxial $\sigma-\varepsilon$ curves obtained from reverse analysis (symbols) are compared with the input (true) data, and show excellent agreement.

where P_1 is measured at $h_1/R = 0.13$ during loading, P_2 is measured at maximum penetration of the experiment, $h_2/R = h_m/r = 0.3$, and S is the contact stiffness, measured from unloading at h_2. The representative stresses $\sigma_R^{(1)}$ and $\sigma_R^{(2)}$ are related with the representative strains, $\varepsilon_R^{(1)}$ and $\varepsilon_R^{(2)}$, respectively, whose functional form and values can be determined from numerical analyses in a way that is similar for sharp indentation test.[70] The functions 6-15 to 6-17 are calibrated from extensive numerical forward analysis; next, during the reverse analysis once P_1, P_2, and S are measured from an indentation test, their values will be substituted into Eqs. 6-15 to 6-17, and the set of candidate material properties which could minimize the total error of the three equations is the identified material. During this course, special attention needs to be paid to the identification of global minimum and to avoid local minima, this can be accomplished via an iteration process: first, the potential material elastoplastic properties are varied in a large range, from which several candidate materials who could lead to small total error of Eqs. 6-15 to 6-17 are identified. Next, from the established forward analyses, the shape factors of the numerical $P-h$ curve of these candidate materials can be compared with that of the measured $P-h$ curve: only the true solution can have the minimum error. This iteration process involves no additional FEM simulation, and it is based on the fact that a deep spherical indentation may lead to unique solution of reverse analysis.

As an illustrative example, in Fig. 6-8 we show that using spherical indentation, the pair mystical materials given in Fig. 6-7a can be effectively distinguished using deep spherical indentation (right). The reverse analysis result (symbols in the figure on left) agree very well with the input material properties (true solution) and properties of this pair of mystical materials can be measured separately.

6. Indentation on Thin Films and Composites

In the indentation analyses discussed above, the specimen is assumed to be a bulk with infinite thickness. This is perhaps the simplest model for a solid and the proposed computational modeling of indentation can be readily applied to homogeneous, isotropic, time-independent biological materials. However, when the specimen thickness is finite (as in many cases for small material structures) and bonded to a substrate or a testing platform, the so-called substrate effect strongly affects the indentation measurement and it usually should be avoided. Note that such effect is very common in biological materials, such as indentation on cell membrane where the effect of cytoskeleton also comes into play significantly. Various techniques have been proposed to subtract off the substrate effect from indentation analysis in order to measure the intrinsic film properties.[1,11,84] In addition, film-coating is just one of the simplest form of a composite material, and composite materials are very common in biological materials. The orientation of the reinforce phase (inclusion) can be random or ordered, and the inclusion can be fiber-like or particle-like. Moreover, other microstructural features, such as porosity, often come into play. All these underlying material structures need to be effectively accounted for during the indentation analysis, and the computational modeling must be adapted to the new changes.

We first use the simplest film indentation as an example. Many researchers[1,3,11] propose that the indentation depth must be sufficiently small such that the substrate effect can be avoided and the intrinsic film property can be measured; typically that means the indentation depth needs to be smaller than about 5% of film thickness. For some biological membranes such as lipid bilayer whose thickness is a few nanometer, this requirement may not be feasible. For very thin films, even if very shallow indentation depths as small as several nm can be made by a high-end commercial nanoindenter, there are many experimental issues interfering with measurements at this extremely small length scale. First, when the contact depth is of the same order as the surface roughness, it is very difficult to determine the contact area or depth accurately. Second, adhesion between the indenter tip and specimen may play an important role and hence the nanoindentation force may not be measured properly. Third, even a brand new indenter tip is blunt at the nanoscale, with a radius of curvature typically tens of nanometers; thus, the conventional sharp indentation analysis cannot be directly applied to the blunt tip at a small scale. Fourth, when h is well below one micron, size-dependent indentation effects have been measured[85,86] that are related to size-dependent plasticity and the effects of strain gradients on strength.[47,87] In addition, nanoindentation results may vary depending on whether the experiment is performed within a grain or at a grain boundary. These effects complicate the measurement significantly and make it more difficult to extract intrinsic film properties such as the elastic modulus, yield strength, and work hardening exponent.

In order to circumvent these issues encountered at extremely small length scales, it is desirable to perform microindentation experiments at moderate indentation depths, *e.g.*, depths on the order of a fraction of a micrometer. Thus, depending on film thickness, it may be inevitable that the substrate should affect the indentation measurement. Indeed, as the indenter approaches the interface between film and substrate, the measured hardness and stiffness are in fact combinations of film and substrate properties. The key issue is to understand the substrate effect and then subtract it from the microindentation measurement, so as to obtain the intrinsic film properties. Although the analysis becomes somewhat complicated, the approach is straightforward and can be based on well-established continuum mechanics, without the uncertainties at the nanoscale.[26,88]

If the substrate is stiffer than the film, when h is normalized by the specimen thickness t_f, the dimensionless variable h/t_f of film indentation would play a somewhat opposite role as h/R of spherical indentation. That is, when the indenter is approaching the coating/substrate interface, the increasing "hard" substrate effect (which is directly related with h/t_f) offers a new independent geometrical factor, just like continuously increasing the sharp indenter angle α in the plural indenter method. Therefore, the substrate effect could be utilized to determine the elastoplastic properties from one indentation test, and it complements the spherical indentation in terms of the trend of variation of α. Zhao *et al.* have proposed a framework of measuring (E, σ_y, n) (with $\nu = 0.3$) based on the normalized indentation loading curvatures taken at $h/t_f = 1/3$ and 2/3, and the normalized unloading work taken at $h_m/t_f = 2/3$. This theoretical formulation, which is similar to the spherical indentation introduced above, is proven to work well for soft metal coatings on elastic and stiff substrates[89,90]; an example is given in Fig. 6-9a for a copper film and it is expected that such technique could work well in general for other thin films on substrates, including the biological membranes. It is further shown that this technique works well to distinguish the mystical materials.[65]

When the specimen is porous as many biological materials do, including bones and tissues, indentation induces densification effect beneath the indenter which must be taken into account. Once again, by comparing the indentation characteristics of a porous specimen with that of a reference specimen, the effect of porosity can be deduced and the existing indentation approach can be modified.[19] An example is given in Fig. 6-9b where a modified factor is proposed: using such factor, the hardness can be easily converted to the intrinsic yield strength of the porous material, depending on different porosity and elastic stiffness of the specimen.

For composite materials or material with other types of micro-structure, the same principle can be applied. More refinements can be incorporated to include time-dependent viscoelastic behavior, nonlinear elastic behavior, anisotropic behavior, and inhomogeneous properties. Many of such properties are important for biological materials where computational modeling of indentation is an indispensable tool for establishing effective algorithms to extract the intrinsic mechanical properties of these materials. A comprehensive summary is given below.

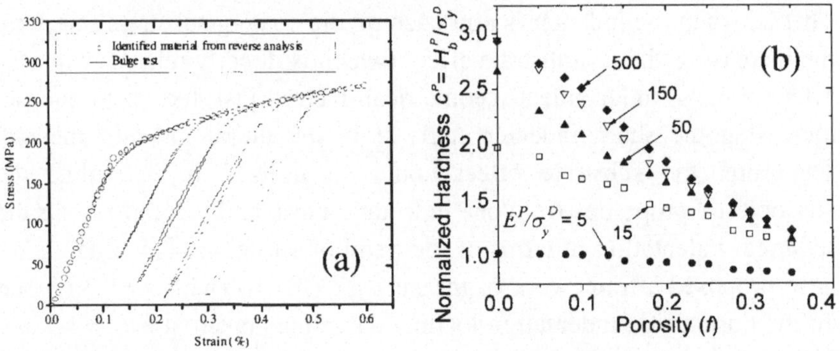

Figure 6-9. (a) Stress-strain curve of a Cu film: reverse analysis result (symbols) obtained from indentation on Cu film deposited on a Si substrate, which agrees well with that measured from a bulge test (line); (b) The hardness correction factor for indentation on porous film.

7. Conclusions and Outlook

In all cases of computational modeling of indentation, based on dimensional analysis, the variables can be grouped into the smallest possible sets needed to characterize the mechanics of indentation. It is straightforward to vary materials parameters over a wide range during the forward analyses, to calibrate the relationships between material variables (*e.g.* constitutive and fracture properties, time-dependent properties, material structure) and indentation responses (*e.g.* shape factors of the indentation load-displacement curve, stiffness, hardness, indenter geometry, crack length), shown in Fig. 6-10. Once indentation data are obtained from experiments, they are substituted into the established relationships and the mechanical properties of the specimen are iterated – the set of material properties that minimizes the total error with respect to the established functions are the parameters identified from the nanoindentation experiment. During such reverse analysis (Fig. 6-10), special attentions are given to ensure that the solution is unique through advanced numerical analyses. In addition, error sensitivity needs to be carried out to evaluate how sensitive the measured material parameters are with respect to common error sources encountered in nanoindentation experiments, as well as to establish appropriate algorithms to alleviate and smooth data fluctuations. Different indenters (*e.g.* pyramidal, conical, spherical) can be used for different systems, and the careful use of plural indenters may help to deduce more material parameters.

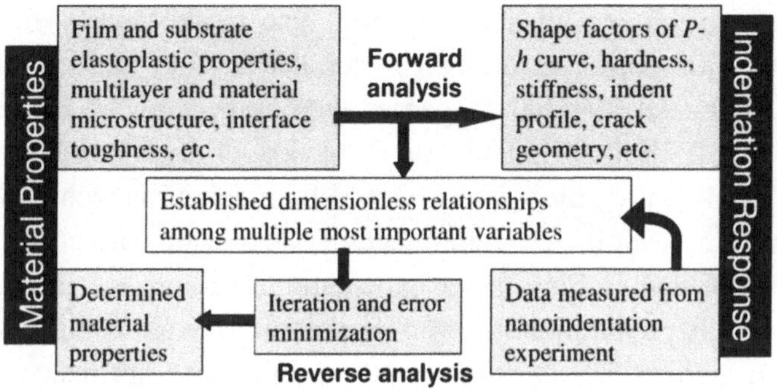

Figure 6-10. Computational modeling of indentation: The general forward (blue) and reverse (red) analysis algorithms, which allow the determination of the mechanical properties of various material structures using instrumented indentation.

The theory established from computational modeling of indentation must be closely tied with experiment, with theory both explaining experimental findings and predicting phenomena for experiments to explore; in turn, the experiment could validate and improve the theory. In use, several issues need to be paid special attention between experimental data and theory, otherwise significant error may emerge.[91]

Most indentation techniques proposed in literature are based on homogeneous, isotropic (and identical tensile and compressive behaviors), stress-free, power-law material, and assuming ideal indenter geometry (and also the specimen surface must be sufficiently flat), which may be referred to as the ideal model. An ideal model is also required for the analysis of experimental indentation data, where phenomenological material parameters may be extracted. If the real specimen does not conform to the five basic assumptions of the ideal model, errors may occur when one tries to use the indentation technique to extract elastoplastic properties from the indentation test. Nevertheless, as shown in earlier examples (*e.g.* Fig. 6-5b) that despite the non-ideal constitutive behavior of the specimen, when the reverse analyses are carried out correctly, the extracted material elastoplastic properties represent an "equivalent" power-law ideal material, whose indentation force-depth curve can well reproduce that of experimental data. The same principle applies to other non-power-law materials and time-dependent materials as long as an effective model can be chosen to effectively describe the constitutive behaviors of such material. The model always has certain deviation with respect to the real material behavior, and it is pertinent that the extracted material properties with respect to the model, could reproduce the similar indentation behavior of the "black box" real material. This is the criterion of a successful indentation technique.

There are several important issues related with correctly using the experimental data.[91] First, since indentation is a surface measurement, during sample preparation it is very important to avoid using mechanical forces to harden or alter the surface; otherwise intrinsic specimen properties are difficult to obtain. Second, the indentation depth should be carefully chosen such that it should not be too small so as to interact with surface roughness and other small-scale uncertainties, and at finite indentation depth boundary effect may come into play, which needs to be modeled using computational analysis. Third, the measurement of contact stiffness or unloading properties are sometimes strongly affected by instrument or indenter tip stiffness, which must be subtracted off. In the establishment of indentation theory via computational modeling (including those introduced above), a rigid indenter is assumed and thus in order to apply such theory, the experimental data must be corrected to remove any contribution from the instrument frame.

Many commercial indenters output the contact stiffness S_0 automatically (which includes the contribution of the indenter tip), using the continuous stiffness measurement, CSM. By employing the widely used diamond properties,[2] Young's modulus (E_i) of 1.141 TPa and Poisson's ratio (ν_i) of 0.07, the intrinsic contact stiffness S (which corresponds to that produced by a rigid indenter) can be obtained by deducting the indenter tip contribution from S_0. There are two approaches of deriving S in practice: (1) If the specimen modulus E is already given, $S = \bar{E}S_0/(\bar{E}^*)$ for both conical and spherical indentation, where $\bar{E} = E/(1-\nu^2)$ is the plane strain modulus of the specimen, and $E_R = \left(1/\bar{E} + 1/\bar{E}_i\right)^{-1}$ is the reduced modulus with $\bar{E}_i$ the plane strain modulus of the diamond tip. (2) In case E is not known *a priori*, based on the knowledge of S_0, E_R can be computed first via the Oliver-Pharr approach for conical indenter[2] or the Fischer-Cripps solution for spherical indenter.[92] Next, $\bar{E}$ can be obtained as $\bar{E} = \left(1/E_R - 1/\bar{E}_i\right)^{-1}$, and the Oliver-Pharr[2] or Fischer-Cripps[92] solution is reversed such that the intrinsic S can be derived for conical or spherical indentation, respectively.

Another important issue is to alleviate the data fluctuation measured in an experiment by imposing proper theory as the guide. For example, during a sharp indentation experiment, very often there is significant data scattering of S_0 even during the holding at the maximum load,[93] and such fluctuation must be removed in order to obtain more reliable material properties from the reverse analysis. Two approaches may be undertaken: (1) Average the S_0 reported by CSM during the entire period of holding at maximum load. (2) According to the Oliver-Pharr method,[2] $S_0/\sqrt{A}$ should be independent of h (where A is the projected contact area), and thus S_0/h should be roughly a constant during sharp indentation (especially when h gets beyond a certain depth). Thus, the averaged S_0/h (when the depth is above a certain threshold) can be used in the reverse analysis. The second method may lead to more stable results of the identified material properties from the reverse analysis. In case of spherical indentation, from Ref. 92, for a given material $S_0/\sqrt{h}$ should be almost a constant during indentation. Therefore, the averaged $S_0/\sqrt{h}$ can be used to derive S_0 at a particular depth.

In this chapter, we focus on the computational modeling of indentation, with an emphasis of extracting the constitutive properties of materials including biological materials. Presently, there are many open questions regarding the factors contributing to the constitutive response and cracking induced by indentation. Answers to these questions are of fundamental value and can be greatly assisted by computational modeling. A significant effort of computational modeling is to establish a reliable reverse analysis approach, which will be employed to determine various unknown material parameters from the experimental data obtained by indentation tests. Once determined, the material properties can be assembled to assess the system integrity, and for biomaterials, key issues related with mechanobiology (such as mechanotransduction[94]) can be addressed. Extensive experimental investigations need to be conducted in parallel with computational modeling, to inspire and verify the established theories. Computational modeling of indentation is still a wide open area thanks to its unlimited possibilities and applications, from which new techniques for probing mechanical behavior of materials with various microstructures are expected.

References

1. Y. T. Cheng and C. M. Cheng, *Materials Science and Engineering* **R44**, 91–149 (2004).
2. W. C. Oliver and G. M. Pharr, *J. Mater. Res.* **7**, 1564–1583 (1992).
3. G. M. Pharr, *Materials Science and Engineering A* **253**, 151–159 (1998).
4. M. Dao, N. Chollacoop, K. J. Van Vliet, T. A. Venkatesh, and S. Suresh, *Acta Mater.* **49**, 3899–3918 (2001).
5. N. Ogasawara, N. Chiba and X. Chen, *Journal of Materials Research* **20**, 2225–2234 (2005).
6. N. Ogasawara, N. Chiba and X. Chen, *Scripta Materialia* **54**, 65–70 (2006).
7. N. Ogasawara, N. Chiba and X. Chen, *Journal of Materials Research* **21**, 947–958 (2006).
8. F. Doerner and W. D. Nix, *Journal of Materials Research* **1**, 601–609 (1986).
9. E. T. Lilleodden, J. A. Zimmerman, S. M. Foiles and W. D. Nix, *Materials Research Society Symposium Proceeding* **673**, 131–136 (2001).
10. N. Huber, W. D. Nix and H. Gao, *Proc. Roy. Soc. London A* **458**, 1593–1620 (2002).
11. R. Saha and W. D. Nix, *Acta Materialia* **50**, 23–38 (2002).
12. T. Y. Tsui and G. M. Pharr, *Journal of Materials Research* **14**, 292–301 (1999).
13. W. C. Oliver, C. J. McHargue and S. J. Zinkle, *Thin Solid Films* **153**, 185–196 (1987).
14. W. D. Nix, *Materials Science and Engineering A* **234**, 37–44 (1997).
15. M. F. Doerner and W. D. Nix, *CRC Critical Reviews in Solid State and Materials Sciences* **14**, 225–268 (1988).
16. A. K. Bhattacharya and W. D. Nix, *Int. J. Solids Structures* **24**, 1287 –1298 (1988).
17. A. M. Korsunsky, M. R. McGurk, S. J. Bull and T. F. Page, *Surface and Coatings Technology* **99**, 171–183 (1998).
18. X. Chen, J. W. Hutchinson and A. G. Evans, *Acta Materialia* **52**, 565–571 (2004).
19. X. Chen, Y. Xiang and J. J. Vlassak, *Journal of Materials Research* **21**, 715–724 (2006).
20. J. Yan, T. Leist, M. Bartsch and A. M. Karlsson, *Acta Materialia*, in press (2008).

21. M. D. Drory and J. W. Hutchinson, *Proc. R. Soc. Lond. A* **452**, 2319–2341 (1996).
22. A. Vasinonta and J. L. Beuth, *Engineering Fracture Mechanics* **68**, 843–860 (2001).
23. J. G. Swadener, B. Taljat and G. M. Pharr, *J. Mater. Res.* **16**, 2091–2102 (2001).
24. W. Y. Ni, Y. T. Cheng and D. S. Grummon, *Surface Coatings Technology* **80**, in press (2003).
25. W. Y. Ni, Y. T. Cheng and D. S. Grummon, *Appl. Phys. Lett.* **80**, 3310–3312 (2003).
26. W. D. Nix and H. Gao, *J. Mech. Phys. Solids* **46**, 411–425 (1998).
27. Q. Ma and D. R. Clarke, *Journal of Materials Research* **10**, 853–863 (1995).
28. S. G. Corcoran, R. J. Colton, E. T. Lilleodden and W. W. Gerberich, *Phys. Rev. B* **55** (1997).
29. D. S. Harding, W. C. Oliver and G. M. Pharr, *Mater. Res. Soc. Symp. Proc.* **356**, 663–668 (1995).
30. D. R. Clarke and R. Tandon, *Mater. Sci. Eng. A* **195**, 207–214 (1995).
31. X. Chen, J. W. Hutchinson and A. G. Evans, *Journal of the American Ceramic Society* **88**, 1233–1238 (2005).
32. Z. Fan, J. G. Swadener, J. Y. Rho, M. E. Roy and G. M. Pharr, *J. Orthopaed. Res.* **20**, 806–810 (2002).
33. S. A. S. Asif and J. B. Pethica, *J. Adhesion* **67**, 153–165 (1998).
34. S. A. S. Asif and J. B. Pethica, *Phil. Mag. A* **76**, 119–122 (1997).
35. B. N. Lucas and W. C. Oliver, *Metall. Mater. Trans. A* **30**, 601–610 (1999).
36. B. J. Briscoe, L. Fiori and E. Pelillo, *J. Appl. Phys.* **31**, 2395–2405 (1998).
37. L. Cheng, X. Xia, W. Yu, L. E. Scriven and W. W. Gerberich, *J. Poly. Sci.* **38**, 10–22 (2000).
38. J. L. Loubet, W. C. Oliver and B. N. Lucas, *J. Mater. Res.* **15**, 1195–1198 (2000).
39. S. Shimizu, T. Yanagimoto and M. Sakai, *J. Mater. Res.* **14**, 4075–4086 (1999).
40. M. L. Oyen and R. F. Cook, *J. Mater. Res.* **18**, 139–150 (2003).
41. V. Gordon, X. Chen, J. W. Hutchinson, A. R. Bausch, M. Marquez and D. A. Weitz, *Journal of the American Chemical Society* **126**, 14117–14122 (2004).
42. P. J. d. Pablo, M. T. Martinez, J. Colchero, J. Gomez-Herrero, W. K. Maser, A. M. d. Benito, E. Muoz and A. M. Baro, *Materials Science and Engineering C* **15**, 149–151 (2001).
43. H. J. Qi, K. B. K. Teo, K. K. S. Lau, M. C. Boyce, W. I. Milne, J. Robertson and K. K. Gleason, *J. Mech. Phys. Solids* **51**, 2213–2237 (2004).
44. M. F. Ashby, *Materials Selection in Mechanical Design, 2nd edition*, Elsevier (1999).
45. ANSYS, *Ansys Release 8.0 Documentation*, ANSYS Inc., Canonsburg, PA (2003).
46. ABAQUS, *ABAQUS 6.4 User's Manual*, ABAQUS Inc., Pawtucket, Rhode Island (2004).
47. N. A. Fleck and J. W. Hutchinson, *Advances in Applied Mechanics* **33**, 295–361 (1997).
48. S. D. Mesarovic and N. A. Fleck, *Proc. Roy. Soc. Lond. A* **455**, 2707–2728 (1999).

49. J. L. Bucaille, S. Stauss, E. Felder and J. Michler, *Acta Mater.* **51**, 1663–1678 (2003).
50. L. Wang, M. Ganor and S. I. Rokhlin, *J. Mater. Res.* **20**, 987–1001 (2005).
51. L. Wang and S. I. Rokhlin, *Int. J. Solids Struct.* **42**, 3807–3832 (2005).
52. M. Li, W. M. Chen, N. G. Liang and L. D. Wang, *J. Mater. Res.* **18**, 73–78 (2004).
53. N. Chollacoop, M. Dao and S. Suresh, *Acta Materialia* **51**, 3713–3729 (2003).
54. M. Futakawa, T. Wakui, Y. Tanabe and I. Ioka, *J. Mater. Res.* **16**, 2283–2292 (2001).
55. A. DiCarlo, H. T. Y. Yang and S. Chandrasekar, *J. Mater. Res.* **18**, 2068–2078 (2003).
56. Y. P. Cao and J. Lu, *Acta Mater.* **52**, 1143–1153 (2004).
57. R. F. Cook and G. M. Pharr, *J. Am. Ceram. Soc.* **73**, 787–817 (1990).
58. M. Zhao, X. Chen, J. Yan and A. M. Karlsson, *Acta Materialia* **54**, 2823–2832 (2006).
59. K. L. Johnson, *Contact Mechanics*, Cambridge University Press (1985).
60. X. Chen and J. J. Vlassak, *J. Mater. Res.* **16**, 2974–2982 (2001).
61. J. C. Hay, A. Bolshakov and G. M. Pharr, *J. Mater. Res.* **14**, 2296–2305 (1999).
62. R. B. King, *Int. J. Solids Struct.* **23**, 1657–1664 (1987).
63. G. M. Pharr and A. Bolshakov, *Journal of Materials Research* **17**, 2660–2671 (2002).
64. Y. T. Cheng and C. M. Cheng, *J. Appl. Phys.* **84**, 1284–1291 (1998).
65. X. Chen, N. Ogasawara, M. Zhao and N. Chiba, *Journal of the Mechanics and Physics of Solids* **55**, 1618–1660 (2007).
66. A. G. Atkins and D. Tabor, *J. Mech. Phys. Solids* **13**, 149–164 (1965).
67. Y. P. Cao and J. Lu, *Acta Mater.* **52**, 4023–4032 (2004).
68. N. Ogasawara, N. Chiba, M. Zhao and X. Chen, *Journal of Solid Mechanics and Materials Engineering* **1**, 895–906 (2007).
69. Y. P. Cao and N. Huber, *J. Mater. Res.* **21**, 1810–1821 (2006).
70. M. Zhao, N. Ogasawara, N. Chiba and X. Chen, *Acta Materialia* **54**, 23–32 (2006).
71. I. N. Sneddon, *Int. J. Engng. Sci.* **3**, 47–57 (1965).
72. N. Ogasawara, N. Chiba, M. Zhao and X. Chen, *Journal of the Mechanics of Materials and Structures*, in review (2006).
73. S. Swaddiwudhiponga, K. K. Tho, Z. S. Liu and K. Zeng, *International Journal of Solids and Structures* **42**, 69–83 (2005).
74. N. Ogasawara, N. Chiba, M. Zhao and X. Chen, *Journal of Materials Research* **22**, 858–868 (2007).
75. Y. P. Cao, X. Q. Qian and J. Lu, *J. Mater. Res.* **21**, 215–224 (2006).
76. J. Alkorta, J. M. Martinez-Esnaola and J. G. Sevillano, *J. Mater. Res.* **20**, 432–437 (2005).
77. K. K. Tho, S. Swaddiwudhipong, Z. S. Liu, K. Zeng and J. Hua, *Journal of Materials Research* **19**, 2498–2502 (2004).
78. O. Casals and J. Alcala, *Acta Mater.* **53**, 3545–3561 (2005).
79. T. W. Capehart and Y.-T. Cheng, *J. Mater. Res.* **18**, 827–832 (2003).

80. J.-H. Ahn and D. Kwon, *J. Mater. Res.* **16**, 3170–3178 (2001).
81. B. Taljat, T. Zacharia and F. Kosel, *International Journal of Solids and Structures* **35**, 4411–4420 (1998).
82. H. Lee, J. H. Lee and G. M. Pharr, *Journal of the Mechanics and Physics of Solids* **53**, 2037–2069 (2005).
83. X. Chen, J. W. Hutchinson and A. G. Evans, *Journal of the American Ceramic Society* **88**, 1233–1238 (2005).
84. X. Chen and J. J. Vlassak, *Journal of Materials Research* **16**, 2974–2982 (2001).
85. R. Saha, Z. Xue, Y. Huang and W. D. Nix, *J. Mech. Phys. Solids* **49**, 1997–2014 (2001).
86. J. G. Swadener, E. P. George and G. M. Pharr, *Journal of the Mechanics and Physics of Solids* **50**, 681–694 (2002).
87. H. Gao, Y. Huang, W. D. Nix and J. W. Hutchinson, *J. Mech. Phys. Solids* **47**, 1239–1263 (1999).
88. W. D. Nix, *Metall. Mater. Trans. A* **20**, 2217–2245 (1989).
89. M. Zhao, X. Chen, N. Ogasawara, A. C. Razvan, N. Chiba, D. Lee and Y. X. Gan, *Journal of Materials Research* **21**, 3134–3151 (2006).
90. M. Zhao, X. Chen, Y. Xiang, J. J. Vlassak, D. Lee, N. Ogasawara, N. Chiba and Y. X. Gan, *Acta Materialia* **55**, 6260–6274 (2007).
91. N. Ogasawara, M. Zhao, N. Chiba and X. Chen, *Journal of Materials Research* **23**, 598–608 (2008).
92. A. C. Fischer-Cripps, *J. Mater. Res.* **16**, 3050–3052 (2001).
93. B. Guelorget, M. Francois, C. Liu and J. Lu, *J. Mater. Res.* **22**, 1512–1519 (2007).
94. Y. Tang, G. Cao, X. Chen, J. Yoo, A. Yethiraj and Q. Cui, *Biophysical Journal* **91**, 1248–1263 (2006).

CHAPTER 7

NANOINDENTATION OF BONE

Virginia L. Ferguson and Sara E. Olesiak

Department of Mechanical Engineering, University of Colorado
E-mail: virginia.ferguson@colorado.edu

While nanoindentation is an ideal tool for mechanically testing at the "tissue-level" of bone, unique challenges exist in both testing methods and analytical approaches. Bone is heterogeneous in composition and structure. Bone's hierarchical organization, typical of many biological tissues, implies that multiple structural levels may influence a single nanoindentation site. Mineral forms a complicated 3-D structure within and around a fibrillar collagen matrix. The degree of organization and mineral infilling varies with factors including the age, type, and location of bone material. Pores sized from nanometers to millimeters contain interstitial fluid that interacts electrochemically and structurally with bone's material constituents, adding further complexities to bone's mechanical response. This chapter reviews practices for the collection of high quality data at small scales where the organization of bone's three phases complicate our ability to prepare samples, visualize 3-D structure, and obtain exact measurements of material properties at small length scales. Nanoindentation, in combination with additional independent measures of the tissue, enable the detailed study of bone's poorly understood 3-D construction, material properties, and mechanical behavior at sub-micrometer scales. This chapter addresses the use of nanoindentation to study bone through consideration of the interactions within and between the collagen, mineral and water phases.

1. Introduction

Bone is a complex biocomposite that serves a wide range of structural and biological functions such as mechanical support, protection of vital organs, bone marrow storage, and calcium homeostasis. Through cellular activities, bone possesses traits that are desired by many materials scientists and engineers. Bone adapts to alterations in mechanical loading by changing its shape, size and location (or moment of inertia). Bone is self-assembling, self-maintaining, and self-repairing. It serves as the major structural framework of the body and possesses extraordinary mechanical properties. However, much is unknown about how the small-scale mechanical behavior of bone relates to its material construction and the physiological processes involved in its dynamic function. A full understanding of bone's mechanical properties, at multiple length scales, is essential for a comprehensive understanding of the structure-property-function relationships within healthy and diseased bone as well as for the ultimate success of bone implants and replacement materials.

Recent advances in depth-sensing indentation have permitted the study of bone mechanics at nanometer and micrometer length-scales that were previously accessible only to those using advanced microscopy and spectroscopic techniques. However, "nano"-mechanically probing such a tissue that possesses material and geometrical heterogeneity, viscoelasticity, and dependence on hydration is far from trivial. This chapter therefore details recent advances and challenges presented in the nanoindentation of bone.

1.1. *The structure of bone*

Bone has a hierarchical structure that requires interpretation of mechanical behavior at multiple length-scales (Fig. 7-1). At macroscopic length-scales, one can readily see that bone exists in different shapes and sizes. Different structural levels of bone also exist that include visibly dense, cortical (or compact) bone and trabecular (or spongy or cancellous) bone. Long bones, such as the "leg bones": the femur and tibia, provide a classical representation of bone. In long bones, dense cortical bone tissue surrounds the hollow mid-shaft region (that houses marrow and fatty tissue in adult mammals) and spongy trabecular bone fills in the regions close to the joint surfaces underneath the articular cartilage. Cortical bone provides the bulk of structural support to the skeleton and contains mostly microscopic pores. In contrast, trabecular bone is made of a network of mostly interconnected plates and rods that reside within the cortical shell or underneath articular joint surfaces. The millimeter-sized spaces in between trabeculae house bone marrow and are much more metabolically active than cortical bone due to their high surface-to-volume ratio. As such, loading patterns within trabecular bone are often visible to the naked eye where trabeculae align along lines of high mechanical loads. It is often the structural integrity of trabecular, not cortical, bone that is lost in disuse or osteoporosis and results in vertebral or hip fractures.

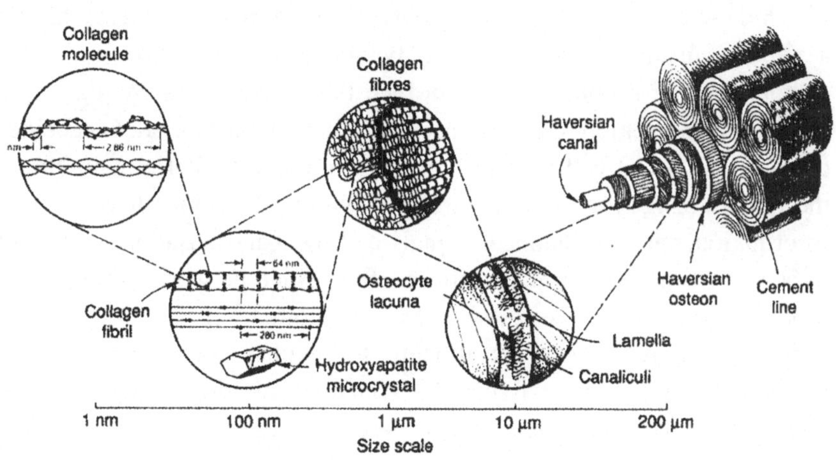

Figure 7-1. Hierarchical structure of human compact bone showing organization from the nanometer scale to the micrometer scale. The bone material is made up of distinct phases: organic (predominately Type I collagen) and mineral (predominately hydroxyapatite with additional forms of calcium phosphates). Water, both bound to the organic molecules and unbound within the interstitial fluid, forms the third constituent phase that acts to plasticize bone and impart viscoelasticity at the tissue (micrometer) level. (Figure reprinted with permission from Macmillan Publishers Ltd: *Nature*,[1] copyright 1993.)

1.2. *Bone cells*

Bone cells perform the key functions of self-assembly, self-maintenance, and self-repair. There exists a complex interplay, one that occurs through both local and systemic communication, between the main types of bone cells: osteoclasts, osteoblasts, and osteocytes. While many decades of research have been devoted to studying the communication among these cells, significant gaps remain in our knowledge of the specific molecular and biochemical mechanisms underlying their behavior.

The principal role of each cell type is generally understood and is covered in detail elsewhere.[2] In brief, osteoclasts are giant (20-100 micrometers in diameter), multinucleated cells that resorb, or remove, bone by solubilizing the mineral and releasing the matrix proteins to the surrounding interstitial fluid. Osteoblasts are small (≈15-30 micrometers in diameter), single nucleus bone-forming cells. They lay down the unmineralized osteoid, an organic extracellular matrix (ECM) onto which these cells subsequently regulate the deposition of mineral through a cellular mediated and/or thermodynamically driven process.[3,4] The osteoblasts line a bone-forming surface that moves through space over time, leaving a region of continual mineralizing bone behind. Osteocytes, the most abundant cell type in bone, are osteoblasts that remain behind the mineralizing front and are 'trapped' within the mineralized osteoid. Information, such as chemical signals and mechanosensation, is thought to be communicated through a porous, fluid-filled network of canalicular processes that connect osteocytes to one another.[5] In addition to other critical roles, osteocytes are thought to regulate local mineralization levels by maintaining extracellular ion concentrations. Through their collective actions and interactions, these three cell types determine and regulate the material properties of bone.

1.3. *Bone as a material*

Bone consists of an organic, mineral, and water phase where each constituent contributes to the mechanical properties at multiple length scales.[6] Fully mineralized, healthy bone is composed of approximately by volume 50% mineral, 20% water, and 30% organic (Fig. 7-2).[7,8] While the macro-scale structure and composition of bone has been widely studied and is fully understood, its nano-scale structure is much more elusive.

At the microscopic scale, bone is formed of distinct types, woven or lamellar, that each contributes to bone's material heterogeneity and

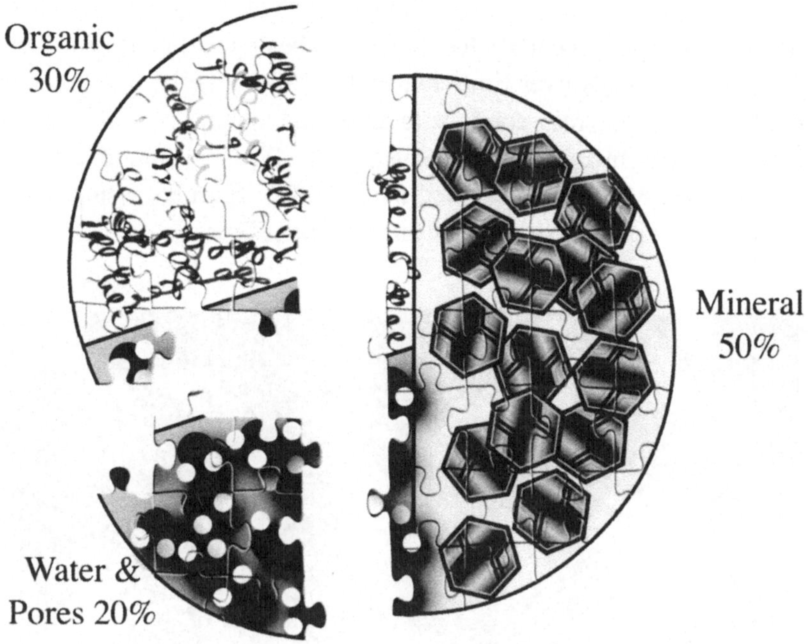

Figure 7-2. The nanometer length scale components and composition of bone are well known, yet how these structures combine to create bone's various microstructures is still under debate. The volume percent of each component is reported as in Gong (1964).[7]

mechanical anisotropy. Woven bone is rapidly deposited and poorly organized at the molecular level. It is formed during periods of rapid growth, such as in the developing embryo, or rapid bone apposition, such as in the initial stages of fracture repair. Woven bone contains a lower mineral content and is later replaced by more highly organized lamellar bone. Lamellar bone, in contrast, consists of highly organized layers (≈1 – 7 μm thick lamellae) of parallel mineralized fibers.[6,9] The lamellae are often arranged in alternating layers of differening collagen orientations (0 - 90°), which appear as "bright" or "dark" layers using circular polarized light microscopy (PLM).[10,11] A transition region between each layer creates the "rotated plywood" structure of lamellar bone that is visible in Fig. 7-1 (far right image).[12,13] In the simplest scenario, these layers form into osteons (≈200 – 250 μm in diameter): concentric lamellae that surround a vascular channel (*i.e.*, Haversian canal) and generally run parallel to the long-axis of the bone[14] (Fig. 7-3). Following the replacement of older bone and the deposition of many new osteons by bone cells over time, remnants of old lamellar structures or possibly woven bone may continue to exist (and mineralize with time) as interstitial bone. Imaging of bone using PLM[10] or scanning electron microscopy (SEM) in backscattered mode (BSE)[15] permits visualization of osteons, individual lamellae, and different bone types.

Figure 7-3. Circularly polarized light showing the lamellar structure in osteons from human femoral midshaft. Differing collagen orientations are visualized as "dark" and "bright" lamellae. Average osteon diameter measures 240 μm (×175; 70 μm thick transverse section).[10,11] Image used with permission from Ascenzi *et al.*, (2003).[10]

1.3.1. *The mineral phase*

Bone mineral is composed of extremely impure form of hydroxyapatite, $Ca_{10}(PO_4)_6(OH)_2$[16,17] which forms nanometer-sized crystals. A substantial amount of carbonate (CO_3^{2-}) is commonly substituted into the apatite lattice for OH^- or PO_4.[3-16] Additionally, apatite bone mineral contains a common Ca^{2+} ion vacancy which often contains sodium, potassium, magnesium, and zinc substitutions.[16] The mineral phase of bone has been widely studied by a variety of techniques: x-ray diffraction, infrared and Raman spectroscopy, neutron diffraction, electron microscopy, nuclear magnetic resonance (NMR), and atomic force microscopy (AFM).[16-23] Despite this wide range of analysis techniques, the collagen-mineral relationship is not fully understood. The nano-sized minerals in bone are smaller and contain more substitutions that those in enamel.[16] These extremely small crystals have the majority of ions occupying surface or near surface positions in the lattice, leading to further uncertainties about the structure of bone mineral. Most bone mineral (~98%) appears to exist as small plate-like crystals that measure $\approx 1 \times 10 \times 15$ nm.[23-25] Additionally, AFM techniques have determined that a few larger crystals exist that measure $\approx 40 \times 60 \times 90$ nm.[23] The size and shape of bone minerals may be a product of the initial formation of crystals between the collagen fibrils, in the hole region, as originally proposed by Petruska and Hodge in 1964.[26] However, the larger crystals are too large to fit within the fibril or hole region and have been observed, via AFM and transmission electron microscopy (TEM), to exist in the interfibrillar region.[23,27,28] Further, TEM observations have shown that the majority of bone crystal exists within and on the surface of the collagen fibers.[29]

1.3.2. *The organic phase*

The organic phase of bone is primarily composed of Type I collagen fibers (90%) and ground substance (proteoglycans, glycoproteins, small proteings such as osteopontin, and lipids). The collagen preferentially aligns and creates a three dimensional network, or scaffold, into and onto which mineral crystals form.[26] The collagen fibrils self-assemble from individual tropocollagen molecules (triple helices) and exhibit characteristic banding every 67 nm due to overlap (Fig. 7-1). Fibrils then form larger collagen fibers that are highly organized and cross-linked.[30] The geometrically complex collagen fibers contribute to the viscoelastic response seen in bone and provide toughness to the brittle mineral phase.[31] The non-fibrillar organic phase is suggested to act as a glue that holds the mineralized collagen fibers together.[32] Additionally, this matrix may serve as a source of sacrificial bonds, thus increasing the energy to fracture.[32]

1.3.3. *Water and pore space in bone*

Porosity and bone fluid play an important structural and mechanical role as fluid-filled pore space makes up approximately 20% of bone's volume. Pore sizes in bone range from millimeters to nanometers.[5] The largest pores exist in trabecular bone, vascular porosity includes Haversian canals (≈ 20 μm), followed by canaliculi and lacunae (≈ 0.1 μm), and finally nanometer-sized pore spaces (or 'matrix micropores'[33]) that exist between and within the collagen and mineral crystals. Within these pore spaces, interstitial fluid transports nutrients and wastes throughout the tissue and likely plays a mechanosensory role, whereby physiological strains result in a shear fluid flow that is thought to be detectable by osteocytes.[34]

Bone contains two types of water: unbound and bound. Unbound water, which exists mostly in the interstitial fluid within larger (≈ 0.1 μm to millimeter-sized) pore spaces, contributes significantly to the time-dependent mechanical properties of bone.[5] Water may also exist in smaller pore spaces between the mineral crystals and collagen fibrils.[35] The collagens and ground substance of the organic matrix contain many charged sites that facilitate interactions with unbound water. Bound water may stabilize the mineral crystallites by occupying OH^- and Ca^{2+} vacancy sites.[17] Overall, the permeability of bone tissue relies on the degree to which tissue is vascularized and the mineral density of the tissue.

1.4. *Motivation for nanomechanical analysis of bone*

Bone's mechanical properties vary widely at different length scales. Large hydrated cortical tensile specimens produce modulus values ranging from 14-20 GPa,[36,37] micrometer-sized specimens yield values of 5.4 GPa,[38] and tensile specimens from dissected osteons measure at 12 GPa.[39] While microhardness testing provides a direct measurement of the resistance to plastic deformation via Vickers hardness, residual indentation impressions are large and often extend far beyond fifty micrometers in width.[40] Nanoindentation provides an advantage that none of these techniques possess: the ability to avoid the influence of bone's complicated three-dimensional geometry on measured properties and to probe bone at the level of the tissue itself.

In order to understand bone's mechanical properties at multiple scales, the basic structural building blocks must also be investigated. Nanoindentation is the ideal tool as it allows testing of small volumes of bone and excludes the effect of large-scale porosity on mechanical property measurements. Additionally, nanoindentation can be used to determine the nanomechanical properties of bone's various microstructures (*e.g.*, individual lamellae), which are on the order of micrometers and are thus ideally suited for indentation testing. Nanomechanical data provide insight into tissue-level alterations in bone that are caused by factors including bone diseases, disuse or weightlessness, and even fossilization.

At any scale, the mechanical response of bone depends heavily on extrinsic factors that include preparation and preservation techniques, sample geometry (*e.g.*, testing a whole bone versus a machined section), hydration state, and mechanical test type and parameters (*e.g.*, strain rate). Additionally, a wide variety of testing parameters such as loading rate and indenter tip shape must be considered in order to extract a specific set of mechanical properties. The Oliver-Pharr formulation used to commonly extract elastic modulus from an indentation unloading curve assumes that the test material is homogenous, isotropic, and elastic-plastic.[41] Instead, bone is heterogeneous, anisotropic, and viscoelastic. Caution must therefore be taken when testing and interpreting nanoindentation results to understand bone's unique structural and mechanical response.

2. Testing Methodologies and Considerations

Nanomechanical testing of bone necessitates the careful selection of test parameters, well-controlled sample collection and preparation, and consideration of alterations within the tissue before, during, and after indentation testing is performed. A key concern lies in maintaining strict control during cutting, polishing, and storage to avoid unintentional modifications of the bone material. Factors ranging from alterations within the tissue itself to changes at the sample's surface are of critical importance. Although alterations in bone with preparation and storage have been addressed for testing bulk, macroscopic-scale specimens, little literature is available regarding the tissue-level changes that would likely affect bone's nanomechanical response.

2.1. *Indentation test parameters*

Every aspect of the nanoindentation test can have an effect on the measured mechanical properties. Nanoindentation of bone has been used by numerous investigators with a large range of testing procedures and sample preparation methods (key published data are shown in Table 7-1). Tip selection, loading rate, maximum load and other variables can play a significant role in the type of data that can be extracted from the test. This section provides an overview of nanoindentation test parameters as they relate to and influence the measured properties of bone tissue.

2.1.1. *Tip geometry*

The selection of the optimal tip geometry is critical for obtaining high quality indentation data. The biggest limitation in tip selection lies in our limited understanding of what occurs within the collagen, mineral and pore spaces within the bone material, at nanometer length scales, during an indentation event. For example, plastic deformation under an indenter tip likely results in compaction of the mineral platelets, possible generation of microcracks within the material of bone, pore collapse, and permanent displacement of water (for bone in a hydrated state).[42] The mechanics of how each tip contacts and indents into the bone material is therefore of critical importance and should be considered when determining the experimental approach to testing elastic, "plastic", or viscous behaviors. However, nanoindentation measurements in bone sample a three-dimensional volume of material, where Berkovich and spherical tips produce roughly comparable values of modulus for bone

(Table 7-2).[43] Thus the selection of tip geometry may be more important to the underlying contact mechanics and subsequent deformation mechanisms than to the measured material properties.

Berkovich: The majority of bone indentation experiments have been carried out with a Berkovich tip, which is supplied with most commercially available testing systems. The typical Berkovich tip, possessing a contact radii that is similar in size to a lamella ($R \approx$ 100-200 nm), has been effectively used to determine mechanical properties of such individual structural features in bone.[44,45] Berkovich tips are often preferred for indentation testing as they are geometrically similar and thus the indentation process is length-scale independent for a monolithic material. However, Berkovich tips are "sharp" and cause significant plastic deformation at the onset of indentation. While the contribution of plastic (and elastic and viscous) deformation to the overall indentation depth can be analytically determined,[46] the nature of this plastic deformation in bone remains unclear. Consideration should be thus given to the effects of indenting the viscoelastic material of bone with a sharp indenter tip.

Spherical: Spherical tips of various radii serve as a powerful tool to study both elastic and viscoelastic properties in bone tissue.[43,47-50] Unlike Berkovich tips, the geometry of spherical tips is less uniform from one tip to another and therefore require time-consuming and careful characterization of the tip area function. For example, spherical tips may flatten slightly or deviate from a perfect spherical geometry.[51] The radius of a spherical tip remains the major factor in producing high quality data.[51]

Spherical tips possess the advantage of producing an initial, primarily elastic response with a detectable transition to plastic deformation.[52] In contrast, Berkovich tips have an immediate onset of plasticity at very small displacements due to the high stress concentration near the material's surface. The use of multiple spherical tips of varying radii, which also differ in geometric similarity, enables a range of microstructural features in bone to be explored within a limited load and depth range.[43,52]

Cube corner: Fracture toughness of human tibial cortical bone has been investigated by using a cube corner tip to induce fracture.[53] Crack lengths likely vary with material anisotropy, compositional variations, and microstructural organization. Scratch testing, also using a cube corner tip, may also prove to be a useful tool for studying variations in bone quality.[54]

2.1.2. *Testing methods*

A variety of testing procedures have been used for the nanoindentation of bone. Three of the most commonly used methods include ramp-and-hold, constant stain-rate testing with continuous stiffness mode (CSM), and a partial-unloading technique. Because bone exhibits time-dependent behavior, properties are sensitive to strain rate and dynamic testing frequency.

Ramp-and-Hold: While the monotonic ramp-and-hold approach is likely the most commonly used test for nanoindentation of bone, variations in its implementation have produced varied results (Table 7-1). The viscoelastic nature of bone creates a significant dependence of the modulus values on the unloading rate and the creep hold at maximum load. Not permitting the creep displacement to saturate enables further displacement to occur during the unloading period, causing the unloading curve to demonstrate a characteristic "bowing out". In this case, the tip continues to sink deeper into the test material even while during unloading.[43] Such behavior increases the elastic recovery rate (dP/dh) and the apparent depth of contact (h_c), thereby overestimating the calculated value for indentation modulus.[41]

Approaches to eliminate time-dependence include a fast unloading rate, at least 10 times of the creep rate in order to elicit a primarily elastic response on unloading,[43] and the presence of a sufficiently long hold time at maximum load to allow creep behavior to minimize under the applied load.[43] The effects of varying the unloading rate and hold time at maximum load are demonstrated within the broader literature. A wide range of loading rates (from 0.01 to 10000 μN/s) and hold times (0 to 1000 s) have been reported (Table 7-1).[43,45,48,50,55-65] The measured modulus increases with the loading rate using a triangular loading-unloading profile (0 s creep hold) when using various loading rates ranging from 10 to 1000 μN/s, indicating the significant presence of

Table 7-1. Comparison of testing procedures used on various types of bone samples where "Inter" = Interstitial and "Trab" = Trabecular.

Reference	**Sample**	**Sample Condition**	**Inter. *E* (STD)**	**Osteon *E* (STD)**	**Trab. *E* (STD)**	**Rate (μN/s)**	**Tip**	**Test Type**
Vanleene *et al.*, 2006[55]	human femur	dry	24.34 (1.58)			0.05 1/s	Berk.	constant strain rate, hold (5 s)
Fan and Rho, 2002[56]	human tibia	dry in epoxy		18.6 to 32.2		10 to 1000	Berk.	precondition & hold (0 to 1000 s)
Rho *et al.*, 1997[57]	human vert. and tibia	dry in epoxy	25.8 (0.7) tibia	22.5 (1.3) tibia	13.4 (2.0) vert.	750	Berk.	ramp-and-hold (10 s)
Rho *et al.*, 2002[58]	human femur	dry in epoxy	23.8 (1.8)	21.8 (2.1)		400	Berk.	ramp-and-hold (200 s)
Hoffler *et al.*, 2005[59]	human femur	wet, dry, or plastic embedded		~20 to 24		0.005, 0.01, 0.02	Berk.	multiple loading
Bushby *et al.*, 2004[43] (data in Table 7-2)	equine third metacarpal	wet, dry, or PMMA		9.9 to 19.5 (E`)		10000	Berk. & *R*=5, 21 μm	ramp-and-hold (120 s) or multiple loading
Gupta *et al.*, 2006[60]	human femur	PMMA		~24 to 27 (Er)		200	Berk.	ramp-and-hold (60 s)
Mulder *et al.*, 2007[61]	newborn pig mandible	PMMA			9.4 (0.3)	80	Berk.	precondition & hold (50 s)

Table 7-1. Continued

Reference	**Sample**	**Sample Condition**	**Inter. *E* (STD)**	**Osteon *E* (STD)**	**Trab. *E* (STD)**	**Rate (μN/s)**	**Tip**	**Test Type**
Silva *et al.*, 2004[62]	mouse femur	PMMA		31.4 (2.7)		300	Berk.	ramp-and-hold (100 s)
Turner *et al.*, 1999[63]	Human femur	PMMA		23.45 (0.21)	18.14 (1.7)	12.5 to 750	Berk.	ramp-and-hold (10 s)
Ferguson *et al.*, 2003[50]	human femur	PMMA			18.8 (0.37) (E`)	N/A	Berk, & *R*=5 μm	partial unload (120 s)
Hengsberger *et al.*, 2002[64]	human femur	PMMA.in Ringer soln. 37°C		~9.5 to 20.5	~8 to 20.5	30 to 300	Berk.	ramp-and-hold (10 s)
Zysset *et al.*, 1999[45]	human femur	epoxy in saline	21.2 (5.3)	19.1 (5.4)	11.4 (5.6)	0.01	Berk.	ramp-and-hold (60 s)
Hoc *et al.*, 2006[65]	bovine femur	submerged in saline		19.3		0.05 1/s	Berk.	constant strain rate. CSM
Bembey *et al.*, 2006[48]	equine third metacarpal	submerged in ethanol		5.46 (0.17) (E`)		333 to 2000	R=21 μm	ramp-and-hold (120 s)

Note: *R* = spherical tip radius and "Berk" = Berkovich. For the column titled "Rate", the values listed are specified as constant strain rate loading (1/s) or loading rate (μN/s).

viscoelasticity and time-dependent plasticity.[66] In practice, the loading rate and creep hold time can, and should, be determined experimentally as the viscoelastic response of bone varies tremendously with many factors.[43] For example, Berkovich indentations in equine cortical bone, dehydrated in 100% ethanol and tested to a 5 mN maximum applied load, show creep on unloading with no hold (0 s) at maximum applied load. However a complete superposition of the unloading responses occurs following unloading after both 120 and 240 s creep hold times – thus indicating that a 120 s hold is sufficient for these samples and testing conditions.[43]

A limitation of many commercially available nanoindentation systems is that they permit only creep, and not stress-relaxation, testing. Because many systems are designed for testing stiff materials with no time-dependence, they are instrumented to control for load and not displacement. Advancements in instrumentation would assist in a sophisticated analysis of viscoelastic, biological materials.

Partial Unloading: Unlike the ramp-and-hold approach, the indentation modulus can be determined at a discrete number of contact depths via a multiple partial unloading technique that is described in detail elsewhere.[50-52] In brief, the indenter tip is incrementally loaded and unloaded into a material over a preselected number of intervals. As the initial unloading response of most materials is elastic, each loading event (to a force, P_1) is followed by a partial unloading to a fraction of that force (P_2) and enables a calculation of modulus at a specific contact depth. Successive loading continues, to a higher value of P_1, and is again followed by partially unloading to the same fraction of P_1. The net effect is that each successive loading event causes the applied load to increase until the maximum desired applied load is attained. A new value for modulus is collected for each pair of load-unload increments, thus permitting modulus to be collected as a function of contact depth. Typical values for partial unloading of bone using a small ($R \approx 5$ μm spherical indenter tip) might be a 5 mN maximum load with unloading to 75% of the applied load at each of 40 incremental loading events.[50]

The partial unloading approach is advantageous in that the separation of the loading and corresponding unloading points can be used to indicate the yield point of the material. This method readily permits the generation of an indentation stress-strain curve and analysis of modulus as a function of contact depth (for each incremental loading step). However, due to the viscoelastic nature of bone, the rates of loading and unloading are of critical importance to once again avoid a creep response from being superimposed on every partial unloading event. Similar to the ramp-and-hold method, such creep behavior results in an overestimation of the measured indentation modulus at each incremental depth.

Constant Strain Rate with continuous stiffness measurement: The continuous stiffness measurement technique utilizes a small sinusoidal oscillation that is superimposed over the main loading function. Continuous stiffness measurement, a standard feature on many nanoindentation systems, enables the continuous determination of stiffness from before making contact with a surface to the maximum contact depth at each indent site. Measurements of continuous stiffness enable the collection and analysis of modulus data as a function of contact depth at a resolution that is much higher than in multiple partial unloading.

In the literature covering nanoindentation of bone, the effect of using various oscillation amplitudes and frequencies is generally unexplored. With continuous stiffness measurement, an increase in mechanical properties seen with increased strain rate is similar to the increase seen with increased loading rate.[55] Even for identical experimental procedures, continuous stiffness measurement tests performed at a constant strain rate take variable amounts of time during the loading portion of the test and depend on the exact material response at each specific indent site, and consequently may vary in loading rates. Bone varies significantly in its inherent structural and material heterogeneity; therefore, continuous stiffness measurement testing will naturally vary from site to site. In addition, variable loading times introduce greater complications when testing a viscoelastic material. The use of continuous stiffness measurement for extracting nanomechanical property information from bone is also complicated due to bone's natural resonant frequencies and its viscoelasticity.[67]

2.1.3. *Preconditioning and repeated loading at single indent sites*

Mechanical preconditioning, a technique that is commonly used in biomechanical testing of viscoelastic soft tissues and elastomers, utilizes multiple loading cycles to establish a repeatable loading history.[68] Over successive loading cycles in soft tissues, the stress versus strain responses shift to the right and eventually converge.[69] However, preconditioning is seldom performed in large-scale mechanical testing of bone as it would likely result in micro-fractures and other permanent structural and material alterations prior to the onset of data collection. Most nanoindentation studies of bone also avoid preconditioning. However, preconditioning is considered here as several studies have employed what amounts to a preconditioning protocol, through repeated loading cycles at a single indentation site. Collection of nanomechanical property measurements is then made on the final loading cycle.[61,66,70,71]

In soft tissues, preconditioning is used to establish a reference state that has a significant effect on the resulting calculations of stress and strain.[68] In large or small-scale mechanical testing of bone, a reference state is naturally set by the rigidity of the material through the organization of collagen and mineral, three-dimensional tissue structure, mineral content, and other factors. Specific to nanoindentation, the reference state of a bone sample is set by the physical boundaries of the half-space – that is, the prepared surface of the bone sample itself. It is this generally stiff surface that provides a consistent definition of the reference state for nanoindentation through a consistent reference point for subsequent data calculations (*e.g.*, indenter contact depth).

The effects of nanometer-scale preconditioning, or repeated loading, on the material bone are unclear. Fan and Rho (2002) hypothesized that, during a plastic indentation, the bone material underneath the indenter tip undergoes radial expansion parallel to the surface and permanent compression perpendicular to the surface.[56] It is not unreasonable to surmise that preconditioning may compact the mineral crystals and, in a hydrated specimen, force water out of the indented region prior to the onset of data collection. While these phenomena have not been explicitly considered in the literature, they likely occur and have a profound influence on the measured nanomechanical properties. Given the difficulties of standardizing preconditioning protocols in areas of soft tissue research,[68] the implementation of such procedures for nanoindentation of bone would also require standardization. Further, it is unclear what is being tested, and consequently what the data indicates, on the final loading cycle of repeated indentations at a single site.

In summary, preconditioning or repeated loading at a single indent site, prior to nanomechanical testing of bone, has a high likelihood of altering bone's tissue-level structure at a scale that is easily resolved with nanoindentation measurements. Most bone samples possess a stiff surface on which nanoindenter systems tend to adequately detect contact. Therefore bone's reference state for nanoindentation is not in question. Preconditioning of bone will more likely complicate, rather than facilitate, the collection of useful and relevant nanoindentation data.

2.1.4. *Influence of contact depth*

The volume of material that contributes heavily to a material's overall nanomechanical response depends on factors that include the indenter contact depth and the size of the tip. The effective volume contributing to elastic modulus measurement can be described by a paraboloid of revolution with radius $3a$ and depth $5a$ (where a = radius of circle of contact between the tip and the surface).[43] It is a simple matter, therefore, to optimize the tip selection and experimental parameters to enable testing of the microstructures or regions of tissue within bone and, conversely, to avoid testing volumes that extend beyond the regions of interest.

The length-scale of bone's microstructural features of interest, for example a lamellae $\approx$5 micrometers thick, is critical to selecting the appropriate tip shape, tip size, maximum indenter penetration depth and corresponding maximum applied load. A wide range of indentation depths, from 100 nm[64] to 3000 nm[55] (Table 7-1), have been used to extract bone's modulus and hardness. The volume of material deformed is laterally seven times the depth of the indent,[72] thus the deeper indents test a larger volume of bone (Table 7-2). Larger testing volumes include the influence of heterogeneities such as pore spaces of various length-scales, changes in the degree of mineralization, increased numbers of lamellae, and multiple types of bone or other structural features. In addition, the maximum depth on loading may produce an effective volume that lies within the boundaries of the region of interest, yet further penetration upon a creep hold at maximum load may further expand the contact volume. Finally, interpretation of nanoindentation results in bone may differ from indents made to a constant maximum applied load, rather than a constant maximum contact depth, due to the inherent heterogeneity of the bone material.

Table 7-2. Typical values for the radius of contact circle: a, contact area: A_c, mean pressure: p_m, and an estimate of the test volume contributing to elastic modulus measurement: V_e; and [mean (SD), number n =] are presented for plane strain modulus, E', for a Berkovich tip (5 mN maximum applied load), an R = 5 μm spherical tip (15 mN), and an R = 21 μm spherical tip (100 mN) in both wet and PMMA embedded equine cortical bone. Data from Bushby *et al.* (2004).[43]

	Berkovich (5 mN)	**$R = 5$ μm (15 mN)**	**$R = 21$ μm (100 mN)**
Wet			
a (μm)	2.54	3.05	5.85
A_c (μm^2)	20.3	29.2	107.5
p_m (MPa)	248	513	931
V_e (μm^3)	1150	2000	14100
E' (GPa), n	11.2 (2.8), 308	9.9 (1.9), 117	11.7 (1.7), 326
Embedded			
a (μm)	1.61	1.81	4.54
A_c (μm^2)	8.14	10.3	64.8
p_m (MPa)	612	1460	1540
V_e (μm^3)	296	419	6630
E' (GPa), n	19.5 (2.7), 349	18.1 (2.4), 548	19.4 (2.1), 283

In general, the measured modulus of bone has been shown to decrease with increased max load ranging from 17.3 GPa measured at 100 μN to 12.3 GPa at 1 N.[73] As the contact area and plastic deformation increase at higher loads, it is likely that more pores and regions high in organic volume fraction are incorporated into a test volume to create a "softening" influence.[73] The consideration of the maximum depth of penetration with regards to the 3-D microstructure is clearly of great importance. Indentation to various depths can be used to probe bone's hierarchical structure. For example, shallow indents (<1000 nm) have been commonly applied to investigate bone's microstructural features – such as individual lamella.[44,60,64] These shallow indents, often performed with an AFM-based indentation system, are more sensitive to heterogeneities as less material volume contributes to the overall mechanical response within a single indentation. Surface roughness, porosity, polishing relief, and other near-surface characteristics may cause a high degree of variability in measured indentation properties.[74] Deeper indents sample large volumes of material and measured mechanical properties are averaged properties of the multiple heterogeneities. Using both experimental results and finite element models, nanoindentation modulus values have been shown to converge to a single value as the depth of the indent, and consequently the effective volume of measurement, increases.[75]

2.2. *Sample preparation*

Small-scale mechanical testing is complicated by the need to prepare samples for analysis. The mere process of removing tissues from *in vivo* physiological conditions triggers a series of events that potentially alter tissue properties. These include altered pH and ionic environment, dehydration, temperature, enzymatic degradation, and cell death. Further processing to prepare samples for analysis may involve storage over time; handling in the form of cleaning, sectioning, and polishing; decalcification (in some cases); and mechanically stabilizing the sample via embedding or through the use of a fixture. Throughout the collection, sample preparation, and testing processes, the maintenance of physiologically relevant conditions is often desired. However, more important is the uniform treatment, or control, of variables that may skew results in ways that are often difficult to recognize or interpret.

While some have attempted to mechanically probe[76,77] or measure strains[78-80] *in vivo*, the bulk of mechanical bone measurements are performed *ex vivo*. Depth-sensing indentation testing of bone in live human subjects is under development and may have great clinical potential[76,77]; however, the challenges of obtaining high quality and repeatable data are immense. Testing bone samples *ex vivo* provides such advantages as the ability to selectively alter bone properties, test specific orientations within the 3-D structure of bone, and couple indentation testing with other analysis methods to enable a multi-factor interpretation of the nanomechanical results.[50]

The key to success lies in tight experimental control of potentially influencing factors. That is, consistently ensuring the preservation of relevant properties during tissue collection, attending to conditions during subsequent storage, and selection of optimal techniques to preserve the condition of the tissue prior to and during nanomechanical testing. Because nanoindentation occurs at very small length scales, consideration must be given to alterations in the surface and near-surface properties of the sample. Further, the influence of collection, storage, and sample preparation on the three phases of bone are critical. Each of bone's components contributes to the mechanical response of the bone, and thus alteration of any single phase will affect the nanoindentation response.

2.2.1. *Sample collection and postmortem changes*

Preservation of properties for a study of tissue-level properties starts with well-controlled collection methods. Bone cell death and cessation of fluid movement through pore spaces likely causes rapid alterations in the biochemical environment. For example, live osteocytes are known to locally regulate mineralization.[81] Without cellular influence, a thermodynamically driven mineralization process may run unchecked. Over hours, catalytic enzymes that are present within the interstitial fluid or that result from bacterial presence may facilitate degradation of proteins that make up the organic matrix. Last, dehydration of the bone tissue may cause permanent changes in the mineral packing, similar to that seen in dehydration of nano-clays, and reorganization of molecular bonds in the organic phase through the removal of the polar water molecules and consequent increase in ionic concentrations in the remaining interstitial fluid.[47,82]

In summary, sample collection from an *in vivo* condition must be rapid. Should preservation of specific cells, proteins or genetic material be desired, immediate tissue processing is mandatory. Even in preparing for mechanical analysis alone, bone samples must be kept cool and hydrated. Tissues obtained from an abattoir are acceptable upon assurance of their history. Irrigation with an isotonic, buffered saline solution and storage on ice are typically sufficient during dissection, cleaning (*i.e.*, removal of non-osseous tissue), and short-term storage (on the order of hours) to prevent significant degradation and dehydration.

2.2.2. *Storage and preservation methods*

While storage and preservation of bone samples beyond the day of collection is often necessary, it is critical to consider the benefits and limitations of each storage method on tissue-level properties. Other more comprehensive reviews of storage and preservation methods exist.[83] A brief overview here examines preservation methods from the perspective of tissue-level alterations that may influence nanomechanical testing results.

Storage in isotonic, calcium-based solutions is often used to maintain hydration and conditions that are similar to those experienced *in vivo*. Significant decreases in bending stiffness with storage in isotonic saline for ten days were prevented by buffering with $CaCl_2$.[84] However, storage of machined equine cortical bone samples in a $CaCl_2$ buffered solution has been observed to result in the deposition of amorphous calcium phosphate crystals onto the sample surface.[85,86] Interactions between the mineral phase of bone and the solution in which samples are stored are also demonstrated by the energetically favorable replacement of hydroxyl groups with fluorine to form fluoroapatite.[87] Mineral deposits have been shown to form on samples following storage in phosphate-buffered fixative solutions.[88] In dental tissues, elastic modulus and hardness decreased significantly following storage in deionized water[89] but showed little change after storage in a balanced salt solution[89,90] or in 70% ethanol.[90] The mineral composition, and possibly the nanomechanical behavior, of a bone sample may therefore be significantly affected by the ionic environment of its storage solution.

Dehydration is a cheap and simple approach that can be employed to store bone samples over long durations. Following removal of non-osseous (non-bony) tissue, bone can be dried in ambient air or under an applied vacuum and stored for extended durations. Mechanical properties of dry bone are significantly different than for hydrated samples due to the formation of microfractures, decreased elasticity of the collagen, and possibly increased packing density of the mineral platelets.[91,92] Water may also serve to plasticize bone tissue.[43,47,48,93] Subsequent mechanical testing performed on rehydrated samples demonstrated little measureable effect at the scale of the whole bone.[94] However, the effects of dehydration in air followed by rehydration on nanomechanical behavior at the tissue-level are unknown.

Air-dried surfaces are readily polished and easy to test, yet substantial microstructural damage occurs through shrinking of the high water-containing organic phase. This is readily seen in tooth tissues, where large cracks typically separate the dental enamel from the underlying dentin in air-dried samples. Volumetric shrinkage measured by dimensional changes has been reported to be ≈6-7%.[95,96] However true volumetric changes may be significantly larger as shrinkage measured by helium pycnometer and the Archimedes principle decrease by 15-16%.[97]

Water is a main constituent of bone, thus the hydration state of the sample plays a vital role in the measured mechanical properties. Dehydration can cause increased modulus of elasticity, tensile and bending strengths along with reduced values of fracture toughness.[59,94,98] Nanoindentation experiments have shown that dehydration increased modulus values up to 50%, yet most studies found an average increase of 15-25%.[43,64,99] While the magnitude of mechanical properties is increased with dehydration, the inherent material relationships (*e.g.*, variations in type of bone) are maintained.

Chemical Fixation: Samples can also be dehydrated via chemical means through submersion in solvents, such as ethyl alcohol (ethanol, C_2H_5OH), which are commonly used to dehydrate tissues in preparation for histology.[47] Ethanol is pH neutral and avoids mineral dissolution, and may be most effective at storage temperatures of 4 to 7°C.[100] Such solvents simultaneously displace the water molecules and preserve the tissue from bacterial invasion and significant enzymatic degradation. It is thought that recovery from this form of dehydration is possible,[101] after rehydrating in an isotonic saline solution. However microcracks often form with such histological preservation methods. The extent of permanent alterations within the bone tissue that may occur following chemical dehydration and that affect nanomechanical behavior remain unknown.

Also, the replacement of unbound water by ethanol has been shown to increase the modulus of bone by more than 20%.[43] This substantial increase is more than expected of a simple fluid substitution and emphasizes the importance of water as a key component of bone. Unbound water may fill fine pore spaces and contribute to the stability of the collagen, while dehydration can cause stiffening of the fiber.[102]

Histological fixation with formalin (typically a 10% buffered solution of formaldehyde in water) or glutaraldehyde will also preserve bone tissue. If not buffered properly, formalin can become acidic, through the generation of formic acid, and dissolve bone mineral.[83] Formalin fixation has been shown to produce a significant decrease in hardness[83] with no concomitant change in stiffness.[103] Storage of dental tissues in ethanol resulted in no significant change in indentation modulus or hardness.[90] Overall, chemical fixatives alter bone's tissue-level properties through precipitation (ethanol) of proteins and the formation of cross-links between proteins (aldehydes). While the ability of bone tissue to fully recover from ethanol dehydration is unclear, chemical fixation in formalin or glutaraldehyde is irreversible.

Freezing is one of the most common approaches to preservation of bone tissue and has the potential to preserve water content, bone structure, cells, proteins, and even genetic material. However, freezing and thawing is known to cause significant damage when done improperly. A rapid transition to the frozen state minimizes the damage to the tissue by ice crystals. Use of a cryoprotectant, to replace water and reduce the amount of ice formed, also prevents destruction by the formation of large ice crystals.[104] However the cryoprotectant is only as effective as its ability to penetrate into the tissue—a process that is likely inhibited by small pore spaces and complicated microstructural geometries within the bone tissue.

Maintaining water content during freezing and thawing may prevent inadvertent damage from even slight dehydration. A common practice is to wrap bones in saline-soaked gauze and freeze in water-tight containers. Alternately, one can wrap a bone sample in cling film (to prevent escape of water) prior to freezing. Use of excess water is not ideal, and the study of cryoprotectants for preserving bone allografts may provide insight into optimal methods for cryopreservation and subsequent thawing and rehydration.[105]

The rate and temperature at which tissue samples are frozen will determine, in part, their quality of preservation. The rate at which the tissue passes through temperatures of ≈ 0°C to −10°C is critical, as this is the temperature range where large, destructive ice crystals may form if given a sufficient period of time. A rapid transition through this region is achieved by the use of small samples and a freezing temperature that is much less than −10°C. The ultimate storage temperature, for temperatures less than −10°C, has little effect on the size of the ice crystals. Rather it is the time that it takes to undergo the transition through this critical temperature range that is the key factor. One complication is that uncharacterized changes often occur within the tissue, and cause micro- or macroscale fractures, during transition from room temperature to very cold temperatures – such as during immersion in liquid nitrogen at −196°C. Attention must therefore be paid to artifacts caused by the freezing process in any experiment.

In general, a conventional refrigerator operates at +4°C and its companion freezer unit at –4°C. Damage to hard and soft tissues following freezing at –4°C is extensive. Further, such units typically include a self-defrosting feature, which cycles the temperature to eliminate the build-up of excess ice, and will rapidly destroy the microstructure of any tissue contained within. Bone structure and cells are reasonably well maintained when frozen at –20°C (the temperature of a typical 'chest' freezer). However the significant thermal mass of bone, especially for large sections, causes the exterior of the sample to freeze much more rapidly than the internal region. It is not uncommon to find small, microscopic fractures within a bone sample that has been frozen at –20°C. Storage at −85°C, the temperature of a typical lab freezer used to preserve proteins and genetic material, will also preserve bone samples for a long period of time.

Mechanical consequences of freezing and thawing on properties of whole or large machined sections of bones are not profound. In a survey of seventeen studies, bone structural variables, rather than material properties, were the most affected.[83] Overall, strength tended to increase and stiffness decrease. It was concluded from the survey that freezing bone samples at a temperature of −20°C, while maintaining moisture content, was sufficient to preserve mechanical properties of whole bones or large, machined bone specimens. Nanomechanical properties of dental tissues weaken significantly with freezing.[90] However, the effects of freezing at −20°C, or any other temperature, on the tissue-level properties of bone measured via nanoindentation remain unclear.

2.2.3. *Mechanical stabilization of samples*

The form of mechanical stabilization needed depends on the intended testing condition (*e.g.*, wet, dry, or embedded) as well as the size and 3-D structure of the bone sample. Bone samples can be machined using conventional machining tools, polished in manners similar to conventional ceramics or metals, and glued to glass slides or indenter 'stubs' to prevent movement during nanomechanical testing. During sample preparation, it is critical to exercise caution in preserving the surface of the bone by avoiding high temperatures, due to excess friction during cutting and polishing, and through modification by exposure to lubricants (including various forms of water) that might alter the bone material. In general, a band saw can be used to cut large blocks, followed by a low-speed diamond blade saw for machining surfaces with minimal friction. The form of lubrication during cutting or polishing must also be considered to avoid significant leaching or deposition of minerals (see Section 2.2.4).

Dehydrated samples are generally infiltrated with, or included within, various types of epoxies. These epoxies are highly viscous, prior to curing, and only fill large void spaces but not smaller pores (*e.g.*, Haversian canals or canaliculi). As the epoxy does not infiltrate into small pore spaces within the bone material, the measured mechanical properties remain unaltered for indent sites that are sufficiently far away from the epoxy bone interface.[56] Inclusion in epoxy resin permits the testing of dry or rehydrated bone samples following surface preparation using conventional (*e.g.*, metallographic) sectioning and surface polishing methods.

A commonly used alternative to including dry bone in epoxy resin is to, instead, embed bone samples in a low viscosity, infiltrating resin or polymer (*e.g.*, PMMA). Embedding facilitates complete infiltration of the embedding medium into pore spaces to preserve cells, maintains tissue morphology, and permits excellent surface preparation. High-quality tissue embedding is time-consuming and typically requires that the bone samples are dehydrated in a series of ethanol, 'cleared' with a solvent such as acetone or xylene that is miscible with the monomer or epoxy, and then slowly infiltrated with the embedding medium (unpolymerized monomer). As rapid curing or polymerization often results in incomplete infiltration into smaller pore spaces, the infiltration process may be prolonged over many days or weeks prior to encouraging polymerization with the addition of a chemical initiator (*e.g.*, benzoyl peroxide) and/or low heat (≈37-40°C).

Tissue embedding is useful for stabilizing the porous 3-D architecture of bone. Additionally, PMMA can stabilize delicate bone samples such as fossilized, demineralized, or deproteinated specimens. PMMA, with a low modulus of ≈ 5 GPa, is thought to minimally contribute to the measured mechanical properties of PMMA-embedded bone.[106] This is demonstrated by the similarity between measured values between dry bone and PMMA-embedded bone.[59]

2.2.4. *Surface preparation*

Surface preparation for nanoindentation is generally achieved by mechanical polishing or with microtome sectioning. Preparation of dried, included, or embedded samples may follow conventional polishing methods whereby the surface is smoothed by progressively finer grit silicon carbide paper and then polished with diamond suspension paste.[58] Between each step, samples are ultrasonically cleaned in an appropriate solution. Preparation of wet samples is possible, yet the fine polish afforded by pastes or slurries finer then ≈1200 grit paper appears to add little benefit. A simple approach to preparing surfaces of wet bone tissue is cryopolishing via placing polishing paper over a frozen metal surface.[43]

Soaking thin bone sections in distilled water for only 6 minutes has been shown to cause demineralization.[107] Boothroyd suggested using water saturated with calcium and phosphate ions,[107] however these ions might precipitate onto the bone structure. An anhydrous preparation using ethylene glycol as a lubricant and ultrasonic cleaning solution has been shown to produce excellent smooth surfaces with no demineralization or precipitation.[74] Alternatively, microtome prepared[60] or micromilled[50,108-111] bone surfaces are optimal in that they possess significantly less surface roughness than polished samples.

When comparing small structural features (*e.g.*, lamellae) at low depths the surface roughness of the sample can have significant effect on the measured mechanical properties. Mechanical polishing of bone selectively removes thin lamellae more quickly and creates a topographical profile of high and low regions which have been observed to vary by about 200 nm.[60] In the valleys created by the thin lamellae the contact area function A_c is underestimated, thus the modulus is over estimated.[60] When comparing the difference in modulus of thick and thin lamellae, a greater difference was seen with samples prepared via microtome due the superiorly flat surface preparation.[112]

Nanoindentation studies of bone have been shown to generate standard deviation values upward of 50%.[45,64] This large variability is partially a reflection of bone's natural tissue structure and heterogeneous nature, yet some sample preparation and experimental variation must also contribute to high standard deviation values. In order to probe individual bone microstructures (*e.g.*, individual lamellae) indentations must be sufficiently shallow to only probe the microstructure of interest, yet sufficiently deep as to avoid influence of surface roughness.

2.2.5. *Nanoindentation of wet bone*

Significant challenges are posed by small-scale mechanical testing of mineralized tissues under physiological conditions: *e.g.*, hydrated with a physiological buffered saline solution at 37°C. However the alternative of testing dry tissue in epoxy or histologically prepared, ethanol dehydrated and embedded, tissues are apparent from the preceding discussions in this chapter. A further complication enters in that testing wet tissues may be limited by the physical set-up of the indentation system, where some instruments lend themselves better to testing in fluid than others, and the challenge of maintaining stable surface characteristics over time.[43,64,85] The selection and validation of an appropriate storage solution and associated conditions, for example, is critical in preventing the appearance of a softer layer on the bulk sample, as was seen by Bushby *et al.* (2004)[43] or altered near-surface properties seen with storage in solutions that cause mineral dissolution or deposition and consequential alterations in nanomechanical behavior.[89,90] Finally, the surface roughness is greater in wet tissues due to difficulties in obtaining a fine polish.[43,74] Overall, the value of testing in wet conditions must be evaluated since obtaining high quality data, especially at shallow contact depths, is full of many challenges.

2.2.6. *Summary*

Physiological conditions are ideal for testing any biological material. Bone is no exception. Many researchers have attempted to determine the relevance of properties of bone *in vivo* and live bone *ex vivo*. However the benefits of testing under the ideal conditions must be weighed with the practicality and the benefits of physiologically relevant testing, versus the collection of high-quality data from well-controlled tests that lack degrees of physiological relevance. On macroscopic samples, mechanical testing of bone is often performed in an environment that mimics critical physiological conditions of hydration, ionic solution, and temperature. The accessibility of some nanoindentation systems to controlling such variables is limited, but adaptations to the systems can be made to enable testing under wet or temperature-controlled conditions.

3. Nanoindentation Behavior of Bone

Conventional analytical techniques for nanoindentation assume an elastic, isotropic, and homogenous material; yet bone fits none of these criteria. Historically, bone nanoindentation results have been analyzed using the Oliver-Pharr method using a Berkovich tip[41] or Hertzian contact mechanics for spherical tips.[51,52,72] Further, bone's complex construction produces time-dependent, or viscoelastic, behavior. While conventional analysis techniques enable a first order approximation of indentation modulus, advanced approaches are needed to accurately quantify the behavior of bone. This section provides an overview of the advances, and current limitations, in our understanding of bone at the tissue-level that have been enabled by nanoindentation.

3.1. *Anisotropy*

The elastic response of bone is anisotropic. Cortical bone, found within the mid-shaft of a long bone, is transversely isotropic[113] where the osteons and Haversian canals generally align with the long axis of the bone.[14] In one study of anisotropy in machined, millimeter-dimensioned sections of human cortical Haversian bone tested in compression and tension, elastic moduli were ≈40% greater in the longitudinal versus the transverse direction in tension and ≈55% greater in compression.[113] Other bone types also possess anisotropy where the properties vary with direction and with the directionality of lamellae, blood vessel networks, and laminae, as well as with anatomical site and age.[113,114]

At smaller scales, bone also possesses anisotropy. The organization of collagen fibrils and mineral crystals within the osteons and lamellar structure create an elastic anisotropy where a large percentage of mineral lies outside of the collagen fibrils.[115,116] Microhardness indentations, sampling areas containing many lamellae and other structures (*e.g.*, osteocyte lacunae), clearly demonstrate anisotropy using a Knoop indenter tip on selected bone surfaces.[117,118]

One key benefit of nanoindentation lies in the ability to measure properties at the level of individual structures within bone, such as within individual lamellae. By conducting nanoindentation testing in different directions, the anisotropy of individual microstructural components can be assessed. Similar to results in bulk sections of bone in compression,[113] Berkovich nanoindentation measurements on dry, cortical bone revealed an elastic modulus for longitudinally-tested interstitial lamellae (25.7 ± 1.0 GPa) that was ≈55% greater than that measured in the transverse direction (16.6 ± 1.1 GPa).[119] Nanoindentation measurements in osteonal bone yielded elastic modulus values (22.4 ± 1.2 GPa) that were ≈35% greater that that of the transversely-tested bone.

Anisotropy is measureable in both mineral and organic phases of bone. Equine cortical bone samples, experimentally manipulated from a wet, baseline condition to produce five different states: wet, dehydrated (in 100% ethanol), embedded (ethanol dehydration followed by PMMA embedding), decalcified, and deproteinated, showed characteristic anisotropic behavior in all but the wet state (Table 7-3).[93] With the exception of nanoindentation in wet bone, the modulus was significantly elevated in the longitudinal direction for all other conditions. Due to difficulties in sample preparation and testing of wet bone (as previously discussed in this chapter), subtle differences in the longitudinal versus the transverse indentation moduli may have been difficult to discern. That anisotropy was measureable even in PMMA-embedded decalcified bone demonstrates that the organization of the organic matrix may contribute to the overall behavior of bone in nanoindentation.

Table 7-3. Values [mean [SD], number n =] of plane strain modulus (E') in GPa from nanoindentation on planes that lie transverse and longitudinal to the long axis of the bone. Student's t-test between transverse and longitudinal directions – results reported as not significantly different (N.S.), significant (*), or highly significant (**). Data are from Bembey *et al.* (2005).[93]

Condition	*E'* Transverse (GPa)	*E'* Longitudinal (GPa)
Wet	11.7 (1.7), 326	11.8 (1.9), 68 (N.S.)
Ethanol Dehydrated	15.0 (2.2), 309	24.8 (2.2), 68 **
PMMA Embedded	19.4 (2.1), 283	25.8 (2.1), 74 **
Deproteinated (Embedded)	13.4 (1.0), 62	20.4 (2.2), 62 **
Decalcified (Embedded)	4.9 (0.3), 59	5.5 (0.3), 49 *

Values for modulus collected for specific anatomically referenced surface orientations can be compared to anisotropic elastic constants.[56,120] Nanoindentation is a multiaxial test where the measured modulus is a combination of moduli in all directions and is relatively insensitive to anisotropy (perhaps another reason for the lack of difference in wet bone samples).[93] The measured indentation elastic modulus will vary with anisotropy but will be weighted toward the properties in the direction of the applied load. Swadener *et al.* (2001) demonstrated that, through integration of measured elastic constants over the indented plane, an anisotropy ratio (based on elastic modulus *E*) can be determined that accounts for the indentation modulus (*M*) ratio in anisotropic materials.[56,120] For bone, the *E* ratio of 1.75 (ratio of longitudinal to transverse moduli) corresponds to a *M* ratio of $\approx$1.4.[120] This approach, in combination with testing bone of experimentally varied compositions, was used to demonstrate a potential mechanism for anisotropic material behavior.[106] Bone mineral therefore was implied to have greater connectivity in the longitudinal versus the transverse directions.

3.2. *Heterogeneity*

The three-dimensional structure and composition of bone is impressively complex. Nanoindentation is an ideal tool for studying the mechanical behavior within individual millimeter-scale structures, such as in trabecular bone, or micrometer-sized tissue-level features, such as lamellae. Further, probing at the nanometer scale permits the study of bone's compositional heterogeneity at the tissue-level and the relationships between the three phases of bone material.

Nanoindentation overcomes many of the difficulties that exist in accurately mechanically testing small volumes of bony structures. Attempts have been made to isolate individual osteons, lamellae, and single trabeculae for micromechanical testing.[121-124] Such examinations are complicated by our inability to accurately dissect tissues at very small scales without causing damage to the test specimen. Testing is also complicated by the geometry of the structural features of interest, such as in the case of a single trabeculae, where the cross-section varies in diameter and geometry along its length.[125] In addition to sample preservation and preparation, micromechanical testing is complicated by adequately gripping the sample, slippage, edge effects, and high measurement uncertainty—especially for custom-built mechanical testing systems. Such factors may profoundly influence testing results. Nanoindentation avoids such complications by providing a means of testing the bone material *in situ*.

Site-specific measurements of material properties can be collected on a scale (<1 micron) that is relevant to the tissue-level organization of bone. Structural features at the tissue-level are commonly at least several microns in width and ideally suited for nanoindentation testing (Section 4.3 below). However, the power of nanoindentation in studying bone's heterogeneity comes from the ease at sampling many different sites and combining mechanical measurements with adjunctive techniques.

Nanoindentation within a heterogeneous material enables statistical sampling of variability within the tissue's properties. Statistical sampling can be used to predict composition-based mechanical performance. For example, a large grid of indents on a well-characterized composite material, *i.e.* titanium-titanium monoboride (Ti-TiB), allowed for identification of the morphological arrangement and volumetric proportions of multiple, discrete, mechanically dissimilar phases.[126] A similar approach has been used to study bone and other mineralized tissues,[50,127,128] however, the complexity of quantifying the contributions of each component leaves more questions than answers. While some basic models have been able to relate mineral content to mechanical property variations,[42] more advanced paradigms are required to fully incorporate all sources of heterogeneity within bone that include mineral content, composition, and orientation, crystallinity, collagen content and organization, porosity, and the level of hydration as dictated by the chemical properties of bone's underlying extracellular matrix.

Depending on factors that include age,[129] type,[45,57,119,129-131] and genetics or disease state,[50] the properties of bone can vary significantly. Simple composites theory has been used to show that no simple composites model can describe the relationship between the elastic modulus and mineral content of bone.[132] Composite bounds, for example, merely predict bounds on bone's elastic behavior. Although many studies have shown a positive correlation between modulus and mineral content, the relationship is not exact and is influenced by degree of anisotropy, organization, and porosity.[133]

Nanoindentation, to examine the tissue-level of bone, has helped to better elucidate the relationship between modulus and mineral volume fraction (mineral V_f).[127] Until recently, this relationship was only studied on large volumes of bone. Similar to the work of Katz on macroscopic sized bone specimens,[132] no simple relationship exists between nanoindentation modulus and mineral content at the tissue-level (Fig. 7-4).[75] The modulus-mineral content relationship falls within the Hashin-Shtrikman (H-S) composite bounds, which describe the continuity between stiff particles in a continuous compliant phase (lower

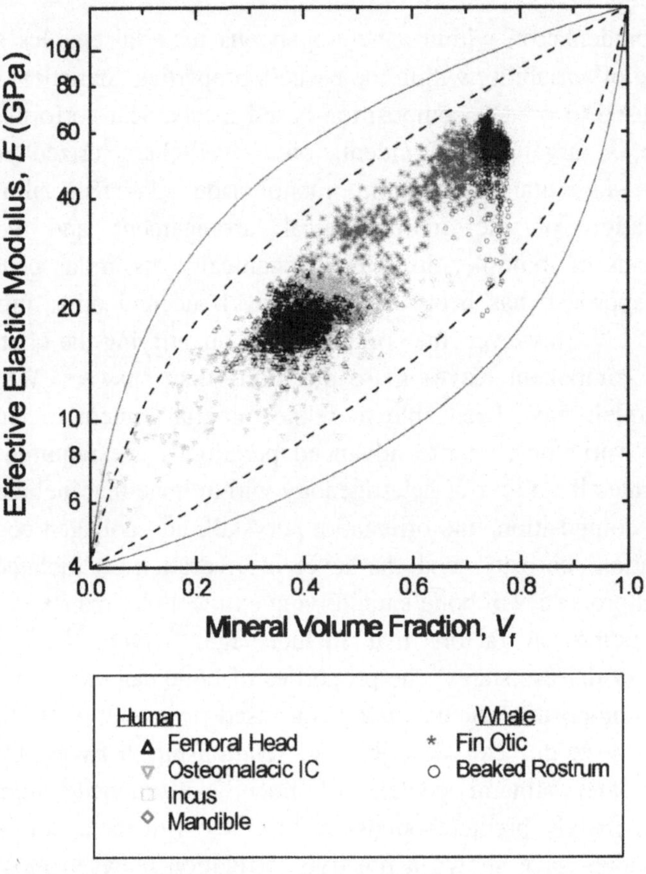

Figure 7-4. Elastic modulus (E) versus mineral volume fraction (V_f) of PMMA-embedded bone samples: human femoral head; human osteomalacic iliac crest (IC); human incus; human mandible; fin whale otic bone; and dense beaked whale rostrum. The solid lines are Voigt-Reuss bounds and dashed lines are the Hashin-Shtrikman bounds. From Oyen *et al.* (2008)[127] with permission.

bound) and compliant particles in a continuous stiff phase (upper bound).[134] In contrast to bone, site-matched measurements of indentation modulus and mineral volume fraction for dental enamel fall outside of the H-S bounds but lie entirely within the Voigt-Reuss bounds (unpublished data).

Elastic indentation modulus increases as a function of mineral volume fraction (as calculated from calibrated quantitative backscattered electron analysis at each indentation site).[50,127] The heterogeneity in factors that include the material composition (*i.e.*, the makeup of the mineral phase) and crystallinity, collagen orientation, and micrometer or nanometer-scale porosity result in the diffuse relationship that exists between modulus and mineral V_f throughout the range of bone types. Further, variance in measurements of modulus and mineralization decreases in tissues containing collagen fibrils that are, in general, uniformly well-aligned such as articular calcified cartilage[50] and in regions of uniform enamel prism orientation.[49]

3.3. *Viscoelasticity*

Bone behaves in a time-dependent manner due in part to its organic phase[135,136] and its water content with a corresponding poroelastic flow through the bone material.[5] Viscoelasticity in bone has been shown to correlate with hydration state[47,135,137] and mineral content.[136] Historically, the general approach for nanoindentation of bone has been to employ a range of techniques to eliminate creep effects, rather than to complicate nanoindentation data analysis with time-dependent behavior. A long creep hold at maximum load can eliminate time-dependent behavior on unloading and thus permits measurement of elastic properties.[138,139] Rapid unloading following no hold at peak load enables determination of the unrelaxed modulus.[140] A third method eliminates viscoelastic effects by correcting the unloading stiffness[41] with the creep rate and contact depth during creep (measured during a hold at max load).[141,142]

Rather than eliminate viscous effects during indentation, more recent works have embraced nanoindentation as a tool to explore the viscous behavior of bone. The use of indentation to study the elastic and viscous responses of biological materials is well established. Further, while nanoindentation was initially developed to study stiff, elastic-plastic engineering materials, its use for examining polymers is becoming widespread.[46,138,143-146]

Early work on the viscoelasticity of dry bone, included in a non-infiltrating epoxy resin, examined the role of loading rates or holding times (at max applied load).[66] The use of various loading rates (with a constant 180 s creep hold) demonstrated that elastic modulus was overestimated at "fast" indentation rates and that an extended hold time (at least 30 s) was required to eliminate the time-dependent plastic effect. Additional testing on dry bone specimens used a viscoelastic correction procedure[141] to produce a corrected elastic modulus. Both of these studies utilized a pointed, Berkovich tip and employed a multiple loading protocol at each indentation site prior to collecting data. The pointed tip may complicate the interpretation of the bone material's response by causing an immediate plastic deformation upon loading.[46] Single loading indentations made into a bone-dental implant interface using a pointed tip, analyzed using a viscous-elastic-plastic model,[147] produced modulus and hardness results that were comparable to those obtained by using traditional elastic-plastic analysis techniques. Further, the variations in the local resistance to plastic deformation dominated the overall response.

It can be argued that a spherical indenter tip provides significant advantages in probing bone – a material possessing elastic, plastic, and viscous behavior. Several studies utilizing a spherical tip have evaluated the viscous behavior of bone via creep testing following a finite-rate ramp loading.[47,48,148] Analytical expressions, based on curve-fits to the load-time or displacement-time data during a creep hold at maximum load, were used to calculate measurements of the creep or stress relaxation functions. This approach was used to examine the role of hydration in the viscous behavior of bone, where cortical bone soaked in varying concentrations of ethanol demonstrated a loss of viscosity with increasing dehydration.[48] Similarly, bone soaked in solvents possessing varying degrees of polarity (and Hansen's solubility parameter for Hydrogen bonding) revealed a similar dependence on hydration and a distinct role of hydrogen bonding in their viscous response.[47] While elastic stiffness, as measured by the values of time-zero shear modulus

(G_0), increased with decreased Hansen's solubility parameter, the viscous creep increased. A poroelastic analysis of the same data set revealed that bone's intrinsic permeability increased with increasing values of the solubility parameter.[148] Correlation of these results to the fundamental porosity length scale indicated that a characteristic pore size of ≈1.6 nanometers describes the intrinsic permeability of water in bone at the scale of the collagen–apatite interactions. While such analyses are simplified and preliminary in nature, they demonstrate the applicability of nanoindentation for the exploration of time-dependent and molecular-level interactions in bone.

Analysis of viscoelasticity in bone is well within the reach of available analytical tools. In particular, approaches for analysis of polymers with viscous behavior are well-developed[46,141,143-146,149] and applicable to the study of biological tissues – including bone.[47,48,66,147] Such analytical formulations are ideal for the analysis of local variations in tissue properties. These analytical methods are easily applied and can provide insight into the viscous behavior at the tissue-level of bone. Finally, such analyses may add to our ability to better understand many research and clinical problems.

3.4. *Finite element modeling*

Finite element analysis (FEA) has been commonly used to analyze nanoindentation of various engineered materials, however application to bone is complicated by its heterogeneous composition and structure. Most finite element models assume bone is elastic-perfectly plastic, homogenous, and isotropic, and thus fail to capture bone's true mechanical response at the tissue-level. Alternatively, several models have investigated bone's complicated structure by modeling bone as a two-phase composite: organic and inorganic. These analyses have employed standard composites theory and modeled bone as a particle or rod-reinforced matrix.[132,150] The mineral volume fraction of normal bone generally ranges from 0.35 to 0.5, yet over this small range the modulus of bone ranges widely.[127,148] As such, the modulus of bone cannot be simply predicted by knowledge of the mineral volume fraction of bone.[132]

Recently, more complicated models have investigated the size and shape of mineral crystals as an elongated, staggered arrangement that is embedded within a compliant organic phase.[151,152] As the removal of the mineral or organic phases leaves a physically intact specimen, both phases form continuous networks and should be modeled as such. One model, using energy arguments, has suggested a continuous mineral phase as a self-supporting foam reinforced with collagen molecules.[153] An additional model of bone that explored material heterogeneities and assumed a continuous mineral phase produced results that compared well to experimentally measured values for elastic modulus.[106] While approximate indentation modulus and hardness values have been predicted using such composites-based models, it is clear that more complex models, for example ones which incorporate bone's porous water phase, are needed to accurately predict the mechanical response of bone at the tissue-level.

4. Experimental Research Questions

The remainder of the chapter will highlight several applications of nanoindentation to investigate the structure-function relationship of bone. Nanoindentation of bone can be a very powerful technique, and especially when it is combined with adjunctive methods of analysis that provide surface, structural, or compositional information. Surface imaging techniques, such as quantitative backscattered electron (qBSE) imaging and polarized or plain light microscopy, are used to image indent sites for location or composition and to visualize pile-up, sink-in, or cracking. Quantitative histomorphometry is used to quantitatively evaluate the amount of bone formation between treatment groups. Other surface characterization techniques, such as infrared (IR) and Raman spectroscopy, allow nondestructive investigation of mineral and organic tissues to determine chemical compositional and structural information. The correlation of nanoindentation data with any such assay enables an advanced interpretation of bone tissue-level properties.

4.1. *Nanoindentation, quantitative histomorphometry, and rodent models*

Histology and histomorphometry provide valuable information about the location and quantity of bone. Quantitative histomorphometry of bone, with a well-defined nomenclature,[154] provides quantitative analysis at the microscopic level of cellular activities (*e.g.*, mineral apposition rate) and bone geometry. Dynamic histomorphometry involves the use of florescent labels, *e.g.*, tetracycline or calcein green, to provide *in vivo* labeling of the skeleton at specific time-points for quantitative measurements of new bone formation (Fig. 7-5).

In mice and rats, the use of florescent labels is common as it enables the assessment of bone metabolism within times ranging from hours to weeks. Mice offer multiple advantages such as rapid bone growth, low cost, small size which allows large sample numbers, and a wide range of transgenic/knockout possibilities to study health and disease.[155] The use

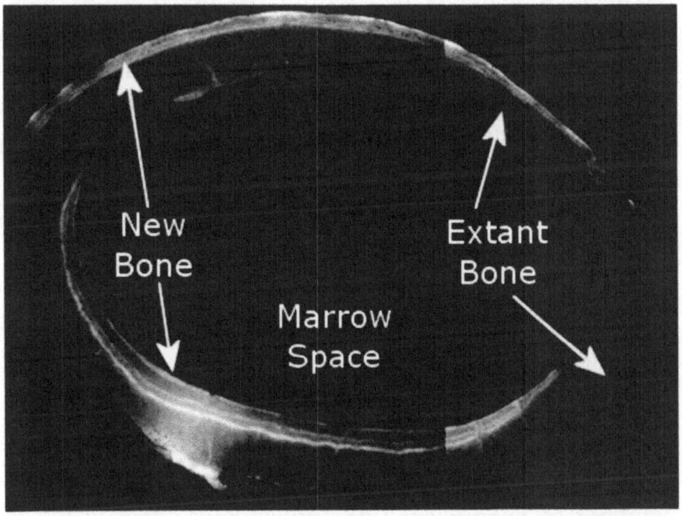

Figure 7-5. Cross-section of a mouse femur treated with bone anabolic therapies.[158] Florescent labels (showing as bright lines within the bone tissue and parallel to the bone surfaces) were administered on specific days during the study to mark new bone formation. Extant, or existing, bone lies "outside" of the labeled areas. Indentations were placed across regions of extant and newly-formed bone to enable comparison of untreated and treated tissue. Field width = 2 mm.

of nanoindentation or microindentation to quantify intrinsic, as opposed to whole bone, mechanical properties enables the examination of the bone tissue.[40,62,156-163] The evaluation of the mechanical properties of both the whole bone and the intrinsic microscale properties are vital to the understanding of the many bone disorders and their potential treatments.

For example one study using mice that were genetically manipulated to exhibit senile osteoporosis showed decreased whole bone bending modulus and an increased indentation modulus.[62] While mechanical properties of bone cannot be directly compared across the macroscopic and microscopic scales, a similar trend might be expected as the whole bone test inherently depends on the intrinsic microscale modulus of bone as well as the bone hierarchical architecture. Such comparisons warrant further investigation and demonstrate the need to probe bone's mechanical properties at multiple hierarchical levels.

The use of labeling associated with nanoindentation provides a practically interesting model that allows for site-specific tests located on extant or newly-formed bone. Fluorescent makers, such as calcein green or oxytetracycline, can be administered at specific time-points during an experiment to label bone apposition at the time of administration.[40,155,162,164,165] Microhardness testing on tail-suspended mice, a common model for osteopenia and space-flight, showed significantly reduced microhardness values in comparison with control values.[166] Within the same bone samples, florescent labels enable identification of extant versus newly formed bone to quantify the relative bone changes that occur with a treatment.[158] Such techniques have been minimally applied with nanoindentation,[158,167] yet have great potential.

4.2. *Microstructural mechanical variation in human bone*

A large number of nanoindentation studies have investigated the wide range of mechanical properties found in various types and regions of human bone. Nanoindentation's high load and displacement resolution is ideally suited to investigate various microstructures in precise anatomical locations. For identical testing methods and anatomical sites, the indentation modulus of interstitial bone is the highest followed by osteonal and then trabecular bone. The nanoindentation modulus of dry trabecular bone samples measured by nanoindentation ranged from 8.2 to 19.4 GPa, whereas cortical (both interstitial and osteonal) bone was significantly higher (18 – 26 GPa).[45,57,59,119] Osteonal bone is generally reported to have a lower modulus value than the higher mineral content, older interstitial bone.[60] The measured moduli for human bone depend greatly on a vast number of experimental, compositional, structural, and anatomical parameters. Additionally, the values reported above are for dry bone and thus do not represent true *in vivo* values. Despite the variability, general property relationships between types of bone are maintained and thus may be investigated using nanoindentation.

Nanoindentation tests to shallow depths can be used to test the mechanical properties of smaller structural features such as individual lamellae. For example, using a ramp-and-hold test to 500 μN (≈100 nm), Gupta *et al.* found alternating regions of high and low stiffness due to the lamellar structure.[60] Less mineral dense regions (in thin lamellae) showed decreased stiffness.[60] Rho *et al.* found a decrease in mean modulus of the thick lamellae with radial distance outward from the center of the osteons.[44] Combined nanoindentation mapping with 2D Raman mapping, of similar spatial resolution, showed a homogenous distribution of the phosphate band in osteons suggesting a uniform mineral distribution.[168,169,170] Collagen fibers, not the degree of mineralization, may not be the dominating faction in elastic modulus values altering with lamellae.[170] X-ray microprobe analysis demonstrated that the thick lamellae had 10-15% higher calcium and phosphorus content than thin lamellae,[171] thus indicating that the mineral content may have a profound effect on the mechanical properties. Most likely, both collagen orientation and the degree of mineralization contribute to the mechanical variations seen between thick and thin lamella. While further studies are needed to fully understand the relationships between structure, function, and composition; nanoindentation, in combination with adjunctive techniques, is an ideal tool to elucidate the role of microstructural variations in bone.

4.3. *Modulus variation with mineral composition*

The elastic modulus positively, and non-linearly, correlates with the mineral content in bulk samples of bone.[132,133] Nanoindentation provides a unique approach to study this same relationship at the tissue-level, whereby measurements of mineral volume fraction or composition can be collected at a similar spatial, micrometer-scale resolution.

Correlation of mineralization with nanomechanical properties can be achieved through combined analysis using nanoindentation and qBSE imaging.[50,172] The mean grey level in the effective contact area for each indent site within a grid array, mapped over a qBSE image of the indented region (Fig. 7-6a), yields the mineral volume fraction through the use of calibrated standard materials.[50,108,127,173] The qBSE image brightness is dependent on the mean atomic number at a given point, which is representative of mineral content, where more highly mineralized regions appear brighter than more poorly mineralized regions.

Correlation with nanoindentation modulus is achieved by collecting qBSE information in areas that match the effective contact area for each indent site.[50,173] Additional refinements to the analysis are provided by segregating data into populations by the type of tissue on which the indents lay. The example provided here shows data for two grid arrays that are separated, for each matched E – mineral V_f data point, into immediate subchondral bone (B) and articular calcified cartilage (C) (Fig. 7-6b). Additional classifications, such as by tissue type or extant versus newly formed bone (through the use of florescent labels), may improve the fidelity of the analysis. As previously mentioned in Section 3.2, a grid indentation approach provides a statistical sampling of a composite material, this method of analysis lacks bias and serves as an effective tool to randomly sample within a region of interest.[126] Plots of E versus mineral V_f thus reveal information about the relationship between nanomechanical properties and mineralization (Figs. 7-4 and 7-6).

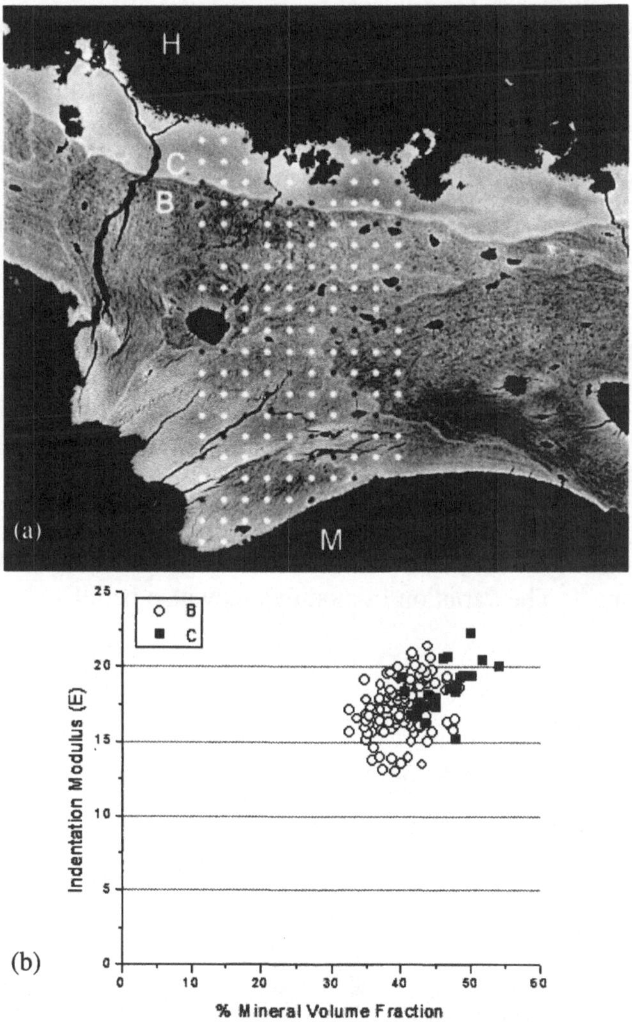

Figure 7-6. (a) qBSE image of normal, post-mortem human femoral head tissue including hyaline articular cartilage (H), articular calcified cartilage (C), immediate subchondral bone (B), and marrow space (M). Calibration standards were used to scale images and calculate mineral volume fraction (black = 0% mineral V_f).[50,108,127] Each indent site was assigned a tissue type (white spots), or discarded from analysis (grey spots), based on overlay of an indentation grid map. Fieldwidth = 594 μm. (b) Indentation modulus versus min V_f plotted for valid B or C sites within the two 5 × 24 indent arrays (20 μm spacing in x- and y-directions) shown in top panel demonstrate the anticipated positive correlation between E and mineral V_f and different ranges of behavior for the more poorly mineralized B versus C. Image (a) used with permission from Ferguson *et al.*, *J. Anat.* 2003).[50]

The positive and non-linear relationship between E and mineral V_f exists over a range of mineralization levels (Fig. 7-4)[127] and tissue types (Fig. 7-6).[50,172] Similar approaches to correlate nanomechanical properties to site-matched measurements of factors that include mineral composition, crystallinity, or organization of the mineral or underlying collagen will provide additional insight into the structure-property-function relationships at the tissue-level.

4.4. *Fossilization*

It is well established that there is a positive correlation between increased mineral volume fraction and the stiffness, or modulus, of mineralized material.[49,132] However, a large range of modulus values for bone is observed at a constant mineral content and thus mechanical properties cannot be explained by the mineral volume fraction and composites theory alone.[132] The variation in modulus may also be related to changes

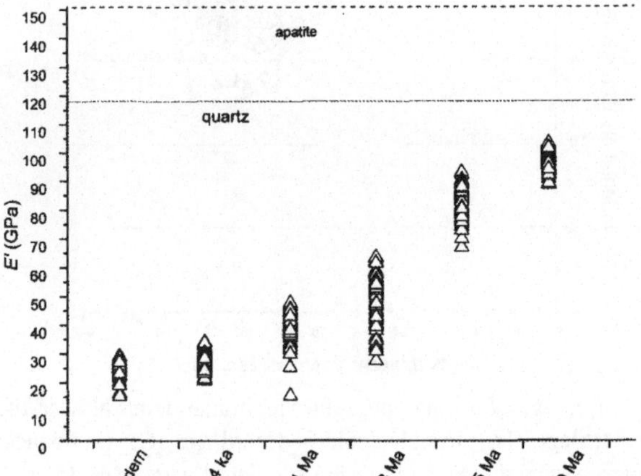

Figure 7-7. Modulus values of fossilized cortical bone samples plotted for each geological age (note: x-axis is not evenly scaled). Data were compared via two-way analysis of variance (ANOVA). A Dunnett's post-hoc analysis shows all fossilized bone moduli values are statistically different from modern bone for $p < 0.05$. Apatite and quartz values are included for comparison to reference samples with relevant composition.[176]

in mineral composition or crystallinity such as increase in crystal size, shape, and perfection. During fossilization, a general increase in mineral volume fraction is commonly observed along with changes in crystallinity and composition. The process of fossilization is extremely complex and can include loss of organic material, partial or complete dissolution of minerals, recrystallization, collagen loss, uptake of trace elements, and changes in porosity.[174,175] This naturally creates a wide range of bio- and geo-mineralized, heterogeneous samples in which to investigate variations of mineral content, composition, and crystallinity.

5. Conclusions and Future Outlook

Nanoindentation of bone and associated mineralized tissues has enabled the mechanical study of bone at the tissue-level. Researchers have long sought ways to evaluate bone at very small length scales, by mechanically testing individual lamellae or a single trabecula or through attempts to correlate non-invasive imaging or high-resolution spectroscopic data with mechanical properties. Nanoindentation has been demonstrated to overcome limitations of larger scale testing methods, where researchers have previously lacked the ability to physically probe bone material at sub-micrometer scales.

Nanoindentation serves as an ideal tool for mechanically testing bone and bone's structural features from tens to thousands of nanometers in depth and indent width. In combination with adjunctive techniques, nanoindentation has and continues to prove its utility in revealing new insight into the microstructural and compositional characteristics of bone across types of mineralized tissues, in health and disease, in response to altered mechanical loading, and in experimentally altered conditions. There is also a tremendous potential for using nanoindentation to evaluate the quality of bone that forms in or adjacent to bone replacement materials and within small volumes of engineered bone tissue.

Because of the potential to collect large, statistically relevant populations of data by indenting many sites within a small region, nanoindentation is an obvious choice for evaluation of precious samples that may exist only in small numbers or volumes of tissue such as human iliac crest biopsies,[108] fossil fragments,[177] or bones from space-flight missions.[162,164] Overall, nanoindentation has and continues to develop into a tool that, when applied to bone samples that have been prepared and tested under well-controlled and thoroughly considered conditions, will continue to reveal insight into many factors ranging from a fundamental study of the elastic, plastic, and viscous contributions of the three constituent phases (*i.e.*, water, mineral and organic) to the development of new bone replacement materials.

References

1. R. Lakes, *Nature* **361**, 511 (1993).
2. R. J. Majeska, Cell biology of bone, in *Bone Mechanics Handbook*, ed. S. C. Cowin, New York, CRC Press, pp. 21–24 (2001).
3. J. M. Gillette and S. M. Nielsen-Preiss, *Journal of Cell Science* **117**, 441 (2004).
4. A. Navrotsky, *Proc. Nat. Acad. Sci. USA* **101**, 12096 (2004).
5. S. C. Cowin, *Journal of Biomechanics* **32**, 217 (1999).
6. J.-Y. Rho, L. Kuhn-Spearing, P. Zioupos, *Medical Engineering & Physics* **20**, 92 (1998).
7. J. K. Gong, J. S. Arnold and S. H. Cohn, *Anatomical Record* **149**, 325 (1964).
8. W. C. Hayes, Biomechanics of cortical and trabecular bone: Implications for the assessment of fracture risk, in *Basic Orthopaedic Biomechanics*, eds. V. C. Mow, W. C. Hayes, New York, Raven Press (1991).
9. S. Weiner and W. Traub, FASEB *Journal* **6**, 879 (1992).
10. M. G. Ascenzi, A. Ascenzi, A. Benvenuti, M. Burghammer, S. Panzavolta and A. Bigi, *J. Struct. Biol.* **141**, 22 (2003).
11. M. G. Ascenzi and A. Lomovtsev, *J. Struct. Biol.* **153**, 14 (2006).
12. W. Gebhardt, *Arch. Entwickl Mech. Org.* **20**, 187 (1906).
13. S. Weiner, T. Arad, I. Sabanay and W. Traub, *Bone* **20**, 509 (1997).
14. M. Petrtyl, J. Hert and P. Fiala, *J. Biomech.* **29**, 161 (1996).
15. A. Boyde and S. J. Jones, *Microscopy Research and Technique* **33**, 92 (1996).
16. J. C. Elliott, Calcium phosphate biominerals, in *Phosphates: Geochemical, Geobiological, and Materials Importance*, eds. M. J. Kohn, J. Rakovan and J. M. Hughes, Washington D.C., Mineralogical Society of America, Vol. 48, pp. 427–453 (2002).
17. E. E. Wilson, A. Awonusi, M. D. Morris, D. H. Kohn, M. M. J. Tecklenburg and L. W. Beck, *Biophysical Journal* **90**, 3722 (2006).
18. K. Sudarsanan and R. A. Young, Acta Crystallographica Section B-Structural Crystallography and Crystal Chemistry **B25**, 1534 (1969).

19. T. I. Ivanova, O. V. Frank-Kamenetskaya, A. B. Kol'tsov and V. L. Ugolkov, *Journal of Solid State Chemistry* **160**, 340 (2001).
20. C. K. Loong, C. Rey, L. T. Kuhn, C. Combes, Y. Wu, S. H. Chen and M. J. Glimcher, *Bone* **26**, 599 (2000).
21. G. Penel, C. Delfosse, M. Descamps and G. Leroy, *Bone* **36**, 893 (2005).
22. A. L. Arsenault, *Bone and Mineral* **6**, 165 (1989).
23. S. J. Eppell, W. D. Tong, J. L. Katz, L. Kuhn and M. J. Glimcher, *Journal of Orthopaedic Research* **19**, 1027 (2001).
24. V. Ziv and S. Weiner, *Connective Tissue Research* **30**, 165 (1994).
25. R. M. Wilson, S. E. P. Dowker and J. C. Elliott, *Biomaterials* **27**, 4682 (2006).
26. J. A. Petruska and A. J. Hodge, Proceedings of the National Academy of Sciences of the United States of America **51**, 871 (1964).
27. W. J. Landis, K. J. Hodgens, J. Arena, M. J. Song and B. F. McEwen, *Microscopy Research and Technique* **33**, 192 (1996).
28. E. P. Katz and S. Li, *Journal of Molecular Biology* **80**, 1 (1973).
29. M. D. McKee, A. Nanci, W. J. Landis, Y. Gotoh, L. C. Gerstenfeld and M. J. Glimcher, *Journal of Bone and Mineral Research* **6**, 937 (1991).
30. B. Alberts, D. Bray, J. Lewis, M. Raff, K. Roberts and J. D. Watson, in *Molecular Biology of the Cell, 3rd edition*, Garland Publishing, New York (1994).
31. S. M. Bowman, J. Zeind, L. J. Gibson, W. C. Hayes and T. A. McMahon, *J. Biomech.* **29**, 1497 (1996).
32. G. E. Fantner, T. Hassenkam, J. H. Kindt, J. C. Weaver, H. Birkedal, L. Pechenik, J. A. Cutroni, G. A. G. Cidade, G. D. Stucky, D. E. Morse and P. K. Hansma, *Nature Materials* **4**, 612 (2005).
33. M. L. Knothe-Tate (ed.), *Interstitial Fluid Flow, 2nd edition*, CRC Press, New York, 22.1–29 (2001).
34. S. C. Cowin, *Journal of Biomechanics* **40**, S105 (2007).
35. W. F. Neuman, T. Y. Toribara and B. J. Mulryan, *Journal of the American Chemical Society* **78**, 4263 (1956).
36. D. T. Reilly, A. H. Burstein and V. H. Frankel, *Journal of Biomechanics* **7**, 271 (1974).
37. H. H. Bayraktar, E. F. Morgan, G. L. Niebur, G. E. Morris, E. K. Wong and T. M. Keaveny, *Journal of Biomechanics* **37**, 27 (2004).
38. K. Choi, J. L. Kuhn, M. J. Ciarelli and S. A. Goldstein, *Journal of Biomechanics* **23**, 1103 (1990).
39. A. Ascenzi and E. Bonucci, *The Anatomical Record* **158**, 375 (1967).
40. V. L. Ferguson, R. A. Ayers, T. A. Bateman and S. J. Simske, *Bone* **33**, 387 (2003).
41. W. C. Oliver and G. M. Pharr, *Journal of Materials Research* **7**, 1564 (1992).
42. K. S. Tai, H. J. Qi and C. Ortiz, *Journal of Materials Science-Materials in Medicine* **16**, 947 (2005).
43. A. J. Bushby, V. L. Ferguson and A. Boyde, *Journal of Materials Research* **19**, 249 (2004).

44. J. Y. Rho, P. Zioupos, J. D. Currey and G. M. Pharr, *Bone* **25**, 295 (1999).
45. P. K. Zysset, X. E. Guo, C. E. Hoffler, K. E. Moore and S. A. Goldstein, *Journal of Biomechanics* **32**, 1005 (1999).
46. M. L. Oyen and R. F. Cook, *Journal of Materials Research* **18**, 139 (2003).
47. A. K. Bembey, A. J. Bushby, A. Boyde, V. L. Ferguson and M. L. Oyen, *Journal of Materials Research* **21**, 1962 (2006).
48. A. K. Bembey, M. L. Oyen, A. J. Bushby and A. Boyde, *Philosophical Magazine* **86**, 5691 (2006).
49. V. L. Ferguson, A. Boyde and A. J. Bushby, *Mater. Res. Soc. Symp. Proc.* **841**, R2.7.1 (2005).
50. V. L. Ferguson, A. J. Bushby and A. Boyde, *Journal of Anatomy* **203**, 191 (2003).
51. J. S. Field and M. V. Swain, *Journal of Materials Research* **8**, 297 (1993).
52. A. J. Bushby, Nondestructive Testing and Evaluation **17**, 213 (2001).
53. L. P. Mullins, M. S. Bruzzi and P. E. McHugh, *Journal of Biomechanics* **40**, 3285 (2007).
54. X. Wang, Y. J. Yoon and H. Ji, *Journal of Biomechanics* **40**, 1401 (2007).
55. M. Vanleene, P. E. Mazeran and M. Tho, *Journal of Materials Research* **21**, 2093 (2006).
56. Z. Fan, J. G. Swadener, J. Y. Rho, M. E. Roy and G. M. Pharr, *Journal of Orthopaedic Research* **20**, 806 (2002).
57. J. Y. Rho, T. Y. Tsui and G. M. Pharr, *Biomaterials* **18**, 1325 (1997).
58. J. Y. Rho, P. Zioupos, J. D. Currey and G. M. Pharr, *Journal of Biomechanics* **35**, 189 (2002).
59. C. E. Hoffler, X. E. Guo, P. K. Zysset and S. A. Goldstein, *Journal of Biomechanical Engineering* **127**, 1046 (2005).
60. H. S. Gupta, U. Stachewicz, W. Wagermaier, P. Roschger, H. D. Wagner and P. Fratzl, *Journal of Materials Research* **21**, 1913 (2006).
61. L. Mulder, J. H. Koolstra, J. M. J. den Toonder and T. M. G. J. van Eijden, *Bone* **41**, 256 (2007).
62. M. J. Silva, M. D. Brodt, Z. F. Fan and J.-Y. Rho, *Journal of Biomechanics* **37**, 1639 (2004).
63. C. H. Turner, J. Rho, Y. Takano, T. Y. Tsui and G. M. Pharr, *Journal of Biomechanics* **32**, 437 (1999).
64. S. Hengsberger, A. Kulik and P. Zysset, *Bone* **30**, 178 (2002).
65. T. Hoc, L. Henry, M. Verdier, D. Aubry, L. Sedel and A. Meunier, *Bone* **38**, 466 (2006).
66. Z. F. Fan and J. Y. Rho, *Journal of Biomedical Materials Research Part A* **67A**, 208 (2003).
67. J. Mencik, G. Rauchs, J. Bardon and A. Riche, *Journal of Materials Research* **20**, 2660 (2005).
68. J. M. Giles, A. E. Black and J. E. Bischoff, *Journal of Biomechanics* **40**, 777 (2007).

69. Y. C. Fung, in *Biomechanics: Mechanical Properties of Living Tissues*, Springer-Verlag, New York (1993).
70. E. Mittra, S. Akella and Y. X. Qin, *J. Biomed. Mater. Res. A* **79**, 86 (2006).
71. B. Tang, A. H. W. Ngan and W. W. Lu, *Philosophical Magazine* **86**, 5653 (2006).
72. K. L. Johnson, in *Contact Mechanics*, Cambridge University Press (1985).
73. J. Zhang, G. L. Niebur and T. C. Ovaert, *J. Biomech.* **41**, 267 (2008).
74. E. Donnelly, S. P. Baker, A. L. Boskey and M. C. H. van der Meulen, *Journal of Biomedical Materials Research Part A* **77A**, 426 (2006).
75. M. L. Oyen and C. C. Ko, *J. Mat. Res.* **23**, 760 (2008).
76. P. K. Hansma, P. J. Turner and G. E. Fantner, *Review of Scientific Instruments* **77** (2006).
77. P. K. Hansma, P. J. Turner, B. Drake, J. Adams, J. LeLujian, A. Proctor and F. Garza de Leon, presented at the 29th Annual Meeting of the American Society for Bone and Mineral Research, Honolulu, HI (2007).
78. A. E. Goodship, L. E. Lanyon and H. McFie, *J. Bone Joint Surg Am* **61**, 539 (1979).
79. C. T. Rubin and L. E. Lanyon, *J. Bone Joint Surg. Am.* **66**, 397 (1984).
80. C. T. Rubin and L. E. Lanyon, *Calcif. Tissue Int.* **37**, 411 (1985).
81. H. M. Frost, *Henry Ford Hospital Medical Bulletin* **8**, 36 (1960).
82. S. R. Eisenberg and A. J. Grodzinsky, *Journal of Orthopaedic Research* **3**, 148 (1985).
83. R. B. Martin and N. A. Sharkey, Mechanical effects of postmortem changes, preservation, and pllograft bone treatments, in *Bone Mechanics Handbook*, ed. S. C. Cowin, New York, CRC Press, pp. 20.1–24 (2001).
84. M. B. Gustafson, R. B. Martin, V. Gibson, D. H. Storms, S. M. Stover, J. Gibeling and L. Griffin, *Journal of Biomechanics* **29**, 1191 (1996).
85. A. Boyde, A. J. Bushby and V. L. Ferguson, University of London, unpublished (2003).
86. G. Cho, Y. Wu and J. L. Ackerman, *Science* **300**, 1123 (2003).
87. R. I. Stearns, *Journal of Dental Research* **50**, 1575 (1971).
88. G. R. Dickson, Chemical fixation and the preparation of calcified tissues for transmission electron microscopy, in *Methods of Calcified Tissue Preparation*, ed. G. R. Dickson, Elsevier (1984).
89. S. Habelitz, G. W. Marshall, Jr., M. Balooch and S. J. Marshall, *J. Biomech.* **35**, 995 (2002).
90. S. Malek, M. A. Darendeliler, T. Rex, O. P. Kharbanda, P. Srivicharnkul, M. V. Swain and P. Petocz, *Am. J. Orthod. Dentofacial Orthop.* **124**, 561 (2003).
91. D. D. Lee and M. J. Glimcher, *Journal of Molecular Biology* **217**, 487 (1991).
92. W. W. Lu, K. D. K. Luk, K. C. M. Cheung, Q. Gui-Xing, J. X. Shen, L. Yuen, J. Ouyang and J. C. Y. Leong, *Spine* **29**, 1196 (2004).
93. A. K. Bembey, V. Koonjul, A. J. Bushby, V. L. Ferguson and A. Boyde, *Mater. Res. Soc. Symp. Proc.* **841**, R/Y2.8.1 (2005).

94. J. J. Broz, S. J. Simske, A. R. Greenberg and M. W. Luttges, *Journal of Biomechanical Engineering* **115**, 447 (1993).
95. S. Lees, L. C. Bonar and H. A. Mook, *International Journal of Biological Macromolecules* **6**, 321 (1984).
96. J. B. Finlay and W. R. Hardie, Proceedings of the Institution of Mechanical Engineers, Part H: Journal of Engineering in Medicine **208**, 27 (1994).
97. W. B. Lievers, V. Lee, S. M. Arsenault, S. D. Waldman and A. K. Pilkey, *Journal of Biomechanics* **40**, 1903 (2007).
98. F. G. Evans and M. Lebow, *J. Appl. Physiol.* **3**, 563 (1951).
99. J. Y. Rho and G. M. Pharr, Journal of Materials Science-Materials in Medicine **10**, 485 (1999).
100. R. K. Schenk, A. J. Olah and W. Herrmann, Preparation of calcified tissues for light microscopy, in *Methods of Calcified Tissue Preparation*, ed. G. R. Dickson, Elsevier (1984).
101. G. Guidoni, J. Denkmayr, T. Schoberl and I. Jager, *Philosophical Magazine* **86**, 5705 (2006).
102. H. Saito and M. Yokoi, *Journal of Biochemistry* **111**, 376 (1992).
103. S. J. Edmondston, K. P. Singer, R. E. Day, P. D. Breidahl and R. I. Price, *Clinical Biomechanics* **9**, 175 (1994).
104. D. E. Pegg, *Methods in Molecular Biology* **368**, 39 (2007).
105. R. J. Egli, C. Wingenfeld, M. Holzle, A. Hempfing, C. R. Fraitzl, R. Ganz and M. Leunig, *Journal of Investigative Surgery* **19**, 87 (2006).
106. M. L. Oyen, C. C. Ko, A. K. Bembey, A. J. Bushby and A. Boyde, *Mater. Res. Soc. Symp. Proc.* **874**, L1.7.1 (2005).
107. B. Boothroyd, *J. Cell Biol.* **20**, 165 (1964).
108. A. Boyde, R. Travers, F. H. Glorieux and S. J. Jones, *Calcif. Tissue Int.* **64**, 185 (1999).
109. A. Boyde, Scanning electron microscope studies of bone, in *Methods of Calcified Tissue Preparation*, ed. G. R. Dickenson, Elsevier (1984).
110. V. J. Kingsmill and A. Boyde, *J. Anat.* **192**(Pt 2), 245 (1998).
111. P. G. T. Howell and A. Boyde, *Bone* **15**, 285 (1994).
112. J. Xu, J. Y. Rho, S. R. Mishra and Z. Fan, *Journal of Biomedical Materials Research Part A* **67A**, 719 (2003).
113. D. T. Reilly and A. H. Burstein, *J. Biomech.* **8**, 393 (1975).
114. J. D. Currey, in *The Mechanical Adaptations of Bone*, Princeton University Press, pp. 294 (1984).
115. R. M. Pidaparti, A. Chandran, Y. Takano and C. H. Turner, *J. Biomech.* **29**, 909 (1996).
116. C. H. Turner, A. Chandran and R. M. Pidaparti, *Bone* **17**, 85 (1995).
117. R. Amprino, *Acta Anat. (Basel)* **34**, 161 (1958).
118. P. E. Riches, N. M. Everitt, A. R. Heggie and D. S. McNally, *J. Biomech.* **30**, 1059 (1997).

119. J. Y. Rho, M. E. Roy, T. Y. Tsui and G. M. Pharr, *Journal of Biomedical Materials Research* **45**, 48 (1999).
120. J. G. Swadener, J. Y. Rho and G. M. Pharr, *Journal of Biomedical Materials Research* **57**, 108 (2001).
121. A. Ascenzi, *J. Biomech. Eng.* **110**, 357 (1988).
122. A. Ascenzi, A. Boyde, M. Portigliatti Barbos and S. Carando, *J. Biomech.* **20**, 1045 (1987).
123. P. L. Mente, Micromechanical testing of single trabeculae, in *Mechanical Testing of Bone and the Bone-Implant Interface*, eds. Y. H. An, R. A. Draughn, Boca Raton, FL, CRC Press, pp. 291–304 (2000).
124. J. G. Ramasamy and O. Akkus, *J. Biomech.* **40**, 910 (2007).
125. P. L. Mente and J. L. Lewis, *J. Orthop. Res.* **7**, 456 (1989).
126. G. Constantinides, K. S. Ravi Chandran, F. J. Ulm and K. J. Van Vliet, *Materials Science and Engineering A* **430**, 189 (2006).
127. M. L. Oyen, V. L. Ferguson, A. K. Bembey, A. J. Bushby and A. Boyde, *Journal of Biomechanics* **41**, 2585 (2008).
128. K. Tai, M. Dao, S. Suresh, A. Palazoglu and C. Ortiz, *Nat. Mater.* **6**, 454 (2007).
129. C. E. Hoffler, K. E. Moore, K. Kozloff, P. K. Zysset and S. A. Goldstein, *J. Orthop. Res.* **18**, 432 (2000).
130. C. E. Hoffler, K. E. Moore, K. Kozloff, P. K. Zysset, M. B. Brown and S. A. Goldstein, *Bone* **26**, 603 (2000).
131. P. K. Zysset, X. E. Guo, C. E. Hoffler, K. E. Moore and S. A. Goldstein, *Technol. Health Care* **6**, 429 (1998).
132. J. L. Katz, *J. Biomech.* **4**, 455 (1971).
133. J. D. Currey, in *Bones: Structure and Mechanics*, Princeton University Press, pp. 436 (2002).
134. Z. Hashin and S. Shtrikman, *Journal of the Mechanics and Physics of Solids* **11**, 127 (1963).
135. N. Sasaki and A. Enyo, *J. Biomech.* **28**, 809 (1995).
136. N. Sasaki and M. Yoshikawa, *J. Biomech.* **26**, 77 (1993).
137. J. Yamashita, B. R. Furman, H. R. Rawls, X. Wang, and C. M. Agrawal, *J. Biomed. Mater. Res.* **58**, 47 (2001).
138. B. J. Briscoe, L. Fiori and E. Pelillo, *J. Phys. D: Appl. Phys.* **31**, 2395 (1998).
139. T. Chudoba and F. Richter, *Surface and Coatings Technology* **148**, 191 (2001).
140. Y.-T. Cheng and C.-M. Cheng, *Journal of Materials Research* **20**, 1046 (2005).
141. G. Feng and A. H. W. Ngan, *Journal of Materials Research* **17**, 660 (2002).
142. B. Tang and A. H. W. Ngan, *Journal of Materials Research* **18**, 1141 (2003).
143. M. L. Oyen, *Philosophical Magazine* **86**, 5625 (2006).
144. C. A. Tweedie and K. J. Van Vliet, *Journal of Materials Research* **21**, 1576 (2006).
145. C. Y. Zhang, Y. W. Zhang and K. Y. Zeng, *Journal of Materials Research* **19**, 3053 (2004).

146. C. Y. Zhang, Y. W. Zhang, K. Y. Zeng and L. Shen, *Journal of Materials Research* **20**, 1597 (2005).
147. M. L. Oyen and C. C. Ko, *J. Mater. Sci. Mater. Med.* **18**, 623 (2007).
148. M. L. Oyen, *Journal of Materials Research* **23**, 1307 (2008).
149. M. L. Oyen, *Journal of Materials Research* **20**, 2094 (2005).
150. J. D. Currey, *Biorheology* **2**, 1 (1964).
151. I. Jager and P. Fratzl, *Biophysical Journal* **79**, 1737 (2000).
152. S. P. Kotha and N. Guzelsu, *Journal of Theoretical Biology* **219**, 269 (2002).
153. C. Hellmich and F. J. Ulm, *Journal of Biomechanics* **35**, 1199 (2002).
154. A. M. Parfitt, M. K. Drezner, F. H. Glorieux, J. A. Kanis, H. Malluche, P. J. Meunier, S. M. Ott and R. R. Recker, *Journal of Bone and Mineral Research* **2**, 595 (1987).
155. S. J. Simske, V. L. Ferguson and T. A. Bateman, Mice and osteoporosis research, in *Recent Research Developments in Biotechnology and Bioengineering*, ed. S. G. Pandalai, Research Signpost, Vol. 5, pp. 97–127 (2003).
156. M. Weber, P. Roschger, N. Fratzl-Zelman, T. Schoberl, F. Rauch, F. H. Glorieux, P. Fratzl and K. Klaushofer, *Bone* **39**, 616 (2006).
157. L. M. Miller, W. Little, A. Schirmer, F. Sheik, B. Busa and S. Judex, *J. Bone Miner. Res.* **22**, 1037 (2007).
158. S. A. J. Lloyd, S. J. Simske, L. K. Bogren, S. E. Olesiak, T. A. Bateman and V. L. Ferguson, submitted (2008).
159. Q. Chen, J. Y. Rho, Z. Fan, S. J. Laulederkind and R. Raghow, *Calcif. Tissue Int.* **73**, 387 (2003).
160. M. P. Akhter, Z. Fan and J. Y. Rho, *Calcif. Tissue Int.* **75**, 416 (2004).
161. N. B. Kavukcuoglu, D. T. Denhardt, N. Guzelsu and A. B. Mann, *J. Biomed. Mater. Res. A* **83**, 136 (2007).
162. T. A. Bateman, S. Morony, V. L. Ferguson, S. J. Simske, D. L. Lacey, K. S. Warmington, Z. Geng, H. L. Tan, V. Shalhoub, C. R. Dunstan and P. J. Kostenuik, *Journal of Bone and Mineral Research* **17**, S136 (2002).
163. N. E. Lane, W. Yao, J. H. Kinney, G. Modin, M. Balooch and T. J. Wronski, *Journal of Bone and Mineral Research* **18**, 2105 (2003).
164. T. A. Bateman, R. J. Zimmerman, R. A. Ayers, V. L. Ferguson, S. K. Chapes and S. J. Simske, *Bone* **23**, 527 (1998).
165. T. C. Lee, S. Mohsin, D. Taylor, R. Parkesh, T. Gunnlaugsson, F. J. O'Brien, M. Giehl and W. Gowin, *J. Anat.* **203**, 161 (2003).
166. S. J. Simske, J. J. Broz and M. W. Luttges, *Journal of Materials Science-Materials in Medicine* **6**, 486 (1995).
167. B. Busa, L. Miller, C. Rubin, Y. X. Qin and S. Judex, *Calcif. Tissue Int.* **77**, 386 (2005).
168. M. Kazanci, P. Roschger, E. P. Paschalis, K. Klaushofer and P. Fratzl, *Journal of Structural Biology* **156**, 489 (2006).
169. A. Carden and M. D. Morris, *Journal of Biomedical Optics* **5**, 259 (2000).

170. T. Hofmann, F. Heyroth, H. Meinhard, W. Franzel and K. Raum, *Journal of Biomechanics* **39**, 2282 (2006).
171. G. Marotti, *Calcif. Tissue Int.* **53**, S47 (1993).
172. H. S. Gupta, S. Schratter, W. Tesch, P. Roschger, A. Berzlanovich, T. Schoeberl, K. Klaushofer and P. Fratzl, *J. Struct. Biol.* **149**, 138 (2005).
173. L. Angker, C. Nockolds, M. V. Swain and N. Kilpatrick, *Arch. Oral Biol.* **49**, 369 (2004).
174. E. Badone and R. M. Farquhar, *Journal of Radioanalytical Chemistry* **69**, 291 (1982).
175. R. E. M. Hedges, *Archaeometry* **44**, 319 (2002).
176. M. E. Broz, R. F. Cook and D. L. Whitney, *American Mineralogist* **91**, 135–142 (2006).
177. S. Olesiak, M. Oyen, M. Sponheimer, J. Eberle and V. Ferguson, *Mater. Res. So.c Symp. Proc.* **975**, DD03.09 (2006).

CHAPTER 8

NANOINDENTATION OF TOOTH TISSUES

Li-Hong He and Michael V. Swain

Faculty of Dentistry, the University of Sydney,
Sydney Dental Hospital
E-mail: mswain@mail.usyd.edu.au

In this chapter, the current progress on nanoindentation measurement of dental hard tissues is summarized. With the help of nanoindentation techniques, it is possible to measure the mechanical properties of small volumes of dental hard tissues at the micro- and nano-level with respect to their unique microstructures, namely the rod unit of the enamel and tubule structure of dentin. The results indicate that the microstructure and organic components regulate the mechanical behavior of these natural biocomposites to a great extent. Furthermore, application of nanoindentation to the investigation of diseased or defective states of tooth tissues, dental treatments and dental materials is also introduced and discussed.

1. Introduction

A tooth is composed of vascularized soft connective tissue—dental pulp—covered by two layers of mineralized hard tissues, namely, inner dentin and outer enamel. As the outer cover and functional layer of a tooth, enamel is the hardest and stiffest structure of mammals. The functional requirements of teeth are that they be able to sustain a wide range of imposed loads and consequent contact induced stresses without failure and retain their shape while doing so.[1] In addition to the normal loads, enamel will be exposed to shear forces because of the direct contact with opposite teeth and external objects during mastication. Moreover, unlike other calcified skeletal structures, fracture of dental

Table 8-1. Conditions under which teeth operate.[2]

Rate of chewing	60-80 cycles/min
Forces: maximum biting force overall teeth	640 N
maximum biting force on a single tooth	265 N
normal forces on a single tooth	3-18 N
Contact time: time maximum pressure sustained	0.07 S
Total contact time: normal	10 min
Bruxist (over 24 h)	30 min - 3 h
Mean sliding distance	1.0 mm
Contact area (1st molar region)	15 mm^2
Maximum stress	20 MPa

tissue is not repairable. Some typical physiological conditions of teeth have been listed in Table 8-1 by Waters.[2] Recently, investigations indicate that masticatory loading forces teeth are exposed to range from 28 N to more than 1200 N.[3,4] At the same time, the contact area may be as small as a few square millimeters. It is amazing that enamel can sustain and survive such high forces for millions of cycles. The mechanical properties and the deformation mechanisms in relation to the compositional and hierarchal microstructural characteristics of enamel are an essential factor to understand this natural material.

As the inner layer of tooth hard tissue, dentin is much softer than enamel. This may be appreciated from the following aspects: structurally, as odontoblasts penetrate into dentin tubules, dentin and pulp are usually considered as a vital complex in response to bacterial invasion; compositionally, dentin has much higher organic component than enamel; functionally, because its major role is to protect the pulp rather than mastication, the mechanical property requirements of dentin are for toughness rather than hardness as for enamel.

Hierarchical microstructures and precisely distributed organic components endow dental hard tissues excellent anisotropic mechanical properties, which ensure their life-time survival in the mouth as a load bearing organ. To understand and illustrate the basis for the excellent performance of tooth tissues, many investigators have undertaken various mechanical tests on them. Moreover, knowledge of the mechanical properties and microstructural features of enamel and dentin is important to understanding stress dissipation in the tooth, for developing biomimetic restorative materials and for improving the execution of clinical dental preparations. Although numerous tests have been conducted to explore the mechanical properties of these natural materials from different perspectives since the 19th century, this is still an area of intense research in the biological materials field because the basis of this naturally optimized biocomposite are still poorly understood.

Recently, with the development of nanoindentation techniques, small volume samples such as dental hard tissues can be measured more accurately. The system enables measurement of the mechanical properties, hardness and elastic modulus, at the surface of a material. The nanoindentation technique is a much simpler procedure compared to other conventional mechanical tests such as compressive, tensile, bending, shear strength and punch shear tests, particularly on small complex shaped samples such as enamel, dentin and cementum. Measurements using an indentation system are relatively non-destructive and the specimen preparation is less time consuming as the test can be done on a bulk polished specimen.[5] More importantly, it allows measurement of the mechanical properties of a very small selected region of the specimen,[6] the dimension of which may be at the micrometer or even nanometer scale. This is of importance for measurement of local properties of inhomogeneous structures such as dental calcified tissues. Indentations of enamel rods and interrods, peritubular and intertubular dentin are achievable, and have greatly contributed to our current understanding of the mechanical behavior of dental hard tissues in association with their microstructural compositions.

The aim of this chapter is to review the application of nanoindentation to dental hard tissue studies and discuss the relevant implications in respect to the microstructural and compositional characteristics of these natural biocomposites. Nanoindentation studies on enamel, dentin and cementum are reviewed. Based on recent investigations by the authors, enamel will provide an example to illustrate the importance of microstructure and organic components on the mechanical behavior of these natural biocomposites. Furthermore, the challenges for future investigations are discussed at the end of the chapter.

2. Formation and Structure of Teeth

For better understanding of its compositional and structural characteristics and the relationship to mechanical behavior, it is essential to briefly review the formation of teeth, with an emphasis mainly on enamel and dentin.

2.1. *Formation and structure of enamel*

The formation of enamel includes secretion and maturation stages. During the initial secretion stage, the ameloblasts develop at their apical (secretory) surface, a single fairly large cell process, the Tomes' process, which is a unique feature of the secretory stage of the ameloblast. The presence and the shape of Tomes' processes are significant for the structure of the developing enamel.

When an ameloblast has completed its principal secretory activities and the developing enamel has reached its full thickness locally, the secretion stage comes to an end. The ameloblast next undergoes a series of cytological changes in preparation for enamel maturation. During enamel maturation, there is a sharp increase in matrix degradation and replacement by tissue fluid. Finally, most of the matrix (specifically the amelogenins) and the fluid have been replaced gradually and the mineral volume fraction component increases steeply from 10-20% to 80-90%.

One of the most important steps in the enamel maturation stage is the organization of HAP crystals into bulk enamel. This organization is achieved by extracellular protein matrices, which are secreted from the Tomes' processes of ameloblasts. Amelogenin undergoes self-assembly to form nano-spheres *in vitro* and *in vivo*.[7] These nano-spheres are located adjacent to the *a*- and *b*-surfaces of the HAP crystallites *in vitro*, possibly favoring crystallite elongation along the *c*-axis resulting in long and thin crystals arranged in parallel.[7] As enamel mineralization proceeds, the organic matrix is progressively removed allowing the crystallites to grow in thickness. The resulting mineral phase forms as a consequence of a unique organic matrix that is competent to direct its own replacement by mineral. The interaction between matrix protein and HAP crystals during maturation indicates an intimate relationship of these two phases in mature enamel.

After maturation, some protein fragments are retained primarily at the incisal edges and proximal sides of rod boundaries defining a rod sheath,[8] which help to define the rounded discontinuity seen around the top of the 'fish scale' on rods. These discontinuities have two very important mechanical functions. First, they define three-dimensional cleavage planes that will deflect cracks.[9,10] This prevents cracks from advancing straight through enamel to cause catastrophic macro-mechanical failure, but instead spreads the damage laterally over a larger volume. Second, the presence of minute quantities of protein remnants allows limited differential movement between adjacent rods.[11] Limited slippage also reduces stresses without crack growth. The minor components of enamel, namely the protein remnants and water, have a profound plasticizing effect. Enamel is known to be much more flexible and softer than its major component, crystalline hydroxyapatite.[10,12]

The result of such a highly precise regulation of the biomineralization procedure is the prominent hierarchy of enamel structure (Fig. 8-1).[13] The smallest structural units are needle or plate-like hydroxyapatite crystallites which are roughly of flattened hexagonal form in cross-section[14] with mean width of 68 nm and mean thickness of 26 nm.[15] In healthy human enamel, these hydroxyapatite crystallites are organized and bundled together by organic molecules into larger-scale structures, called 'keyhole-shaped' enamel rods. At this level, the directional arrangement of the hydroxyapatite crystallites varies. Crystallite plates in the central part of the rod are parallel to the rod axis while those near the edge of the rod usually have an angle of near 45 degrees to the longitudinal axis of the rods.[16] The arrangement of rods determines the next level unit, namely enamel types. Human enamel consists of ~5 μm diameter rods encapsulated by thin protein rich sheathes that are arranged parallel in a direction perpendicular to the dentino-enamel junction (DEJ) from dentin to the outer enamel surface. In some areas, rods may twist together or change their direction slightly to reinforce the whole structure.[17] This complex hierarchical structure has elicited a number of attempts to model it,[18-21] however in all instances the emphasis has been placed on the elastic rather than the non-linear deformation and toughness of the structure.

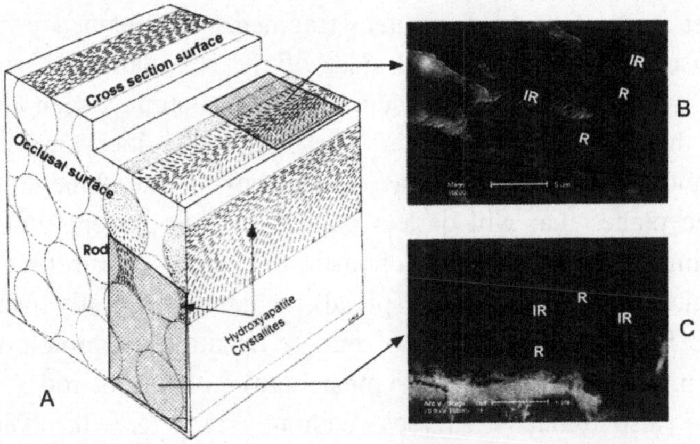

Figure 8-1. Schematic illustration of enamel hierarchical structure (modified from Nanci[22]), and SEM images of different faces. (R: rod; IR: interrod).

2.2. *Formation and structure of dentin*

As with enamel, the formation of dentin starts with the activation of cells, namely odontoblasts. As the dental mesenchyme cells begin to differentiate into odontoblasts, they begin to secrete several molecules that promote mineralization of the extracellular matrix. These proteins include tenascin, bono-1, dental sialophosphoprotein, and alkaline phosphatase.[23] Finally, as the odontoblast phenotype emerges, osteonectin and type I collagen are secreted as components of the extracellular matrix and form an approximately 10–30 μm wide unmineralized zone between the mineralized dentin and odontoblasts, namely predentin. Thus, totally different to that of enamel, the mineralization pattern of dentin is developed layer by layer immediately after the secretion of the protein matrix. Similarly, as described above, the matrix proteins in predentin (newly secreted protein-rich unmineralized dentin) regulate the growth and deposition of the HAP crystals. Moreover, odontoblasts synthesize and secrete type I collagen to the predentin layer. In predentin, collagen molecule fibers aggregate with their long axes in parallel into fibrils, which further arrange to bundles, possibly with the help of proteoglycans,[24] to form the collagen component of mature dentin.

The dentin structurally consists of tubular, peritubular and intertubular dentin (Fig. 8-2). As a result of cell activity of odontoblasts and due to the insertion of the odontoblasts process into the mineralized structure, dentin forms its tubular pattern characteristic feature (Fig. 8-2). From the outer surface of the dentin to the area nearest the pulp, these tubules follow an S-shaped path. The diameter and density of the tubules are greatest near the pulp.[26] Tapering from the inner to the outermost surface, they have a diameter of 2.5 μm near the pulp, 1.2 μm in the middle of the dentin, and 900 nm at the dentino-enamel junction. Their density is 59,000 to 76,000 per square millimeter near the pulp, whereas the density is only half as much near the enamel. Dentinal tubules give high permeability to the dentin. In addition to an odontoblast process, the tubule contains dentinal fluid, a complex mixture of proteins, such as albumin, transferrin, tenascin and proteoglycans.[26] In addition, there are branching canalicular systems that connect to each other to form a complex network. Dentin tubules are surrounded by highly calcified peritubular dentin, which is more radio-opaque and electron-dense than intertubular dentin. The less calcified intertubular structure, which contains more organic material than peritubular dentin, comprises the remaining dentin body and lies between regions of peritubular dentin.

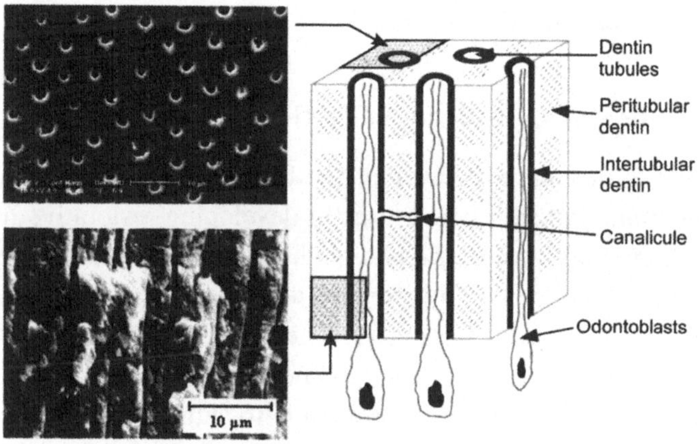

Figure 8-2. Schematic illustration of dentin tubule structure; and SEM images from different directions, modified from Angker *et al.*[25]

2.3. *Structure of cementum*

Cementum is a thin layer of dental calcified tissue that covers the tooth root by firmly adhering to dentin. Similar to bone tissue, it consists of an inorganic component (65 wt%) mainly comprising hydroxyapatite, an organic component of collagen type I (23 wt%) and the remainder is water (12 wt%). Based on the presence or absence of cells, cementum is grouped into cellular and acellular cementum. The cellular cementum develops a relatively thicker layer on the apical third of the root particularly the outer region, while the acellular cementum, which is more highly calcified, but structureless, covers the cervical two-thirds of the root. The fibers found in cementum could be extrinsic (Sharpey) fibers from periodontal ligaments, or intrinsic fibers originating from cementoblasts, or both types of fibers. The cementum facilitates the attachment of collagen fibers of periodontal ligament to tooth root, whereby the tooth is physically maintained and supported in an alveolar socket.[27] Cementum responds to mechanical loadings to which the tooth is subjected, and is continuously modified through life. Excessive forces may lead to a pathological condition resulting in either a resorption or a hyperplasia of the cementum. Dental disease such as caries can affect the cementum as well as other pathologic conditions involving dental pulp and periodontium. Chemical substances from dental procedures may also initiate pathosis in the cementum.

3. Nanoindentation Behavior of Teeth

As enamel is the outmost layer of tooth, its mechanical properties are more important and significant to us in developing restorative materials. Moreover, the implication of its hierarchical microstructure in regulating mechanical behavior of biocomposites has attracted considerable attention of materials scientists. Therefore, more nanoindentation investigations have been done on enamel than dentin. In the following sections, we will take enamel as a sample to summarize and discuss the nanoindentation of teeth.

3.1. *Elastic modulus and hardness of enamel*

Mechanical tests to determine elastic modulus and hardness of enamel have experienced several stages of development from macro-scale to micro- and nano-scale over the last several decades. Due to the size limit of human enamel samples, it has been difficult to perform macro-scale tests. Therefore, most investigations on the mechanical properties of human enamel have been done with nanoindentation. A partial listing of such investigations is listed in Table 8-2. Also, the range of the results obtained has reduced considerably, due to the precise control of the test area.

From the table, we may conclude that, firstly, the large mineral level (about 96 wt%) as well as the unique microstructure of enamel contribute to its large mechanical properties.[28-30] The stiffness of a composite material like enamel is the combination of the stiffness of its multi-phase components and associated microstructures; the inorganic-hydroxyapatite crystals and organic tissues which may be considered as a fibrous composite model.[18]

Secondly, current understanding of the mechanical behavior of enamel at the nano-scale associated with its anisotropic microstructure is possible with nanoindentation. The hardness and elastic modulus of enamel were found to be a function of enamel rod alignment.[30,36] The measured elastic modulus varies with the direction of the load relative to the apatite crystals of enamel. Higher elastic modulus values are found when the applied force is parallel to enamel rods as the deflection will be resisted mostly by the aligned crystals within the rods, while in the perpendicular directions, the displacement will be influenced more by the less stiff organic tissue surrounding the crystals thereby resulting in lower elastic moduli. The rod orientation dependence is also seen in a uniaxial loading three-dimensional finite element analysis of enamel.[18] This also explains the discrepancy in the elasticity data of enamel rods and interrods. The interrod was reported to have hardness and elastic modulus of about only 25 and 50%, respectively of the values of the rod.[6]

Table 8-2. Nanoindentation elastic modulus and hardness values of enamel.

Author(s)	Surface and Site	Test Load	Hardness (GPa)	Elastic Modulus (GPa)
Willems *et al.*[31]	top surface	10 mN	3.39±0.18	90.59±16.13
Cuy *et al.*[29]	cross section: outer enamel DEJ	Depth control (400 and 800 nm)	2.7-6.4 >6 <3	47-120 >115 <70
Zhou *et al.*[32]	top surface	Depth control (100-2000 nm)	5.7-3.6	104-70*
He *et al.*[33]	top surface cross section	1-450 mN	4.9±0.5 4.5±0.4	102.5-53.45* 82.7-40.3*
Ge *et al.*[6]	top surface: rod interrod	1 mN 0.3 mN	4.3±0.8 1.1±0.3	83.4±7.1 39.5±4.1
Mahoney *et al.*[34]	cross section (primary molar)	50 and 150 mN	4.9±0.4	80.4±7.7
Marshall *et al.*[35]	cross section (DEJ area)	Up to 14 mN	3.51±0.13	63.55±1.46
Poolthong[36]	cross section: buccal side cusp tip	50 and 150 mN	4.74±0.2 4.45±0.3	90.79±2.2 83.83±6.3
Fong *et al.*[37]	top surface cross section	0.3-2.5 mN	4.78±0.36 4.53±0.26	98.3±5.9 95.6±4.9
Habelitz *et al.*[30]	top surface cross section head of rod tail of rod interrod	1.5 mN	3.8±0.31 3.3±0.35 4.3±0.4 3.7±0.4 3.9±0.4	87.5±2.1 72.7±4.4 88.0±8.6 80.3±7.2 86.4±11.7
Habelitz *et al.*[38]	cross section	0.75 and 1.5 mN	3.2±0.4 – 3.7±0.5	74±4 – 80±9.1
Barbour *et al.*[39]	top surface	3 mN 5 mN 7 mN	4.81±0.15 4.77±0.13 4.75±0.12	99.6±1.8 101.9±1.6 105.2±1.3

*In these test, authors reported a depth/contact area dependent effect of elastic modulus of enamel, namely the higher the penetration depth/contact area, the lower the elastic modulus.

Thirdly, the mechanical properties of the enamel are intra-tooth site-dependant. The inner enamel near the dentino-enamel junction (DEJ) has on average a reduced hardness and elastic modulus by about 26% compared to the outer enamel.[29] The size of enamel rod decreases with distance toward DEJ, consequently the volume fraction of softer interrods has a two fold increase, which undoubtedly contributes to a decrease in composite hardness and stiffness.[36]

Recently, the authors did a systemic nanoindentation investigation on the elastic modulus of sound enamel of permanent teeth with different tips and various loads.[33] The interesting finding to emerge from such an investigation is: the elastic modulus of enamel was not a constant value and had size effect dependent on the indentation contact areas.

Figure 8-3 shows a consistent decrease in elastic modulus for all indenters in both rod orientations. This illustrates that elastic modulus, in the case of enamel, is a function of contact area and the effective volume of the contact stress field from which the elastic modulus is determined, irrespective of what type of indenter tip is used. This can be rationalized by the unique rod-interrod microstructure of the biocomposite, in which the nanohardness and elastic modulus of interrod region were respectively 73.6% and 52.7% lower than those of the rod.[6] Consideration of the rod structure of enamel, with increasing peak load during nanoindentation, more and more prism-sheath units are involved in the indentation stress field and contribute to the measured nanoindentation elastic modulus. Due to the decreased modulus of the protein rich sheath region, the more the sheath area becomes involved in the deformation field, the lower the stiffness of the measured area. This may also partially assist in explaining the wide range of elastic modulus reported in other investigations and wider standard deviations from AFM based nanoindentation tests.

3.2. *Stress-strain relationship of enamel*

In our previous study, we developed a nanoindentation-based method for the measurement of indentation stress-strain relationship of the material.[33] Based on Tabor's theory, In the case of a conical-spherical tipped indenter, contact stress, σ, and strain, ε, can be estimated as:

$$\sigma \approx H/3 \tag{8-1}$$

$$\varepsilon \approx 0.2 \tan\theta \tag{8-2}$$

in which H is contact pressure (hardness) and θ is contact angle of the spherical tip or cone angle. Therefore, H-tan θ may be regarded as a contact stress-strain relationship. Since this is an empirical relationship and the actual strain and strain about an indentation impression has severe gradients, it is not surprising that some investigations have shown that the nominal stress-strain curve from this approach did not superimpose with standard uniaxial stress-strain curves.[40] Therefore,

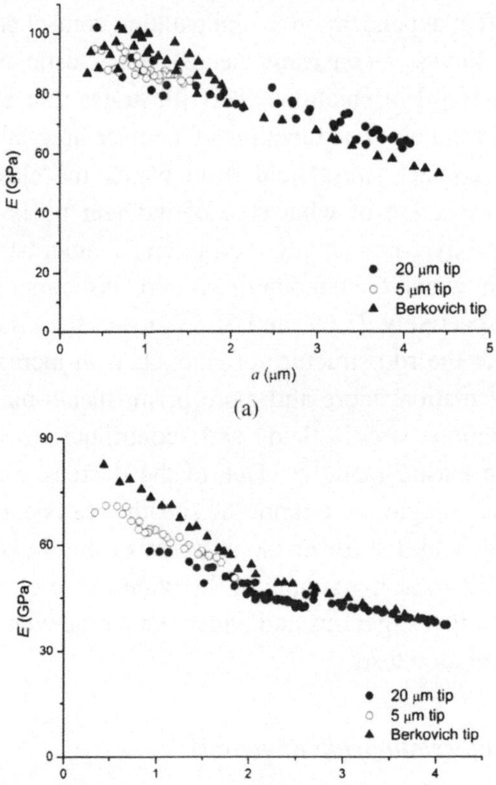

Figure 8-3. Elastic modulus (E) plotted against contact radius (a) for the three different indenters on (a) the top surface and (b) cross section surface of enamel.

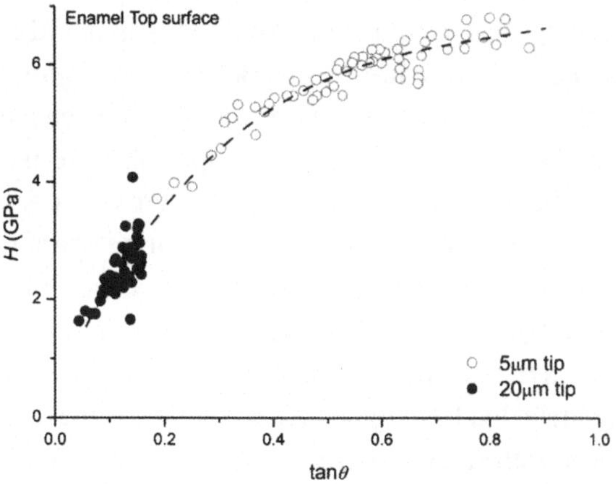

(a)

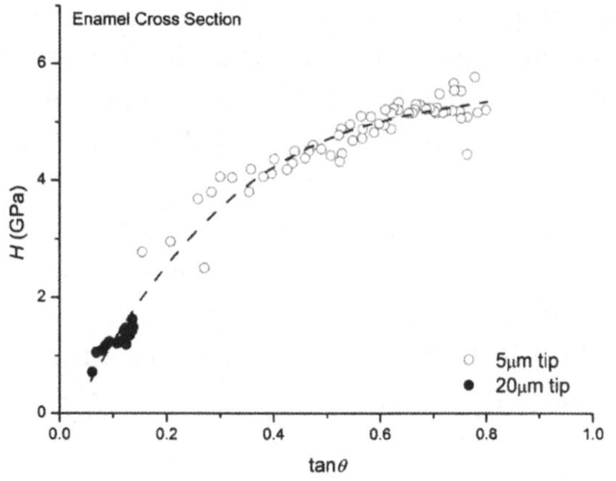

(b)

Figure 8-4. Nanoindentation stress-strain (H-tan θ) curves on (a) top surface and (b) cross section of enamel from two spherical tipped indenters.

we retained the values of H and tan θ in our results to avoid misunderstanding that our plots are standard stress-strain curves. However, using the same method, the H-tan θ curves enable comparison between different materials and are indicative of stress-strain behaviour. With this method, enamel as well as a wide variety of dental materials were tested and compared. Some interesting results were found.

As shown in Fig. 8-4, enamel from directions both parallel and perpendicular to the rod axis showed smooth stress-strain curves without a prominent elastic-plastic transition point. This is mainly caused by the unique composite rod structure of enamel and its organic rich interrod which makes the whole material more deformable than pure inorganic hydroxyapatite. The stress-strain curves, especially the inelastic part for the top surface, lie well above those for the cross section surface. For indentations on the top surface, the force direction is parallel to the prism with the yield behavior primarily determined by shear deformation of the very thin protein "glue" between the individual apatite crystallites. In the case of the cross-section surface, the loaded rods deflects under the force because the contact load is supported not only by the rod structure but also by the more deformable and thicker interrod structure surrounding the rods parallel to the surface.

Anisotropic properties of enamel, which arise from the rod like prism-sheath structure, are well reflected in the stress-strain curves. The top surface is stiffer and has higher stress-strain response than an adjacent cross section surface because of the greater influence of the prism sheaths in the latter behavior.

From a comparison of stress-strain curves for a range of dental materials (Fig. 8-5), enamel was found to behave more like a metallic material rather than a ceramic. A comparison with pure hydroxyapatite disk (HAP), which is the main component of the enamel, indicated that enamel had a much different stress-strain response. This suggests that the minor organic components in enamel influence the mechanical behavior of this highly mineralized biocomposite significantly. From this point of view and the above results, traditional criteria of dental material selection based on elastic modulus and hardness may not be appropriate for good restorative materials which can protect opposing teeth from excessive attrition. The nanoindentation stress-strain evaluation should also be involved in future dental restorative material selection and development studies.

3.3. *Energy absorption of enamel*

Another approach that further reinforces the significant role of the minor organic component on the mechanical properties of enamel is the issue of

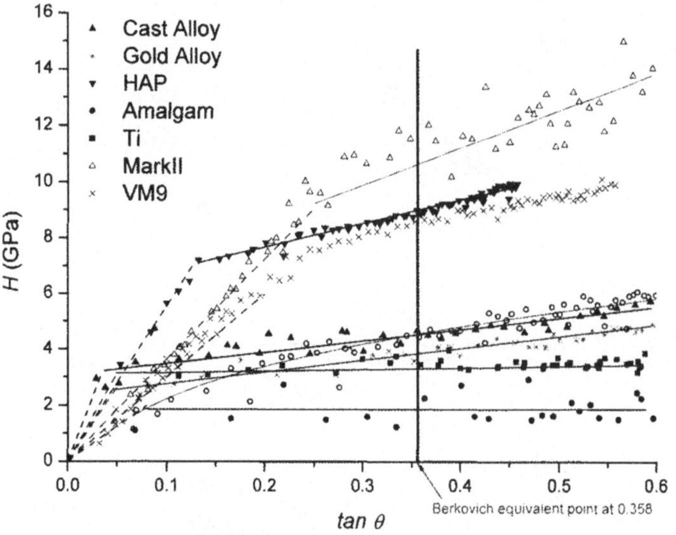

Figure 8-5. Stress-strain (H-tan θ) relationship of different materials. (Solid lines represent plastic part and short dots lines represent elastic part of materials, which was calculated from the Hertz expression, Eq. 5-5.)

energy adsorption, an approach that has received relatively scant attention to date. From nanoindentation load-displacement curves, the elastically recovered energy, U_E, and inelastically absorbed energy, U_I, of one indentation test can be calculated; the total energy $U_T = U_E + U_I$. The ratio of the inelastic energy to the total energy, $U\% = U_I / U_T * 100\%$, is used to normalize the energy expenditure of each test. This value may also be denoted the "plasticity index" to characterize the plasticity of the material. This U_I / U_T ratio ranges from 0 for an ideally elastic material (all energy is returned on unloading) to 1 for an ideally plastic material (no energy is returned).

In a recent study, we investigated the energy absorption ability of enamel by nanoindentation.[41] As shown in Fig. 8-6a, energy expenditure ratio was plotted against the contact strain. In this way, data from different tips, especially spherical tips, are consistent and form a continuous curve. For enamel, energy expenditure ratio, $U\%$, begins at 30% and has a monotonic increase with the contact strain, ε, with no obvious elastic part. In contrast, from Fig. 8-6b, HAP showed a linear

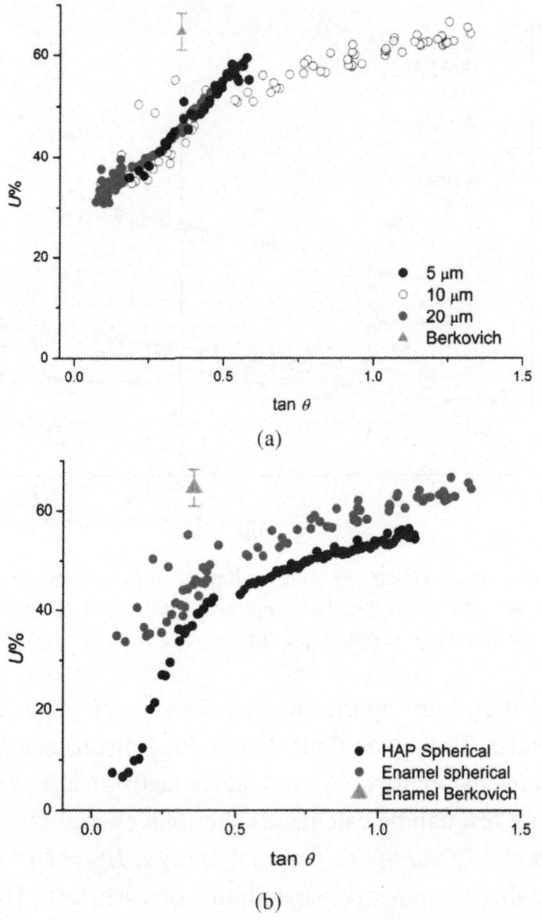

Figure 8-6. (a) Energy expenditure of enamel from different indenters as a function of indentation contact strain; (b) Comparison of energy absorption ability between enamel and HAP disk based on the nominally 10 μm spherical and Berkovich indenters.

$U\%$-ε part for strain values less than 0.5 and the curve started from 0%. Although the major component of the two materials in Fig. 8-6b was similar, enamel had slightly higher energy dissipation ability (~10%) than HAP at the same strain level. The origin of slightly higher energy expenditure from Berkovich indenter may be because the Berkovich indenter has a sharp tip and ridges with higher local stresses in their vicinity and in addition it may cause cracks, which is another source of unrecoverable energy dissipation. Another possible reason may be that the contact strain across the area of contact beneath a spherical tip is not constant but has a maximum at the edge of contact. Whereas for a Berkovich indenter the contact strain is constant across the contact area, that is, the mean contact strain beneath a Berkovich indenter is greater than that for a spherical indenter of nominally equivalent contact strain ($\tan \theta = a/R$).

Where is the extra energy in enamel dissipated? If we initially consider the apatite crystals, their very small size would have near theoretical strength, approximately 10% of the shear modulus of ~10 GPa. The mean contact stresses for enamel do not approach this value and only in the vicinity of the tip in the case of the Berkovich indenter may such stresses be achieved. Hence, it is not considered that energy loss by conventional dislocation based plastic deformation of the inorganic phase is important for enamel. Therefore, the focus of the discussion is centered on the role of the organic protein components between and surrounding hydroxyapatite crystallite rods.

Spears cited the elastic modulus of hydroxyapatite crystals and organic tissues as 114 GPa and 4.3 GPa, respectively, in his FEA model.[18] Moreover, AFM nanoindentation[6] also reinforced the argument that the protein rich sheath has a lower elastic modulus and hardness than crystal containing rods: the nanohardness and elastic modulus of the sheaths as reported above were 73.6% and 52.7% lower than those of the prisms, respectively. The nanoindentation curves shown by Ge *et al.*,[11] although not commented upon, showed that sheaths had more energy absorption than prisms. White *et al.*[10] also established that mineral apatite crystals had a micro-hardness approximately 1.8 times greater than human enamel. These studies indicate that the minor components of enamel, namely protein and water, have a profound softening or plasticizing effect. Other possible mechanisms responsible for the greater energy loss include: fracture of the apatite crystallites, deformation of the very thin protein layers between the crystallites within the prisms or by deformation of the protein and movement of fluid in the nano-scale porous structure.

In conclusion, enamel showed much higher energy absorption capacity than HAP, a ceramic material with similar composition. This may be because of the minor organic component of enamel. For the spherical indenters the energy loss was found to be a function of the contact strain. Energy loss with a Berkovich indenter was greater than a spherical indenter at equivalent contact strain.

Higher energy expenditure of the enamel was primarily related to the minor organic components. Three different mechanisms were considered to contribute to the measured energy absorption, namely fluid flow within the sheath structure as proposed by Fox,[42] protein "sacrificial bond" extension as proposed by Smith *et al.*[43] for nacre and Fantner *et al.*[44] for bone, or nano-scale friction within sheaths associated with the degustation of enamel rods as considered by Macho *et al.*[17] Further work is required to more quantitatively assign the energy loss observations to these or other mechanisms.

3.4. *Creep behavior of enamel*

Viscous responses of natural materials play an important role in their function to bear and distribute force, and especially for bone, and has attracted considerable attention.[45-50] Jäger[46] summarized the characteristics of typical viscous deformation of biological tissues as: (1) the force necessary for deformation depends on the deformation rate; (2) force-displacement hysteresis occurs during loading and unloading; (3) application of a (constant) force leads to transient and possibly permanent creep; (4) after removing the deformation force there is recovery.

Restricted by measurement methods, the limited viscosity and temperature range over which natural biomaterials operate, the viscous response—especially creep behavior—is difficult to quantify. For metallic and polymeric materials, traditional tensile or compressive tests are used to investigate the creep response of a specimen. Two main problems limit the usage of these methods for biological materials such as enamel. Firstly, these tests need multiple identical specimens and are time consuming to conduct. Secondly, the accuracy of these methods depends on the creep response of the material. For small samples with limited creep, such as enamel, it is not very reliable. In the biomaterials field, only limited indentation creep tests have been done on biological hard tissues. Torsion[47,48] and three-point bending tests[45] were used to measure the relaxation elastic modulus and shear modulus of bovine bone. Further more, because of the anisotropic property of natural biomaterials, the loading direction must be controlled. This limitation, along with the size of teeth, severely restricts the usage of bulk methods for the evaluation of dental tissue.

As already mentioned, depth-sensing indentation techniques have become widely used for mechanical property evaluation, and may be used for the determination of creep behavior. Indentation creep tests are a powerful technique for studying the rheological properties and relaxation dynamics of different materials such as metals,[51] polymers[52,53] and other materials.[54,55] In our recent investigations,[56] the creep behaviour of enamel and dental materials were described and compared by a one-step loading-unloading method from nanoindentation with a Berkovich indenter. Samples were indented at the maximum force of 250 mN with a one-step loading method. Holding time at maximum load was 900 seconds for all tests. During one-step unloading, another holding at 5 mN was activated. This unload-holding time was 900 seconds as well. Each group of tests commenced at least two hours after a thermal soak period in the instrument, and thermal drift was assessed before each experiment. Thermal drift during all the tests was deemed to be negligible (<0.02 nm/sec).

In Fig. 8-7, the relative penetration depth, h_t', was calculated by subtracting the initial depth at the beginning of the holding time from the increasing depths during the holding period at maximum and minimum loads. Creep data at maximum-load holding exhibited increasing penetration depth with time for most of the investigated materials (Fig. 8-7a). Amalgam showed the maximum creep, followed by titanium. However, considerable scatter was observed in the data. Enamel, on the other hand, had a relatively stable creep response. In contrast, the indentation creep depths of hydroxyapatite and cast Cr-Co alloy were

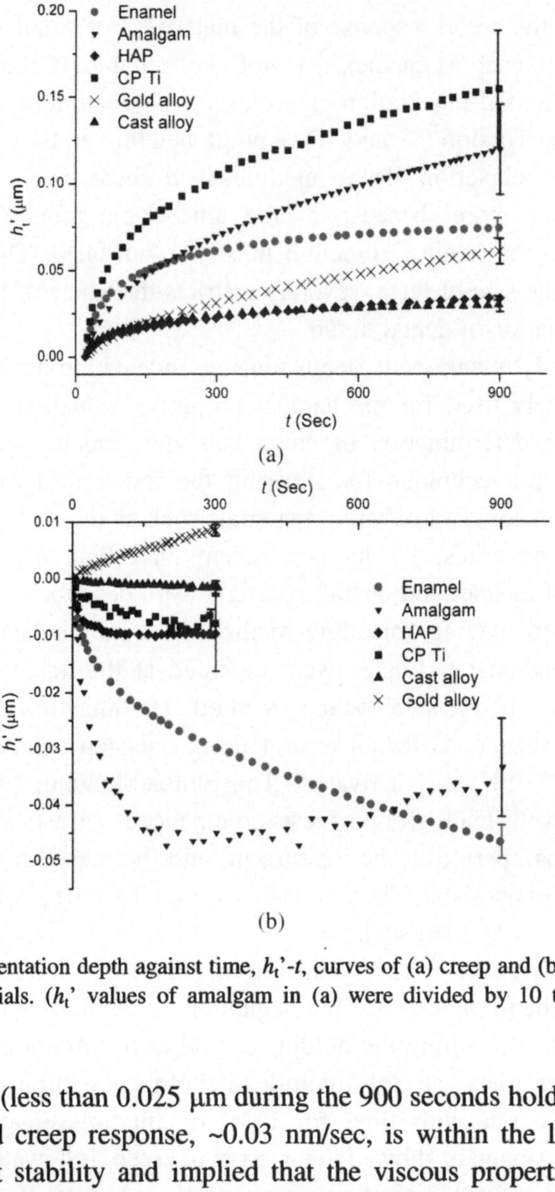

Figure 8-7. Indentation depth against time, h_t'-t, curves of (a) creep and (b) backcreep of different materials. (h_t' values of amalgam in (a) were divided by 10 to fit onto the diagram.)

very limited (less than 0.025 μm during the 900 seconds holding period). Such limited creep response, ~0.03 nm/sec, is within the limit system thermal drift stability and implied that the viscous properties of these materials are negligible compared with their elastic-plastic properties.

At the minimum-load holding period (named "*backcreep*" period), the recovery was prominent only for enamel and amalgam samples but remained within the instrument's stability range (<0.03 nm/sec) for other samples (Fig. 8-7b). The displacement rate and limited backcreep of hydroxyapatite (~0.03 nm/sec) and cast alloy (~0.01 nm/sec) were all within the limit of system thermal drift. Therefore, only enamel and amalgam exhibited significant backcreep response. When considered in more detail, the backcreep curves of enamel and amalgam were totally different. Enamel showed a progressive recovery while amalgam recovered very quickly in the first 150 seconds and reached the limit of nearly -0.05 μm in h_t' values. The latter portion of the amalgam curve only reflects system thermal drift response.

This result implies that enamel and some dental metallic materials all exhibit creep behavior but with different characteristics. This indicates different mechanisms are responsible for this behavior. The visco-elastic property of enamel may be considered as follows: First, the proteins and other organic (and water) components within enamel may be regarded as a biopolymer. Viscoplastic properties of these biopolymers stem from the configurational rearrangements, disposition and interaction among the macromolecules in both their short- and long-range network interactions.[57] Under indentation load, the local re-distribution response of the structure is rapid, while the long-range interaction response is slow due to the time-dependent molecular rearrangement. Upon unloading, due to the memory effects associated with the cross-linked network, the molecules have a tendency to return to their initial form and positions.[57] In the case of indentation, the surrounding elastically deformed hinterland about the residual impression provides an additional driving force for such backcreep deformation. These factors result in the recovery of the indentation impression being also time-dependent. Second, proteins within enamel may behave similarly to the recovery domain bond within many proteins, called a "sacrificial bond".[58] Rief *et al.* noted that the longer the delay before the next application of force, the more reformation of the bond occurred during relaxation; refolding of protein domains was also observed during this process.[58] Similarly, proteins within the organic rich sheath and between the apatite crystallites of enamel may have a similar function and response as those in bone.

In short, this reversible viscoelastic property of enamel may help teeth distribute the localized high contact or impact stresses resulting from chewing forces especially the high localized contacts from asperities or small hard particles trapped between teeth. Also, sliding wear between teeth and food particles may introduce high local tensile, compressive and shear forces around sliding contact area. This reversible, inelastic property may be vital to limit damage accumulation and prevent excessive wear in enamel.

3.5. *Nanoindentation measurements on dentin*

Most nanoindentation investigations on dentin have concentrated on the measurement of elastic modulus and hardness with respect to its microstructure and composition. Nanoindentation allows for the measurement of the highly calcified peritubular dentin and the intertubular dentin separately, demonstrating the structural dependence of dentin mechanical behavior. Data on the mechanical properties of dentin in both permanent and primary teeth based on nanoindentation investigation are summarized in Table 8-3.

From these investigations, we may conclude firstly that dentin is an anisotropic material and nanoindentation elastic modulus and hardness of dentin are microstructurally dependent. As with enamel, the mechanical properties of the dentin are larger when applied loads are parallel to tubule directions compared to orthogonal directions.[59] Intertubular dentin has hardness and elastic modulus of 0.51 ± 0.02 and 21.1 ± 1.3 GPa, respectively, but values decrease with proximity to the pulp.[60] In contrast, the more mineralized structure of peritubular dentin not only has greater hardness and elastic modulus, but also is invariant with respect to intra-dentin position. Overall the outer dentin has higher hardness and elastic modulus compared to the inner dentin,[61,62] which may be due not only to the changes in the mechanical properties of intertubular dentin, but as a consequence of thicker peritubular dentin and fewer tubules.[61]

Secondly, primary dentin is mechanically inferior to the permanent dentin, which has been shown with nanoindentation tests by Mahoney *et al.*[34] and Poolthong.[36] This is because the dentin of primary teeth is less calcified than the dentin of permanent teeth.[63,64] And lastly, changes

Table 8-3. Nanoindentation elastic modulus and hardness of dentin.[25]

Authors	Surface and Site	Test Load	Hardness (GPa)	Elastic Modulus (GPa)
Angker *et al.*[61]	primary molars axial surface: - outer dentin - middle dentin - inner dentin	25 mN	0.91±0.15 0.85±0.19 0.52±0.24	16.33±3.8 17.06±3.09 11.59±3.95
Marshall *et al.*[35]	axial surface: intertubular dentin adjacent to DEJ	6 mN	0.74~0.92*	18.61~20.69*
Tesch *et al.*[65]	axial surface - DEJ region	0.5 mN	0.65	19.5
Fong *et al.*[37]	axial surface	0.4 mN	0.79±0.04	24.8±1.4
Mahoney *et al.*[34]	primary molars transversal section of mid-coronal dentin	50 and 150 mN	0.92±0.11	19.89±1.92
Poolthong *et al.*[59]	position to tubule: - perpendicular - parallel - perpendicular# - parallel#	50 and 150 mN	1.20±0.11 1.03±0.06	27.23±0.63 23.10±1.01 27.20±1.73 21.30±1.01
Kinney *et al.*[60]	peritubular dentin intertubular dentin: outer dentin inner dentin	0.3 mN	2.45±0.14 0.51±0.02 0.15±0.03	29.8±8.9 21.1±1.3 17.7±0.3
Van Meerbeek *et al.*[66]	axial surface	5 mN	0.49	19.26

*Range

#Spherical indenter, others used a Berkovich indenter.

in the mechanical properties of dentin are mineral content-dependant as the collagen makes no significant contribution to the overall hardness and elastic modulus of dentin.[67] The mechanical properties of dentin collagen remain constant across the dentin, of which dehydrated collagen has considerably higher hardness (0.2 GPa) and elastic modulus (2 GPa) compared to the elastic modulus of hydrated collagen (381 kPa).[67] Indeed another study showed that upon the removal of dentin collagen using sodium hypochlorite, the dentin had relatively unchanged hardness and elastic modulus, but the opposite outcome was seen when the dentin was etched beforehand.[68]

In conclusion, considering the microstructural variations of the dentin and their impact on the mechanical properties measured, any attempt to investigate dentin mechanical properties has to take into account sample selections, that is, whether permanent or primary teeth, age of the tooth, young teeth are less mineralized and have bigger tubules, testing sites, dentin near the pulp has decreased mechanical properties due to increased tubule numbers and dimensions,[69] less mineralized intertubular dentin[60] and the decreased dimension of highly mineralized peritubular dentin,[70] direction of load whether it is perpendicular or parallel to tubules[71] and whether the test is conducted on hydrated or dehydrated dentin.[72,73] Furthermore, storage media have been shown to have a detrimental effect on the mechanical properties of dentin,[38] but this may be minimized by preparing the specimens just before testing.

The micromechanical properties of dentin are of relevance to the issue of etch-based pretreatment of dentin prior to use of a composite resin restorative. Bonding between adhesive restorative materials and dentin is normally facilitated through partial ionic bonding of the polymer of the restoration to the calcium of hydroxyapatite and the collagen of the dentin, as well as micro-mechanical retention of the resin into the etched dentin. Toledano *et al.*[74] showed that etched dentin prepared for restorative bonding had a significantly reduced hardness, which could not be fully restored following the impregnation of bonding resin into it, thus the hardness of the resin-dentin interdiffusion zone was considerably lower than that of unaltered dentin. A gradient of moduli of elasticity was observed from the stiffer sound dentin to a less stiff resin-dentin interdiffusion zone, and adhesive resin layer to the restorative composite.[66] This is simply explicable as the restorative materials used have an elastic modulus far inferior to sound dentin, typically ranging from 5.8 to 9.2 GPa. Further investigation may lead to more appropriate technique and material selections that can create a mechanically sustainable dentin-resin bonding.

3.6. *Nanoindentation measurements on cementum*

Nanoindention investigations of cementum generate the baseline knowledge of the mechanical properties of sound cementum and allow for an understanding of abnormal behavior due to various pathologic conditions. It is anticipated that changes in the mechanical properties of cementum may be related to its mineral levels and porosity resulting from cellular activities surrounding the root.

The mechanical properties of cementum have been shown to vary significantly due to the local anatomy and cementum type (Table 8-4). Unlike enamel and dentin, only limited data on cementum are available in the literature. Poolthong[36] reported that cementum has a hardness equivalent to dentin, but lower elastic modulus. The hardness and elastic modulus of the apical third range from 0.8 – 0.9 and 9.2 – 9.4 GPa, respectively, while the cervical two third were 1.4 – 1.6 and 11.5 - 12.1 GPa, respectively.[36] Using a 20 μm spherical indenter, Malek and coworkers[75] were able to investigate the mechanical properties of the unpolished surfaces of the outer cementum. They found that the mechanical behavior of the outer cementum is root site dependant ranging from 0.18 – 0.6 and 1.7 – 10.8 GPa, for hardness and elastic modulus respectively, which are noticeable lower than the previous data. The outer surface has more organic material as a result of the presence of Sharpey fibers compared to the rest of the cementum.[22] Due to the enriched presence of protein and collagen fibers, cementum is highly viscoelastic, especially when hydrated.[76] The cemento-dentinal junction was shown to have large quantities of proteins including collagens that contribute to hydration and significantly reduced the mechanical properties. It has decreased mechanical properties compared to the adjacent hard tissues, cementum and dentin.[77]

The mechanical preparation required for nanoindentation as well as the storage media were shown to affect the mechanical properties of the cementum measured. It was suggested that an ultra-sectioning technique using a microtome generates a flat surface of cementum with the advantage of clear visualization of its microstructure. The later is not evident in polished specimens possibly due to the smear layer generated from polishing procedures.[76] As a result more diverse data are obtained

Table 8-4. Nanoindentation elastic modulus and hardness of cementum.[25]

Authors	Tooth Sample	Sample Site and Sectioning	Hardness (GPa)	Elastic Modulus (GPa)
Srivicharnkul *et al.*[78]	orthodontic-treated premolars	outer surface of cementum#: - cervical third - middle third - apical third	0.25 ± 0.09 0.24 ± 0.07 0.18 ± 0.06	4.4 ± 2.4 3.4 ± 2.0 2.4 ± 1.8
Ho *et al.*[79]	mandibular third molars	transversal section: - ultrasectioned - polished	0.48 – 1.09* 0.79 – 0.89*	14.2 – 25.9* 18.2 – 20.8*
Malek *et al.*[75]	premolars	outer surface of cementum#	0.41 ± 0.14	5.68 ± 2.65
Poolthong[36]	premolars	Longitudinal section: - middle third - apical third	0.63 ± 0.06 0.57 ± 0.48	11.95 ± 1.24 8.55 ± 0.84

*Range
#Spherical indenter, others used a Berkovich indenter.

from an ultra-sectioned cementum, which clearly represent local variations. Purified water was reported to be a better storage media, compared to alcohol or desiccation, in which the mechanical properties of cementum were noted to be stable for up to nine months.[80]

Nanoindentation has also been applied to investigating the impact of orthodontic treatments on tooth root. Ideally an orthodontic force should be able to generate tooth movement without any pathological symptoms. It is believed that excessively large applied forces may initiate resorption in cementum. However, nano-indentations on the cementum of premolar teeth, which had been subjected to 25 g and 250 g orthodontic loads for four weeks showed no significant difference in their hardness and elastic modulus.[78] The latter authors consider that this is possibly due to short treatment period and large inter-individual variations between teeth. The reported values are however, lower than those previously reported by Malek *et al.*[75] for untreated teeth (Table 8-4). Further investigations are necessary to more critically evaluate spatial variations and the role of orthodontic forces and periodontal diseases on the mechanical properties of cementum.

3.7. *Nanoindentation of carious lesions*

Dental caries may be defined as a bacterial disease of the calcified tissues of the teeth characterized by demineralization of the inorganic and destruction of the organic substance of the tooth.[81] Due to its long history and wide prevalence, dental caries has attracted considerable attention of both clinical practitioners and researchers.

From a materials science point of view, caries and pre-caries lesions (*e.g.* white spot lesions in clinic) will have inferior mechanical properties compared to their healthy counterparts. By considering the above mentioned hierarchical microstructures of both enamel and dentin and the natural procedure of acid invasion, the changes of mechanical properties may have defining characteristics and features, which may allow us to understand the formation and development of the lesion such that remineralization treatments may be developed. In this field, nanoindentation is obviously a powerful tool.

Recently, a nano-indentation characterization of artificially generated white spot lesions (WSLs) has been reported.[82] After a period of time soaking in a mild acidic solution, sound teeth samples showed degrees of subsurface demineralization. The partially demineralized lesion areas have lower elastic modulus and hardness than sound areas, due to the loss of mineral crystals. At the surface of the lesion, a less demineralized surface zone was found.

Very recently, our group has undertaken nanoindentation investigations of natural WSLs (Fig. 8-8). The lesions exhibited a number of specific mechanical property features:

- There is a rapid reduction in mechanical properties from sound enamel to the WSL body;
- The body of the lesion appears inhomogeneous with some islands that show higher or lower elastic modulus and hardness;
- The surface layer of the lesion showed higher elastic modulus and hardness than the lesion body.

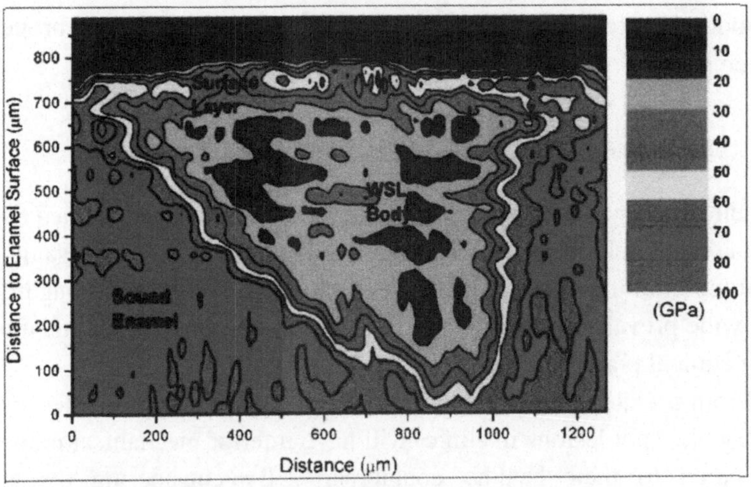

(a)

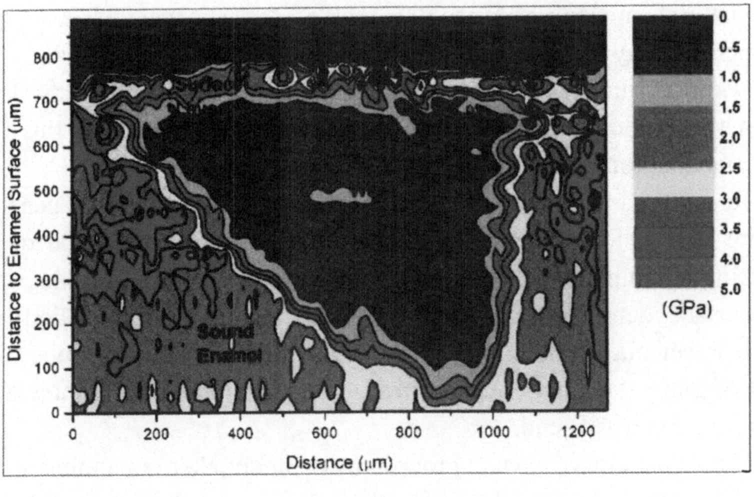

(b)

Figure 8-8. Nanoindentation mapping of a natural WSL on (a) elastic modulus and (b) hardness (load applied was 25 mN and the interval between two indents was 25 μm).

Moreover, the reduction of elastic modulus and hardness of natural WSLs was considered as a consequence of mineral loss. Therefore, a correlation analysis of elastic modulus and mineral density was performed on the natural WSL samples. Mineral density was measured

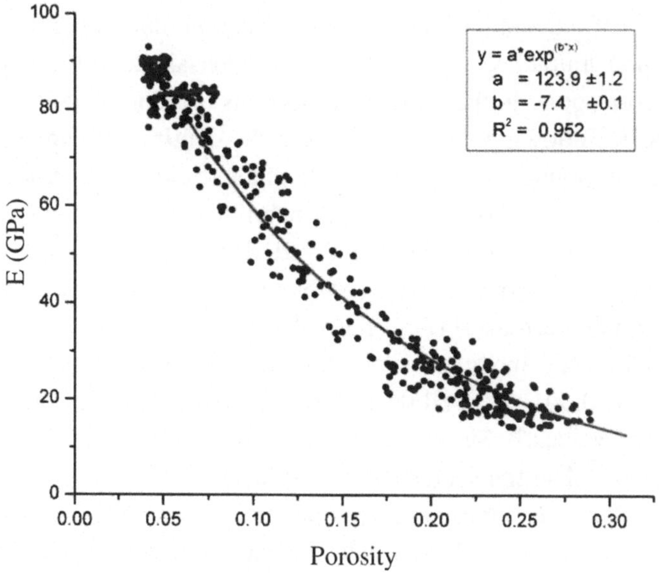

Figure 8-9. Nanoindentation elastic modulus against the mineral porosity of WSL sample.

and calculated from the gray value of micro-CT images.[83] By taking the mineral density of sound enamel ρ_{sound} as 3.0 g/cm^3, the porosity of every indented lesion point can be calculated with the equation: Porosity = $\rho_{lesion}/\rho_{sound}$, and correlated with the elastic modulus at that point (Fig. 8-9). A good exponential relationship was found and the result can be related to Rice's minimum solid area model for porous materials.[84] In addition, recent high resolution electron microscopy observation provided direct evidence on the dissolving and pore formation of enamel HAP crystals.[85]

Deterioration in the mechanical properties of dentin is not only known as an early sign of carious demineralization in dentin, but of clinical relevance for caries assessment and treatment selections. The nanoindentation technique enables testing of carious dentin in a hydrated state, which is of particularly significance as the hardness and elastic modulus of carious dentin have been shown to increase significantly with dehydration, and a hydrated condition is considered more clinically appropriate.[72] This difference was not evident using traditional approaches such as Knoop indentation, which depends on visual measurement of indentation imprint on test material due to highly elastic recovery upon unloading. As a composite material, dentin mechanical properties correspond to the relative proportions of mineral, collagen and water content. Following demineralization, the dentin will contain more water than the sound dentin and the remnant fluid-filled structure of collagen with its high poroelastic deformability will become dominant, particularly in the severely demineralized dentin.[86] The hardness and elastic modulus of carious dentin were found to decrease significantly, of which the outer carious surface normally has the lowest value. The hardness and elastic modulus values for primary carious dentin ranged from as low as 0.001 GPa and 0.015 GPa, respectively to just below the sound dentin values.[86] Similar findings were reported for permanent carious dentin.[73] The transparent, sclerotic layer which is assumed to be hypercalcified, as a result of dental pulp attempts to block potential toxin movement towards the pulp, has mechanical properties slightly lower than the sound dentin.[73] In contrast to the changes in mineral levels through a carious lesion, which appear to indicate a linear function with lesion depth,[72,87] the deterioration in the mechanical properties of carious dentin was found to exhibit a logarithmic function.[72]

3.8. *Nanoindentation on hypomineralized dental hard tissues*

Developmental defects of dental hard tissues mainly include amelogenesis imperfecta (defects of enamel) and dentinogenesis imperfecta (defects of dentin). The etiology of these developmental defects is complex, including genetic, environmental and systemic factors. Amelogenesis imperfecta has two subtypes: hypoplastic type (defective matrix production) and hypomineralized type (defective maturation or mineralization).[81] In general, most of these defective enamel and dentin have less mineral and higher organic contents. Theoretically, from Spears' finite element model,[18] the change of the mineral to organic ratio of the composite will decrease the mechanical properties of the natural biocomposite significantly. In recent nanoindentation investigations, the hardness and elastic modulus of the hypomineralised enamel were reported as 0.53±0.31 and 14.49±7.56 GPa, respectively, which are much lower than the sound counterpart.[88]

Similarly, another nanoindentation mapping showed the hardness and elastic modulus of hypomineralized region ranged from 0.07 – 1.7 GPa and 3.2 – 40.9 GPa, respectively, which are less than a quarter of that of the normal enamel.[89] It was also suggested that this reduction is not only due to variations in mineralization, but more likely from differences in the apatite crystalline and crystal orientations of the hypoplastic enamel. Moreover, a newly published report showed that hypomineralized enamel also showed a size effect on nanoindentation elastic modulus, as described in Section 3.1 of this chapter, but the decreasing trend was more significant than sound enamel.[90] The corresponding TEM observation illustrated thicker interrod structures and higher levels of organic content surrounding the HAP crystals within both rod and interrod regions.[90] By employing Spears' finite element model, the inferior mechanical properties can be reasonably explained in respect of the microstructure of the hypomineralized enamel.

4. Dental Applications of Nanoindentation Techniques

4.1. *Evaluation of dental materials*

With the technique of nanoindentation, small regions of material can be tested accurately. Due to this advantage, the technique has been employed in dental material development, evaluation and selection. A good example is the increasing reports on the measurement of dental composite materials with nanoindentation.[91-98] An investigation on fiber-reinforced composites tested the mechanical properties of reinforcement fibers and polymer matrix separately and reported the influence of water storage.[98] The effects of coupling agents on hydroxyapatite (HA) before incorporating into polymer matrix to form bioactive composite materials were investigated with nanoindentation,[99] in which the hardness and elastic modulus of small HA particles, the polymer matrix and the particle-matrix interface were tested and reported separately. Different surface finish methods were also compared with nanoindentation on dental composite materials.[100] Nanoindentation also helped in the evaluation of different coatings on dental alloys, systems that have a range of potential dental applications.[101]

4.2. *Understanding of soft drinks*

The consumption of soft-, sport-, and energy-drinks has increased significantly over the last few years. The refreshing taste of many soft drinks and fruit juices is based mainly on their content of different acids. Beyond improving taste, acids in soft drinks may exhibit additional functions, such as preservation or stabilization of the drinks. It is known that an acidic environment has potential harmful effects—namely dental erosion—on dental hard tissues, especially the outermost enamel. During erosion, calcium and phosphate ions are dissolved from the enamel, which eventually leads to a collapse of the enamel surface structure and loss of the outermost enamel layers. A key factor for soft drinks on the erosion of dental enamel is related to the early stages of enamel demineralization at the tooth-soft drink interface, as at this stage the demineralization process was assumed to be reversible by saliva-induced remineralization. Nevertheless, evaluation of the early stages of enamel demineralization was not accessible for investigation until recently due to a lack of suitable technology.

The sensitivity of nanoindentation allows the study of the very early stages of enamel erosion. AFM and nanoindentation in combination can be considered the current state of the art standard methods in the quantification of food- and beverage-induced enamel material loss, and has revolutionized the investigation of food-induced enamel erosion. An AFM based nanoindentation test[102] illustrated for the first time that the hardness of eroding enamel is reduced as a function of time and pH. After a minimum hardness value of the eroded enamel was reached (enamel softening) the structure of the enamel collapsed and some enamel material was lost. Also, nanoindentation was used in this field to find suitable concentrations of calcium and phosphate in soft drinks, which may reduce the erosive potential of soft drinks.[103] At the same time, temperature was considered as a confounding influence factor contributing to the demineralization procedure.[104] Furthermore, deciduous and permanent enamel were compared under similar conditions.[105] In short, the atomic force microscope (AFM) and the AFM nanoindenter have been shown to be cutting edge tools to probe the early stages of enamel demineralization with previously unmatched precision on the nanoscale.[106] With these advanced techniques, enamel erosion has been found to occur in three stages: enamel softening, enamel material loss on a microscopic scale, and macroscopic enamel loss. Based on the above findings from nanoindentation, healthy functional soft drinks contain calcium and phosphate were developed and introduced to market.[106]

4.3. *Assessment of dental treatments*

Bleaching of teeth is gaining popularity for cosmetic reasons. Present tooth-bleaching techniques use hydrogen peroxide as the active agent. It is applied directly, or produced in a chemical reaction from sodium perborate-carbamide peroxide. Hydrogen peroxide is a reactive oxygen species, along with superoxide ($O_2^{\cdot}$), hydroxyl radical ($HO^{\cdot}$), peroxyl (ROO) and alkoxyl (RO).[107] These free radicals oxidize a wide variety of organic and inorganic compounds by attacking the long chained, dark colored chromophore molecules and split them into smaller, less intensely colored and more diffusible molecules. Although the bleaching treatments have been exposed to numerous evaluations,[108] the effect these have on teeth is still largely unknown, especially the potential harmful effects of free radicals on organic components of enamel. Recently, nanoindentation was used to evaluate the effects of bleaching agents on the mechanical properties of human teeth.[109] The results illustrated that the mechanical properties of the bleached dentin were significantly decreased. For intertubular dentin, the mean hardness decreased by 29-55% and the mean elastic modulus decreased by 19-43%. For enamel, the mean hardness decreased by 13-32% while the mean elastic modulus decreased by 18-32%.

As a common daily oral hygiene procedure, tooth brushing may have abrasive effects on the surface of enamel, especially acid-softened enamel. AFM nanoindentation was employed to investigate the issue[110] and the results indicated that tooth brushing of surface softened enamel leads to minor changes in the surface morphology and nanomechanical properties; the amount of enamel lost due to tooth brushing was independent of the demineralization time and was lower compared to the mineral loss caused by the demineralization treatments. The exposure to artificial saliva offers minimal protective effect to the softened enamel surface against tooth brushing.

Remineralization of demineralized dental hard tissue has long been considered as a non-invasive option for the prophylaxis and treatment of caries. Some commercial remineralization products are currently available in the market. Unfortunately, most evaluations of the remineralization potential of such treatments have concentrated on mineral density measurements and so far there are no nano-mechanical reports on the effects of these remineralization products. Only one report suggested that human saliva has the function to remineralize a structurally weakened enamel surface due to its high calcium and phosphate concentrations.[111]

Similarly, lasers have also long been utilized in the dental clinic as a cutting method or surface treatment method for caries prophylaxis. Considerable morphological and histological research has been undertaken on laser treated dental hard tissues. Again, the method is lacking comparable micro-mechanical characterization, for instance using nanoindentation to establish its efficacy.

5. Conclusions

The use of nanoindentation in dental hard tissue studies has significantly improved our understanding of the mechanical behavior of enamel, dentin and cementum with respect to their microstructural and compositional characteristics, as well as the consequences of disease and of chemically-induced conditions. At the same time, dental hard tissue diseases and dental materials as well as treatments have begun to be evaluated by nanoindentation, which has led to dramatic progress in these areas. Detailed investigations of the inelastic behavior of enamel by nanoindentation have opened a new window for comparative biomaterial evaluation with advanced techniques now possible, and these investigations may influence the biomaterial and restorative material selection criteria in the future.

However, caution must be expressed as the nanoindentation technique evaluates only the mechanical properties of a very shallow surface region of a specimen, which may have undergone damage associated with mechanical preparation required to achieve a satisfactorily flat sample for testing. Further investigations on this issue of surface preparation are necessary. The technique is also very surface sensitive with a fine polished surface a prerequisite. In addition, the inhomogeneous and anisotropic nature of these viscoelastic materials add further complications which to date have barely been addressed.

Acknowledgements

L. H. He thanks the Australian Department of Education, Science and Training (DEST) the Westmead Hospital and University of Sydney for financial support in the form of IPRS scholarship. Partially financial support by an ARC Linkage grant (LP0561184) and ADRF grants (60/2006 and 29/2007) are also acknowledged.

References

1. J. M. Rensberger, in *Tooth Enamel Microstructure*, eds. W. V. Koenigswald and P. M. Sander, pp. 237–257, Balkema, Rotterdam (1997).
2. N. E. Waters, in *The Mechanical Properties of Biological Materials*, eds. J. Vincent and J. Curry, pp. 99–134, Cambridge University Press (1980).
3. V. F. Ferrario, C. Sforza and G. Zanotti, *J. Dent.* **32**, 451–457 (2004).
4. C. P. Fernandes, P.-O. Glantz and S. A. Svensson, *Dent. Mater.* **19**, 118–126 (2003).
5. J. Mencik and M. V. Swain, *Mater. Forum* **18**, 277–288 (1994).
6. J. Ge, F. Z. Cui, X. M. Wang, *et al.*, *Biomaterials* **26**, 3333–3339 (2005).
7. A. G. Fincham, W. Luo, J. Moradian-Oldak, *et al.*, in *Development, Function and Evolution of Teeth*, eds. M. F. Teaford, M. Meredith-Smith and M. W. J. Ferguson, pp. 37–61, Cambridge University Press (2000).
8. C.-C. Hu, M. Fukae and T. Uchida, *Journal of Dental Research* **76**, 648–657 (1997).
9. A. Boyde, in *Ciba Foundation Symposium 205 Dental Enamel*, eds. D. J. Chadwick and G. Cardew, pp. 18–31, Wiley (1997).
10. S. N. White, W. Luo, M. L. Paine, *et al.*, *J. Dent. Res.* **80**, 321–327 (2001).
11. L. H. He and M. V. Swain, *Appl. Phys. Lett.* **90**, 171916 (2007).
12. H. H. K. Xu, D. T. Smith, S. Jahanmir, *et al.*, *J. Dent. Res.* **77**, 472–480 (1998).
13. D. R. Eisenmann, in *Oral Histology*, ed. R. T. Cate, pp. 218–235, Mosby, St. Louis (1998).
14. P. D. Frazier, *J. Ultrastruct. Res.* **22**, 1–11 (1968).

15. B. Kerebel, G. Daculsi and L. M. Kerebel, *J. Dent. Res.* **57**, 306–312 (1979).
16. D. F. G. Poole and A. W. Brooks, *Arch. Oral. Biol.* **5**, 14–26 (1961).
17. G. A. Macho, Y. Jiang and I. R. Spears, *Journal of Human Evolution* **45**, 81–90 (2003).
18. I. R. Spears, *J. Dent. Res.* **76**, 1690–1697 (1997).
19. D. Shimizu, G. A. Macho and I. R. Spears, *American Journal of Physical Anthropology* **126**, 427–434 (2005).
20. Y. Jiang, I. R. Spears and G. A. Macho, *Arch. Oral. Biol.* **48**, 449–457 (2003).
21. I. R. Spears, R. van Noort, R. H. Crompton, *et al.*, *Journal of Dental Research* **72**, 1526–1531 (1993).
22. A. Nanci, in *Ten Cate's Oral Histology: Development, Structure, and Function*, ed. A. Nanci, Mosby, St. Louis (2003).
23. M. J. James, E. Järvinen and I. Thesleff, *Gene Expr. Pattern* **4**, 595–599 (2004).
24. A. L. Boskey, *Bone Miner* **6**, 111–123 (1989).
25. L. Angker and M. V. Swain, *Journal of Materials Research* **21**, 1893–1905 (2006).
26. C. D. Torneck, in *Oral Histology: Development, Structure and Function*, ed. A. R. Ten Cate, pp. 169–217, Mosby, St. Louis (1994).
27. B. Berkovitz, G. Holland and B. Moxham, *Oral Anatomy, Histology and Embriology*, Mosby Ltd. (2002).
28. P. Anderson and J. C. Elliott, *Caries Res.* **34**, 33–40 (2000).
29. J. L. Cuy, A. B. Manna, K. J. Livi, *et al.*, *Archives of Oral Biology* **47**, 281–291 (2002).
30. S. Habelitz, S. J. Marshall, G. W. Marshall, *et al.*, *Archives of Oral Biology* **46**, 173–183 (2001).
31. G. Willems, J. P. Celis, P. Lambrechts, *et al.*, *J. Biomed. Mater. Res.* **27**, 747–755 (1993).
32. J. Zhou and L. L. Hsiung, *J. Biomed. Mater. Res.*, in Press (2006).
33. L. H. He, N. Fujisawa and M V. Swain, *Biomaterials* **27**, 4388–4398 (2006).
34. E. Mahoney, A. Holt, M. V. Swain, *et al.*, *J. Dent.* **28**, 589–594 (2000).
35. G. W. Marshall, M. Balooch, R. R. Gallagher, *et al.*, *J. Biomed. Mater. Res.* **54**, 87–95 (2000).
36. S. Poolthong, PhD Thesis, *Faculty of Dentistry*, University of Sydney (1998).
37. H. Fong, M. Sarikaya, S. N. White, *et al.*, *Mater. Sci .Eng. C* **7**, 119–128 (2000).
38. S. Habelitz, G. W. Marshall, M. Baloochb, *et al.*, *J. Biomech.* **35**, 995–998 (2002).
39. M. E. Barbour, D. M. Parker and K. D. Jandt, *Journal of Colloid and Interface Science* **265**, 9–14 (2003).
40. E. G. Herbert, G. M. Pharr, W. C. Oliver, *et al.*, *Thin Solid Films* 398–399, 331–335 (2001).
41. L. H. He and M. V. Swain, *J. Biomed. Mater. Res.* **81A**, 484–492 (2007).
42. P. G. Fox, *J. Mater. Sci.* **15**, 3113–3121 (1980).
43. B. Smith, T. Schaffer, M. Viani, *et al.*, *Nature* **399**, 761–763 (1999).

44. G. E. Fantner, T. Hassenkam, J. H. Kindt, *et al.*, *Nature Materials* **4**, 612–616 (2005).
45. T. Goto, N. Sasaki and K. Hikichi, *Journal of Biomechanics* **32**, 93–97 (1999).
46. I. L. Jäger, *Journal of Biomechanics* **38**, 1451–1458 (2005).
47. N. Sasaki and A. Enyo, *Journal of Biomechanics* **28**, 809–815 (1995).
48. N. Sasaki, M. Yoshikawa and A. Enyo, *Journal of Biomechanics* **26**, 1369–1376 (1993).
49. J. H. McElhaney, *Journal of Applied Physiology* **21**, 1231–1236 (1966).
50. I. L. Jäger and P. Fratzl, *Biophysics Journal* **79**, 1737–1746 (2000).
51. B. N. Lucas and W. C. Oliver, *Metallurigical and Materials Transactions* **30A**, 601–610 (1999).
52. S. Yang, Y. W. Zhang and K. A. Zeng, *Journal of Applied Physics* **95**, 3655–3666 (2004).
53. M. L. Oyen and R. F. Cook, *Journal of Materials Research* **18**, 139–149 (2003).
54. B. C. Wei, T. H. Zhang, W. H. Li, *et al.*, *Materials Transactions* **46**, 2959–2962 (2005).
55. A. K. Bembey, M. L. Oyen, A. J. Bushby, *et al.*, *Philosiphical Magazine* 86, 5691–5703 (2006).
56. L. H. He and M. V. Swain, *J. Dent.* **35**, 431–437 (2007).
57. B. Ji and H. Gao, *Journal of the Mechanics and Physics of Solids* **52**, 1963–1990 (2004).
58. M. Rief, M. Gautel, F. Oesterhelt, *et al.*, *Science* **276**, 1109–1112 (1997).
59. S. Poolthong, M. V. Swain, T. Sumii, *et al.*, *J. Dent. Res.* **77**, 847 (1998).
60. J. H. Kinney, M. Balooch, S. J. Marshall, *et al.*, *J. Biomech. Eng.* **118**, 133–135 (1996).
61. L. Angker, M. V. Swain and N. Kilpatrick, *J. Dent.* **31**, 261–267 (2003).
62. Y. Hosoya and G. W. Marshall, *J. Mater. Sci. Mater. Med.* **16**, 1–8 (2005).
63. R. Hickel and J. M. Manhart, in *Advances in Glass-Ionomer Cements*, eds. C. L. Davidson and I. A. Mjor, Quintessence, Chicago (1999).
64. A. Hirayama, *Shikwa Gakuho* **90**, 1019–1036 (1990).
65. W. Tesch, N. Eidelman, P. Roschger, *et al.*, *Calcif. Tissue Int.* **69**, 147–157 (2001).
66. B. Van Meerbeek, G. Willems, J. P. Celis, *et al.*, *J. Dent. Res.* **72**, 1434–1442 (1993).
67. M. Balooch, I. C. Wu-Magidi, A. Balazs, *et al.*, *J. Biomed. Mater. Res.* **40**, 539–44 (1998).
68. G. W. Marshall, N. Yucel, M. Balooch, *et al.*, *Surface Science* **491**, 444–455 (2001).
69. D. H. Pashley, K. A. Agee, R. M. Carvalho, *et al.*, *Dent. Mater.* **19**, 347–352 (2003).
70. D. A. Sumikawa, G. W. Marshall, L. Gee, *et al.*, *Pediat. Den.* **21**, 439–44 (2000).
71. S. Poolthong, D. Low, M. V. Swain, *et al.*, *Journal of Dental Research* **77**, 917 (1998).

72. L. Angker, C. Nockolds and M. V. Swain, *Archives of Oral Biology* **49**, 369–378 (2004).
73. G. W. Marshall, S. Habelitz, R. Gallagher, *et al.*, *J. Dent. Res.* **80**, 1768–1771 (2001).
74. M. Toledano, R. Osorio, E. Osorio, *et al.*, *J. Dent.* **33**, 349–354 (2005).
75. S. Malek, M. A. Darendeliler and M. V. Swain, *Am. J. Orthod. Dentofacial Orthop.* **120**, 198–208 (2001).
76. S. P. Ho, H. Goodis, M. Balooch, *et al.*, *Biomaterials* **25**, 4847–4857 (2004).
77. S. P. Ho, M. Balooch, H. E. Goodis, *et al.*, *J. Biomed. Mater. Res. A* **68**, 343–351 (2004).
78. P. Srivicharnkul, O. P. Kharbanda, M. V. Swain, *et al.*, *Am. J. Orthod. Dentofacial Orthop.* **127**, 168–176; quiz 260 (2005).
79. S. P. Ho, M. Balooch, H. E. Goodis, *et al.*, *Journal of Biomedical Materials Research Part A* **68A**, 343–351 (2004).
80. S. Malek, M. A. Darendeliler, T. Rex, *et al.*, *Am. J. Orthod. Dentofacial Orthop.* **124**, 561–570 (2003).
81. J. V. Soames and J. C. Southam, *Oral Pathology*, Oxford University Press (2005).
82. M. E. Dickinson, K. V. Wolf and A. B. Mann, *Archives of Oral Biology* **52**, 753–760 (2007).
83. T. Huang, A. S. Jones, L H. He, *et al.*, *J. Dent.* **35**, 737–743 (2007).
84. R. W. Rice, *J. Mater. Sci.* **31**, 102–118 (1996).
85. T. Yanagisawa and Y. Miake, *Journal of Electron Microscopy* **52**, 605–613 (2003).
86. L. Angker, M. V. Swain and N. Kilpatrick, *Journal of Biomechanics* **38**, 1535–1542 (2005).
87. A. Tsuda and J. Arends, *J. Dent. Res.* **73**, 1703–1710 (1994).
88. E. K. Mahoney, R. Rohanizadeh, F. S. M. Ismail, *et al.*, *Biomaterials* **25**, 5091–5100 (2004).
89. E. Mahoney, F. S. Ismail, N. Kilpatrick, *et al. Eur. J. Oral. Sci.* **112**, 497–502 (2004).
90. Z. H. Xie, E. K. Mahoney, N. M. Kilpatrick, *et al.*, *Acta Biomaterialia* **3**, 865–872 (2007).
91. L. Ceballos, M. A. Garrido, V. Fuentes, *et al.*, *Dental Materials* **23**, 100–105 (2007).
92. J. L. Drummond, *Journal of Biomedical Materials Research Part B-Applied Biomaterials* **78B**, 27–34 (2006).
93. M. G. Sowa, D. P. Popescu, J. Werner, *et al.*, *Anal. Bioanal. Chem.* **387,** 1613–1619 (2007).
94. Y. J. Wang, J. J. Lee, I. K. Lloyd, *et al.*, *Journal of Biomedical Materials Research Part A* **82A**, 651–657 (2007).
95. D. Mohamad, R. J. Young, A. B. Mann, *et al.*, *Journal of Dental Research* **82**, 525–525 (2003).

96. H. H. K. Xu, D. T. Smith, G. E. Schumacher, *et al.*, *Journal of Biomedical Materials Research* **52**, 812–818 (2000).
97. G. Willems, J. P. Celis, P. Lambrechts, *et al.*, *Journal of Biomedical Materials Research* **27**, 747–755 (1993).
98. S. Bouillaguet, A. Schutt, P. Alander, *et al.*, *Journal of Biomedical Materials Research Part B-Applied Biomaterials* **76B**, 98–105 (2006).
99. E. Ho and M. Marcolongo, *Dental Materials* **21**, 656–664 (2005).
100. S. M. Chung and A. U. Yap, *J. Dental Materials* **21**, 1008–1016 (2005).
101. G. T. Liu, J. G. Duh, K. H. Chung, *et al.*, *Surface & Coatings Technology* **200**, 2100–2105 (2005).
102. M. E. Barbour, D. M. Parker, G. C. Allen, *et al.*, *European Journal of Oral Sciences* **111**, 428–433 (2003).
103. M. E. Barbour, P. M. Parker, G. C. Allen, *et al.*, *European Journal of Oral Sciences* **111**, 258–262 (2003).
104. M. E. Barbour, M. Finke, D. M. Parker, *et al.*, *Journal of Dentistry* **34**, 207–213 (2006).
105. F. Lippert, D. M. Parker and K. D. Jandt, *European Journal of Oral Sciences* **112**, 61–66 (2004).
106. K. D. Jandt, *Trends in Food Science & Technology* **17**, 263–271 (2006).
107. C. Gibbs, P. Mahan, A. Mauderli, *et al. J. Prosth. Dent.* **56**, 226–229 (1986).
108. A. Joiner, *J. Dent.* **34**, 412–419 (2006).
109. B. R. H. Nizam, C. T. Lim, H. K. Chng, *et al.*, *J. Biomech.* **38**, 2204–2211 (2005).
110. F. Lippert, D. M. Parker and K. D. Jandt, *Caries Research* **38**, 464–472 (2004).
111. F. Lippert, D. M. Parker and K. D. Jandt, *Surface Science* **553**, 105–114 (2004).

CHAPTER 9

NANOINDENTATION OF SOFT TISSUES AND OTHER BIOLOGICAL MATERIALS

Donna M. Ebenstein

Biomedical Engineering Department, Bucknell University
E-mail: donna.ebenstein@bucknell.edu

Nanoindentation is a valuable tool for mapping mechanical properties in samples from small animal models and can aid in generation of structure-property relationships in hierarchical biological materials at some of their smallest length scales, which can have important applications in clinical medicine and materials development. Although a young field, nanoindentation has been applied to studies of diverse biological materials. This chapter will focus primarily on soft tissue characterization using nanoindentation, with an emphasis on cartilage studies, but will also briefly review the application of nanoindentation techniques to other non-mineralized biological materials such as wood, plant leaves, and spider silk to highlight the broad range of research questions that can be addressed using this high resolution mechanical characterization tool. The challenges that are faced in adapting the nanoindentation technique to the study of compliant, hydrated soft tissues will also be discussed, and some of the current approaches to overcoming those challenges will be summarized.

1. Introduction

The mechanical characterization of tissues and other biological materials has important applications in clinical medicine and materials development. For example, many of the tissues in the body are load-bearing tissues. Characterization of the mechanical behavior of these tissues can provide valuable information for understanding the function of the tissues in health and disease, developing diagnostic tools and treatments, and designing replacement tissues that can adequately replace the load-bearing function of the native tissue. From the materials development perspective, there is a growing interest in the field of "biomimetics", or studying nature to develop solutions to engineering problems. For example, spider silk fibers have a combination of high strength, high toughness, and low density unmatched by synthetic fibers (silk is stronger by weight than steel), and bone and nacre (abalone shell) are ceramic composites with better fracture toughness than traditional engineering materials. Understanding structure-property relationships in these materials could aid in the development of synthetic materials with comparable properties.

In addressing these medical and materials problems, there is a clear need for submicrometer or micrometer-level characterization of the mechanical properties of biological materials. For example, tissues and other biological materials are typically comprised of hierarchical structures with relevant length scales down to nanometers or micrometers. Characterization of their organization and mechanical behavior at those small length scales is critical to understanding, and ultimately mimicking, their behavior at the macroscale. Small sample sizes are another motivation for the use of high spatial resolution mechanical testing techniques for characterization of biological materials. For example, spider silk has a diameter of ~5 micrometers, and many tissue specimens are acquired from small animal models (such as mice, rats, and rabbits) or biopsy specimens. These specimens may not be suitable for traditional macroscopic testing techniques.

A third argument for mechanical testing on small length scales is the desire to characterize materials on a size scale of interest to cells or insects. For example, neurons have been shown to preferentially grow on less stiff substrates, and many plants are designed to interact with the feet of insects. Since cells are on the order of 10 micrometers, and insect feet can end in hairs on the order of 1 micrometer,[1] the ability to probe the mechanical properties of the substrates using contacts of similar size to the biological entities that will be interacting with the substrate can be quite valuable. Finally, for some applications it is desirable to map the mechanical properties in a material and correlate changes in mechanical behavior with changes in morphology or composition. Mechanical testing techniques with high spatial resolution will allow for more refined mapping of material properties in biological systems.

Although tools have long existed to characterize the structure of materials on small length scales, such as optical, X-ray and scanning electron microscopy (SEM), only in recent years have tools become available to measure mechanical properties on similar length scales. Two of the most versatile commercial approaches to characterizing mechanical properties of biological materials at small length scales are the scanning probe techniques of atomic force microscopy (AFM) and nanoindentation.

As discussed in detail in Chapter 2, the atomic force microscope (AFM) is a tool traditionally used for submicrometer testing. While the AFM was originally developed as a tool for mapping surface topography,[2] it has since been adapted to measure both tribological properties (such as adhesion and friction) and elastic and plastic material properties.[3] A typical AFM probe consists of a sharp tip attached to the end of a cantilever beam. For force measurements, the cantilever tip is pressed into the surface and the deflection is monitored and converted to force values based on contact mechanics theory and a cantilever mechanics model.[4] Due to the low loads applied in AFM and its high sensitivity, atomic force microscopy is most commonly applied to the measurement of atomic and molecular forces. In biological applications, AFM has been effectively used to measure intermolecular forces and mechanical properties of biomolecules and living cells.[4,5] Characterization of tissue-level properties using an AFM is challenging due to the very small displacement and load ranges of the probes, and the use of the cantilever to apply load, and hence AFM has seen limited use in the area of tissue characterization. Instead, nanoindentation has served as the submicrometer resolution mechanical testing tool for tissue-level studies.

In comparison to AFMs, nanoindenter systems apply perfectly normal loads to the sample and can accurately measure loads and depths of higher magnitude. With a typical working force range of 100 nN to

500 mN and displacement range of 1 nm to 200 μm, this technique bridges the gap between atomic force microscopy (AFM) and macroscale mechanical testing, as illustrated in Chapter 1, Fig. 1-2. In typical commercial instruments, displacement is monitored with a capacitance gage, while force actuation is provided through electrostatic force generation or magnetic coils. Load and displacement are monitored continuously, and the elastic modulus and hardness of the substrate material can be calculated from the resulting load-displacement curve using well-established contact mechanics models.[6] Nanoindentation is more readily adapted to the measurement of tissue-level properties than AFM due to its higher load and displacement capabilities, normal sample approach, and well-defined tip shapes. For this reason, nanoindentation techniques have been used to measure local mechanical properties in tissues ranging from bone and teeth to cartilage and arteries,[7] as well as other biological materials such as wood[8] and spider silk.[9]

Despite the recent rise in the use of nanoindentation to characterize biological materials, there are many challenges associated with adapting commercial nanoindentation instruments to the study of biological materials. While all biological materials present some challenges, soft tissues provide the greatest challenge to the field of nanoindentation because their inherent properties are the most different from the traditional indentation substrates for which commercial nanoindentation systems were designed. Hence, soft tissue characterization will be the primary focus of this chapter.

Nanoindentation was developed for application to silicon and other engineering materials. These are materials that exhibit nominally elastic behavior, are stiff (elastic moduli >> 10 GPa), can be polished to present a flat surface, and can be tested in ambient air, without concern for rigorous temperature or humidity control. Soft tissues, on the other hand, generally exhibit time-dependent (viscoelastic or poroelastic) behavior, are compliant (moduli on the order of kPa to MPa), have irregular surfaces, exhibit significant amounts of creep, and have properties that vary with degree of hydration and with temperature. Hence, numerous modifications to traditional indentation techniques have been necessary in order to apply commercial nanoindenters to the characterization of these complex materials. These adaptations include developments in sample preparation, data acquisition, and data analysis techniques to better characterize this class of materials, including methods for sample hydration, selection of appropriate tip geometries, and experimental protocols that account for time-dependent properties, adhesion, and material anisotropy.

This chapter will present a review of much of the recent work in nanoindentation characterization of biological materials. The primary focus of the chapter will be on soft tissue characterization. A review of the literature on cartilage indentation will be used to illustrate the variety of research questions that can be addressed using nanoindentation and some of the challenges faced in applying nanoindentation to the characterization of hydrated, compliant materials. To further demonstrate the diversity of biological, biomedical, and materials design questions that can be addressed through nanomechanical testing, examples will also be presented from nanoindentation studies of a wide range of other biological materials including other soft tissues, plants and acellular materials. This overview will be followed by a discussion of some of the key challenges faced in adapting nanoindentation to biological material characterization, including summaries of some of the approaches that have been taken to address these challenges. Finally, the chapter will conclude with suggested directions of future research to advance the field of indentation of soft tissues.

2. Nanoindentation of Articular Cartilage

Nanoindentation is still a fairly young technique in the field of biological materials characterization. Early work using commercial instrumentation focused primarily on mineralized tissues such as bones and teeth. These materials are stiff and were often tested dry after embedding and polishing, allowing them to approximate more traditional indentation substrates. Since its success with mineralized tissues, nanoindentation has been increasingly applied to the study of a broad range of biological materials, including soft tissues, plants, and acellular biomaterials with biomimetics applications. Among the soft tissues characterized using nanoindentation, the most papers have been published in the area of nanoindentation of articular cartilage.

Articular cartilage is a highly studied tissue for clinical reasons. Articular cartilage is hyaline cartilage that covers the ends of long bones and sesamoid bones within synovial joints.[10] The function of the cartilage is to cushion the ends of the bones that articulate at the joint, provide a low friction sliding interface, and provide lubrication to the joint. Articular cartilage is of clinical importance because it is often the site of osteoarthritis and other damage. If the cartilage on the ends of the long bones deteriorates, joint replacement surgery is required. Hence, there has been much research into understanding structure-property relationships in healthy, diseased, and repair cartilage in an effort to develop better diagnostic tools and treatments for cartilage degeneration.

To understand why cartilage is particularly suited for nanoindentation testing, an understanding of cartilage structure and traditional mechanical testing is necessary. The following sections will briefly describe cartilage structure and the mechanical testing techniques commonly used to characterize cartilage. Then several cartilage nanoindentation studies will be reviewed, concluding with a discussion of the implications of the results of cartilage studies for the future development of soft tissue indentation techniques.

2.1. *Cartilage structure*

The structure of articular cartilage makes it challenging to test using traditional macroscale mechanical testing techniques, such as bulk tension or compression tests. Articular cartilage is found as a thin layer on the ends of long bones, with thicknesses ranging from micrometers to millimeters depending on the joint and animal being studied. The thin cartilage layer is made up of chondrocytes (cartilage cells) in a highly organized matrix of collagen II fibers (to resist tensile and shear loads) and hydrated proteoglycans (to resist compressive loads).[10,11]

Cartilage tissue has a complex microstructure and composition that varies with depth. The cartilage thickness is typically divided into three zones: the superficial zone (10-20% of the cartilage thickness), the middle zone (40-60% of the cartilage thickness), and the deep zone (30% of the cartilage thickness).[10,11] The collagen orientation varies with depth in the cartilage, aligned parallel to the surface in the superficial zone and perpendicular to the surface in the deep zone, with random orientation in the middle zone . The proteoglycan content and chondrocyte distribution also vary between the different zones.

Because of the complex microstructure of cartilage, and the key roles played by both the collagen II and proteoglycan molecules in load support in the joint, it is desirable to measure mechanical properties in cartilage without disrupting the cartilage structure. This means that, ideally, mechanical testing should occur *in situ*, with the cartilage layer still attached to the bone, so that water can only flow from the cartilage along the free surface and the proteoglycan matrix will remain trapped within the collagen network.

2.2. *Mechanical characterization of cartilage*

Macroscale and microscale indentation tests are the most common method for characterizing the mechanical properties of cartilage because they allow for *in situ* testing without disruption of the cartilage microstructure.[10,11] In a typical indentation test, the entire joint is mounted in the testing apparatus, and a porous cylindrical tip with diameter on the order of 1 mm is used to compress different regions of the sample, mapping mechanical properties across the joint surface.

Load-relaxation (or stress-relaxation) tests are frequently used to characterize the mechanical properties of cartilage samples using indentation.[10,11] In a load-relaxation test, a displacement is applied to the cartilage surface using the indenter tip. While maintaining the same displacement, the force (or load) on the tip is monitored. In viscoelastic materials, the load tends to decrease with time under a constant displacement, a phenomenon referred to as relaxation. Eventually, usually after minutes or hours, equilibrium is reached and the load stabilizes. Depending on the assumed material model used to interpret the load-relaxation results, elastic, viscoelastic or biphasic poroelatic properties can be measured from this indentation test.[10,12-15]

Indentation and microindentation tests using a variety of tip geometries and material models have been used to characterize cartilage in animal models *in situ* for the past 30 years.[10,12-15] The correlations observed between mechanical properties measured through indentation experiments and cartilage disease or degeneration have led to the more recent development of commercial hand-held indenters for diagnostic use during arthroscopy surgeries.[16,17] Using either quasi-static[17] or dynamic[16] compression, these hand-held indenters are used to probe joint tissue during surgery and identify regions of cartilage degeneration. These developments demonstrate the value placed on mechanical characterization of cartilage tissue as a marker for disease or degeneration, and the important role of indentation in the field of cartilage biomechanics.

The use of nanoindentation to characterize mechanical properties in cartilage is a natural extension of the popular indentation testing technique. While macroscale and microscale indentation work well for large joints from large animal models, such as dog or cow knee joints, they cannot be used to map mechanical properties of articular cartilage from small animal models (such as murine models) or from small joints (such as the finger joints) because the cartilage is too thin and mapping requires finer spatial resolution. It is therefore not surprising that researchers have turned to nanoindentation to map mechanical properties under conditions that require finer spatial or displacement resolutions.

However, it is important to note that nanoindentation studies of cartilage have not simply replicated the traditional load-relaxation tests at a smaller length scale. Load-relaxation tests require displacement-control for implementation, and typically take many minutes or hours to reach equilibrium. In addition, biphasic poroelastic analysis sometimes requires the use of a porous indenter tip to allow fluid flow. Due to instrumentation and experimental limitations with commercial nanoindentation systems (discussed in Section 4), it can be difficult to exactly replicate the traditional load-relaxation test protocol and biphasic analysis. Instead, most nanoindentation experiments in cartilage have either been based on quasi-static nanoindentation testing of polymers with nominally elastic analysis or have involved development of entirely new experimental protocols. The next section will review the types of experiments that have been performed on cartilage, and the types of research questions that are being addressed. Advantages and limitations of current nanoindentation methods for characterization of cartilage and other soft tissues will also be discussed.

2.3. *Review of major cartilage nanoindentation studies*

To date, articular cartilage is the soft tissue most commonly studied using nanoindentation. Nanoindentation has proven to be a valuable tool for cartilage characterization due to its ability to measure mechanical properties *in situ* without disrupting the complex microstructure and its ability to map mechanical properties with microscale resolution. Applications have included comparing the mechanical properties of control and repair cartilage in rabbit knees[18] and minature pig knees,[19] correlating mechanical properties in different regions of rabbit metacarpophalangeal (finger) joints with morphological parameters,[20] and measuring the fracture toughness of cartilage in bovine patellar

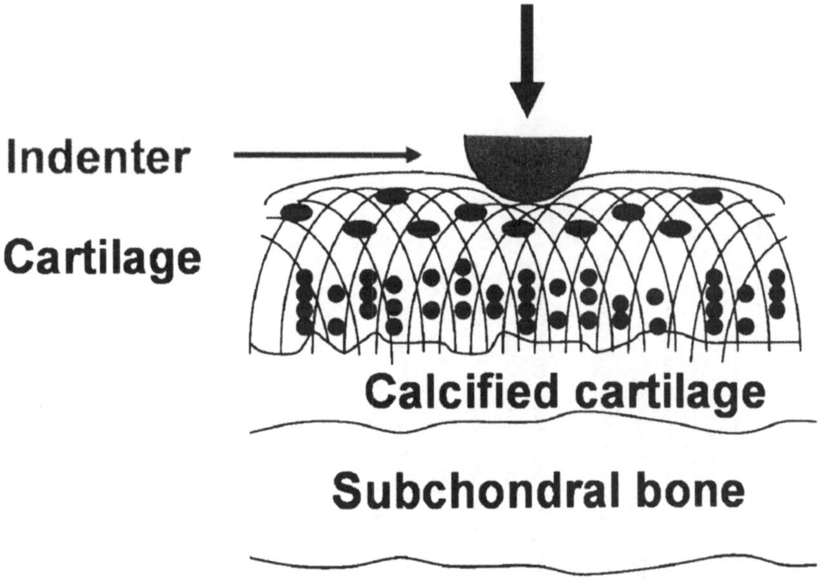

Figure 9-1. Schematic illustration of in situ cartilage nanoindentation using a spherical tip, with collagen fibers intact. The vertical arrow illustrates the applied load or displacement. Reprinted with permission from the nanoindentation of biological materials review by Ebenstein and Pruitt.[7]

Table 9-1. Summary of nanoindentation studies of articular cartilage.

Study	Purpose of Study	Animal Model	Tip Geometry	Mechanical Measures
Ebenstein *et al.* 2004[18]	Compare mechanical properties of repair and control cartilage; Correlate mechanics with histology	rabbit knee	100 μm radius of curvature sphere	Functional parameters (contact stiffness, resistance to penetration (~hardness), and volumetric creep rate)
Franke et al. 2007[19]	Compare mechanical properties of repair and control cartilage; Correlate mechanics with histology; Compare effects of storage fluids; Investigate depth dependence of mechanical properties	miniature pig knee	Berkovich (3-sided pyramidal) tip	Contact stiffness, reduced modulus, and hardness
Li *et al.* 2006[20]	Compare mechanical properties of cartilage in different regions of joint; Correlate mechanics with morphological measures	rabbit metacarpo-phalangeal (finger) joint	100 μm radius of curvature sphere	Functional parameters (contact stiffness, resistance to penetration (~hardness), and volumetric creep rate)
Simha *et al.* 2004[21]	Measure fracture toughness	Bovine patella	10 μm end diameter flat-ended cone (67° included angle)	Fracture toughness
Simha *et al.* 2007[22]	Determine effect of indenter size on mechanical properties	Bovine patella	5 μm to 4 mm end diameter flat-ended cone (60° included angle) and cylindrical tips	Equilibrium elastic modulus

articular cartilage.[21] Another study has also recently been performed to assess the effect of indenter size on the equilibrium modulus of bovine patellar cartilage measured by nanoindention, for future application to measuring mechanical properties in cartilage in murine models.[22] These studies are summarized in Table 9-1, and each will be discussed in more detail below. A schematic illustration of *in situ* cartilage indentation with a spherical tip is shown in Fig. 9-1.

In one of the early cartilage indentation studies performed using a commercial nanoindenter, Ebenstein *et al.*[18] used nanoindentation to compare the mechanical properties of regions of repair cartilage and control cartilage in the rabbit knee, and to correlate those properties with histological composition. Quasi-static indentation was performed with a 100 μm radius of curvature conospherical tip to generate load-displacement curves in each tissue region. As illustrated in Fig. 9-2, indents in tissue regions with different histological features (healthy cartilage vs. hyaline-like repair cartilage vs. fibrocartilage) showed significant variation in load-displacement behavior. Because of concerns over the validity of the traditional indentation contact models for this application, the authors did not use the compliance method[6] to calculate reduced modulus values. Instead, the authors reported functional parameters of contact stiffness, resistance to penetration, and volumetric creep rate to quantify the indentation load-displacement behavior in each tissue region.[18] Using these functional parameters, a correlation was found between the mechanical behavior and the histological classification of each tissue region tested. The results showed that even when repair cartilage had developed some histological features of healthy cartilage (hyaline-like repair), the mechanical behavior was not fully restored. These results demonstrated the importance of mechanical assessment of repaired cartilage in addition to histological analysis for determining the functional behavior of repair tissue.

In a more recent cartilage repair study in miniature pig knees by Franke *et al.*,[19] a correlation was again found between histological markers of tissue composition and mechanical properties measured by nanoindentation. In this study, indentation was performed using the traditional Berkovich (3-sided pyramidal) tip, and the penetration depths

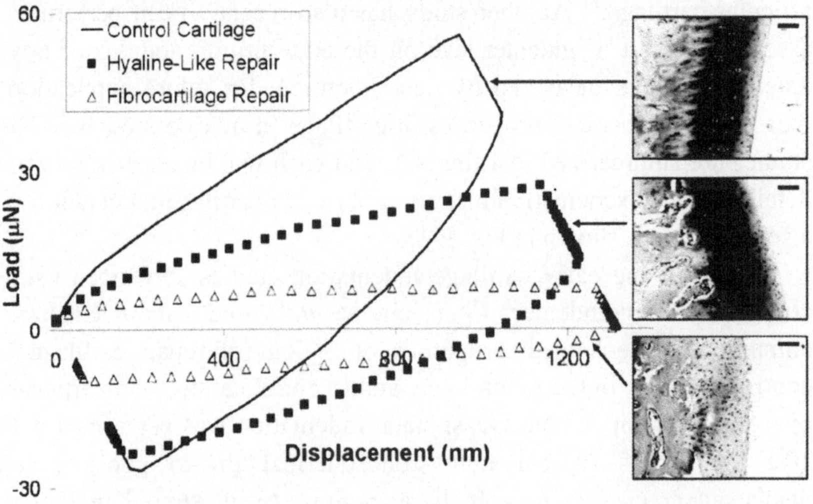

Figure 9-2. Load-displacement curves and histological sections from three different cartilage regions: control cartilage, hyaline-like repair tissue, and fibrocartilage repair tissue.[18] Note the variations in load-displacement behavior between the three tissue regions, with the control cartilage showing the stiffest response and the fibrocartilage repair tissue showing the most compliant behavior. The hyaline-like repair tissue had intermediate mechanical behavior. As illustrated in the histological sections, the hyaline-like repair tissue was histologically more similar to the control cartilage, with intense staining for proteoglycans and depth dependent distribution of cells. In contrast, the fibrocartilage repair tissue did not stain strongly for proteoglycans and had evenly distributed cells as a function of depth. Bars on histological images = 50 μm. (Data adapted from cartilage study by Ebenstein *et al.*[18])

and peak indent loads were much higher than in the previous cartilage repair study (loads up to 10 mN, displacements over 100 μm in some cases). In addition, multiple load-unload cycles were used to investigate changes in mechanical properties as a function of penetration depth, and the effect of fixing cartilage tissue in paraformaldehyde (PFA) on the mechanical properties was investigated.[19] The authors in this study reported contact stiffness, reduced modulus, and hardness, relying on tip calibration in fused silica (E = 72 GPa, ν = 0.18) to determine the tip shape and estimate the contact area between the tip and the sample and using the compliance method[6] of analysis to calculate reduced modulus.

The reduced modulus values reported were approximately 3 MPa for the intact cartilage, 0.5 MPa for repair tissue, and 10 MPa for PFA-fixed cartilage tissue.[19] Interestingly, the intact cartilage exhibited an increase in contact stiffness as a function of depth, while the repair cartilage, which was shown through histological analysis to be fibrocartilage, did not show a dependence on penetration depth. This difference in behavior was attributed to differences in structure in the two tissues, with the intact cartilage having the graded structure typical of articular cartilage with parallel-aligned collagen in the superficial layer, while the fibrocartilage was more uniform in structure and composition.[19] The results of this study reinforced the conclusions of the previous cartilage repair study, again demonstrating the importance of mechanical assessment when interpreting the functionality of repair tissue, and showing that nanoindentation can be used to perform that assessment. In addition, the comparison of PFA-fixed and unfixed healthy cartilage clearly demonstrated the importance of sample hydration and choice of hydrating fluid on mechanical properties.

In a study by Li *et al.*,[20] mechanical parameters measured through nanoindentation in specific regions of rabbit metacarpophalangeal (finger) joint cartilage were correlated with morphological parameters such as cell density and thickness of the superficial zone. This study followed the methodology of the cartilage repair study by Ebenstein *et al.*,[18] performing indents using a 100 micrometer radius of curvature conopherical diamond probe tip and reporting the functional parameters of contact stiffness, resistance to penetration, and volumetric creep strain to avoid assuming a specific contact model.[20] Indents were performed at predetermined regions of interest on the joint surfaces expected to have different properties due to different loading conditions *in vivo*. The functional mechanical properties were compared to morphological parameters obtained with light and polarized light microscopy to correlate local cartilage structure with mechanical properties. Both stiffness and resistance to penetration were found to correlate strongly with superficial zone thickness, suggesting that the superficial zone, where collagen is aligned parallel to the surface, plays a significant role in supporting load.[20]

All three of the above studies could not have been performed using bulk mechanical testing techniques due to the small size of the regions of interest in the tissue samples. In all three cases, the cartilage was tested *in situ*, with the cartilage surface supported by the underlying bone, and the bone being mounted in the indentation apparatus. In addition, sample hydration was maintained in all three cases, either through the use of fluids-soaked gauze,[18] submerging the sample during the entire indent process,[19] or keeping the sample fully submerged in a fluid cell until just before indentation.[20] The choice of tip geometry varied among the studies, with two groups using a 100 μm spherical tip[18,20] while the cartilage repair study by Franke *et al.* used the Berkovich tip typically used for stiff materials.[19] However, all three studies used fairly standard nanoindentation techniques with only slight modifications to the protocols used for traditional nanoindentation substrates. One of the key modifications was the incorporation of the hold period at peak load to allow time for creep to dissipate prior to unloading, a common method for dealing with time-dependent behavior of polymers and soft tissues, but a method which does not allow for rigorous biphasic poroelastic or viscoelastic analysis. These and other methods for dealing with hydration, tip selection, time-dependent properties, and other concerns when indenting soft tissues will all be discussed in more detail in Section 4.

While the three studies described above represent only slight modifications to traditional quasi-static indentation analysis, the fourth study developed an entire new model and methodology for indentation of soft tissues. The goal of this study by Simha *et al.*[21] was to develop a technique for measuring the fracture toughness of articular cartilage using nanoindentation. A truncated diamond conical tip with a 67° included angle and 10 micrometer end diameter was used in the study, with the intent of penetrating the tissue surface rather than elastically deforming it. A method was developed to detect penetration into the surface using power as an indicator, and only indents that exhibited penetration were analyzed. Penetration depths ranged from 100-250 micrometers, much deeper than indents typically performed with nanoindenters (<1 μm), so the authors called the technique "micropenetration".[21] The predicted fracture toughness from indentation testing was not significantly different from the values measured in macroscopic notch tests, and penetration depths predicted by the indentation depth were not significantly different from the values measured in *post-hoc* histological analysis.[21]

This fracture-toughness study differs largely from the previous studies in that it develops a new experimental approach to characterizing a biological material, rather than simply trying to adapt nanoindentation techniques developed for stiff, elastic engineering materials to compliant, time-dependent materials. This is a path that should be pursued more in the future if quantitative materials characterization, with parameters comparable to macroscale properties, are desired. The results of this paper also present some interesting implications for future soft tissue studies. The ability to penetrate the cartilage surface using a blunted indenter tip raises some concerns over the use of sharp (*i.e.* small included angle conical, cube-corner, or Berkovich) indenter tips to indent soft tissues. The authors noted that they chose the truncated cone because it was difficult to detect when penetration occurred using an unflattened 60° included angle conical tip.[21] Presumably, however, penetration would occur more easily with a sharp tip than with the blunter tip used in this study. Hence, in future studies, care should be taken when considering the use of sharp tips to indent soft tissues, particularly if large loads or displacements are used.

The final cartilage indentation study, from the same group, highlights several challenges associated with nanoindentation testing of cartilage,[22] many of which will apply to other biological materials as well. The goal of the study was to investigate the effect of indenter size on the elastic modulus of cartilage measured using nanoindentation techniques. In this study, cartilage samples were indented using a series of flat-ended 60° cone and cylindrical indenter tips with end diameters ranging from 5 μm to 4 mm. An elastic model was used to analyze each load-displacement curve and calculate an equilibrium modulus. The study revealed that there was a size effect when indenting cartilage, with the equilibrium modulus decreasing as the tip diameter increased.[22] The authors speculated that the inhomogeneous material properties of cartilage, such as variations in tissue organization and composition with depth in the tissue or surface characteristics, could have been the source of the size effect.[22] However, no direct correlations were investigated in this study.

Based on their size effect results, the authors further suggested several implications of the study for interpretation of indentation testing of cartilage and design of future experiments.[22] First, the mechanical properties measured by different investigations may vary substantially if different indenter sizes are used, so care must be taken in comparing the results of different studies. Second, to minimize the influence of size effects in an experiment, a single size tip should be used for the entire study; consistent cartilage thickness may also be important. The authors further suggest that a tip should be selected to ensure a large enough contact area to measure tissue level properties rather than cell-level properties, and a penetration depth should be selected to simulate the 20% compression typically experienced during cartilage loading *in vivo*.[22] Similar constraints with regards to sampling tissue-level properties and simulating physiologically relevant loading conditions could be considered for all soft tissues.

The indenter size effect observed in this study[22] and the depth dependent contact stiffness observed by Franke *et al.*[19] both highlight a potential limitation in comparing data between experiments. Both authors postulate that these effects are due to the inhomogenous cartilage structure as a function of depth.[19,22] Even if this is the source of variation, it makes it difficult to compare mechanical properties between different experiments and even different tissue regions, since tissue thickness and/or structural organization may vary from region to region. This concern could translate to other inhomogeneous tissues as well. Another possible contributor to both the observed trends is substrate effect. The cartilage regions tested in these studies have finite thicknesses, typically on the order of 0.5 to 2 mm. Below the cartilage layer is a much stiffer bone layer. As penetration depths increase relative to cartilage thickness, the modulus measured by indentation could become increasingly influenced by the underlying bone layer. In addition, the aspect ratio (the ratio of tip radius to sample thickness) has been shown to influence the mechanical properties measured by indentation, leading to the development of correction factors as a function of aspect ratio for macroscale cartilage indentation.[12,14,17] The substrate effect and its role in influencing tip selection will be discussed in more detail in Section 4.

2.4. *Conclusions*

These cartilage studies show the variety of research questions that can be investigated through cartilage nanoindentation studies. Nanoindentation studies of cartilage repair tissue in rabbit and minature pig knee models illustrated the importance of using mechanical characterization in conjunction with histology to demonstrate the functional behavior of repair tissue. In addition, nanoindentation was successfully used to investigate structure-property relationships in rabbit finger joint cartilage, and to measure fracture properties in bovine knee cartilage. Most of these studies could not have been performed using bulk mechanical testing techniques due to the small size of the regions of interest in the tissue samples, and hence demonstrate the value of having a tool that can measure mechanical properties in soft tissues with high spatial resolution.

In addition, these cartilage studies reveal some valuable insights into the interpretation of past studies and the development of future studies of soft tissues using nanoindentation techniques. Some basic issues were raised for future experimental design, such as the importance of selection of appropriate tip geometries, tip dimensions and loading conditions to measure physiologically relevant properties. In addition, the importance of sample hydration and the choice of hydrating fluid were addressed. Limitations associated with inhomogeneous tissue properties and indenter tip size effects were also raised. Methods for dealing with hydration, tip selection, time-dependent properties, substrate effects, and other challenges when indenting soft tissues will be discussed in more detail in Section 4.

Despite these ongoing challenges, one important insight into data interpretation that can be gained by comparing these studies is that even if the traditional quantitative parameters (modulus, hardness, etc.) are not physically meaningful for every experimental set-up, they or other functional parameters can still provide a valuable qualitative measure of relative mechanical properties. Often, this relative assessment is sufficient to answer research questions such as "Does the repair cartilage have the same mechanical behavior as healthy cartilage?" Material parameters that can translate to the macroscale are not always necessary. While these qualitative comparisons are effective within studies, comparison of quantitative results between studies may not be meaningful if different experimental techniques are used (such as different tip sizes and penetration depths),[22] particularly when the material being tested has a microstructure that varies with depth.

Although functional parameters may be sufficient to answer many research questions, the fracture toughness study[21] shows that there is a potential to develop entirely new experimental methods and models that can take into account the unique properties of a given material and allow accurate measurement of truly quantitative mechanical properties and material parameters that may be comparable to bulk values. The development of new indentation models for soft tissue characterization should continue to be pursued in the future, rather than relying on adapting models that were developed for very different materials.

3. Nanoindentation of Other Biological Materials

Cartilage is only one of many biological materials which have been characterized using nanoindentation. The previous chapters reviewed the substantial literature on nanoindentation of bones and teeth, which were the first biological materials to be studied using nanoindentation. In more recent years, nanoindentation methods have been applied to the study of a variety of other biological materials, ranging from soft tissues to plants to acellular natural materials. Studies of articular cartilage were reviewed above, demonstrating applications ranging from characterizing cartilage repair tissue to correlating mechanical properties with morphology in small animal models. The sections below will provide more examples of nanoindentation studies of soft tissues, plants, and acellular biological materials to illustrate the wide range of biological, biomedical and materials design questions that can be addressed through nanomechanical testing.

3.1. *Soft tissues*

Soft tissues represent a class of hydrated biological materials that often have a complex and hierarchical microstructure. Characterization of soft tissues at small length scales has applications in basic science and clinical studies. Understanding the structure-property relationships in soft tissues at small length scales can help researchers understand the behavior of the tissue at the macroscale, and can also aid in the development of improved tissue replacements with appropriate mechanical behavior. In addition, nanoindentation studies of soft tissues can aid in understanding disease processes and repair mechanisms in tissues.

As was demonstrated in the review of cartilage indentation studies, applications of nanoindentation to cartilage characterization have focused on correlating local mechanical properties in small tissue samples with tissue morphology[20] and investigating tissue damage[21] and repair processes.[18,19] In contrast, nanoindentation studies in arteries have focused on gaining understanding of disease processes rather than repair processes. Several studies have looked at the influence of mineralization in human atherosclerotic plaques on the mechanical behavior of the tissue,[23-25] with the goal of generating input parameters for computational models of diseased arteries. Healthy arteries have also been characterized using both elastic[25] and viscoelastic[26] analyses to serve as a baseline for comparison of diseased artery properties.

Other soft tissue studies have included quasi-static and dynamic mechanical analysis of the 20-40 micrometer thick stratum corneum layer of skin,[27] the characterization of viscoelastic properties of demineralized dentin,[28] and the indentation of live fruit fly larvae at various developmental stages to investigate the role of the developmental stage on cuticle stiffness.[29] All the soft tissue studies required maintenance of tissue hydration during testing, and many highlighted other challenges in sample preparation, tip selection, and data acquisition and analysis that had to be overcome in applying nanoindentation to the study of soft tissues. The challenges associated with soft tissue testing will be discussed in more detail in Section 4.

3.2. *Plants*

Characterization of wood, bamboo and the waxy layers on pitcher plants have also been the subject of many nanoindentation studies. Wood, like many tissues, is a material with a complex hierarchical structure with relevant length scales at the submicrometer level. Hence, nanoindentation has become a valuable tool for understanding structure-property relationships in wood, particularly at the cell and sub-cell level.[8] The primary goals of nanoindentation studies of wood have been to understand how wood maintains such a high strength-to-weight ratio and to develop methods to treat wood to improve its surface properties without sacrificing those unique mechanical properties.

Many papers have investigated the effects of the wood cell microstructure and composition, including fiber orientation within the cell wall and the degree of lignification, on the wood cell's mechanical properties.[8,30,31] Other studies have studied mechanical properties as a function of wood maturity,[32] tree ring,[31] or degree of pyrolization.[33] Another focus of wood research is product development,[8] such as investigating the penetration of coatings and adhesives into the wood cells to determine whether the mechanical properties of the wood will be influenced by the treatments.[8,34-37] Investigations into the anisotropic properties of wood[30,38,39] and bamboo[40] have also been performed, including a study specifically highlighting the concerns of using nanoindentation to characterize highly anisotropic wood cells.[38]

The waxy layers on pitcher plants play a key role in trapping insects in the plant by interfering with their adhesion mechanisms. Because individual setae on insect feet end in spatulae on the order of 1 micrometer in diameter,[1] the nanoindenter provides a valuable method for investigating the behavior of the waxy coatings on the same scale as insect feet are interacting with those coatings. Several studies have investigated the mechanisms by which these waxy layers prevent insect attachment using nanoindentation in conjunction with high resolution imaging and chemical analysis tools.[1,41]

Both the wood review by Moon *et al.*[8] and the pitcher plant studies[1,41] identify numerous limitations of the nanoindentation technique for characterization of plant materials. These limitations and areas for improvement will be summarized in Section 4. These studies of wood and pitcher plants also represent examples of biomimetic applications of nanoindentation. Biomimetics is the field of learning design concepts from nature to solve engineering problems. Studies of wood shed insight into how to achieve high strength and low density, while pitcher plants represent models of surfaces that prevent insect attachment. In the next section, more examples will be provided of biomimetic materials applications of nanoindentation.

3.3. *Acellular biological materials with biomimetic applications*

An increasing interest in biomimetics, designing materials based on natural biological structures, has led to another focus area of biomaterial indentation: acellular biological materials. In order to design materials that mimic nature, characterization of structure-property relationships in natural materials down to the micro- or nano-scales is often necessary. Indentation has become a valuable tool for characterizing mechanical properties to complement the structural analyses also being performed at small length scales.

One example of a biological material with desirable mechanical properties is the nacre (mother-of-pearl) that lines mollusk shells. Nacre is a material with extremely high fracture toughness that researchers want to mimic when designing vehicle armor or other high impact ceramics. Nacre is both strong and tough because it combines a strong ceramic with a polymeric interface to provide toughness. The ceramic plates, made up of aragonite, are typically 5-8 micrometers long and 200-800 nm thick.[42] Because of these small dimensions, nanoindentation has been used in conjunction with computational modeling to determine the mechanical properties of the argonite blocks and the interfacial protein that make up nacre's complex microstructure.[42,43] Spider silk is another biomaterial of interest to researchers, due to its unusual combination of high strength and toughness. Nanoindentation has been used to measure anisotropic mechanical properties in two orthogonal planes in the 5 micrometer diameter spider dragline silk fibers using both quasi-static indentation and dynamic stiffness imaging methods, as illustrated in Fig. 9-3.[9]

Other biomimetics applications of nanoindentation have included characterization of structure-property relationships in specific regions of insect cuticle (exoskeleton),[41,44] mapping the variation in modulus and hardness across the 200–600 micrometer diameter sponge spicule,[45]

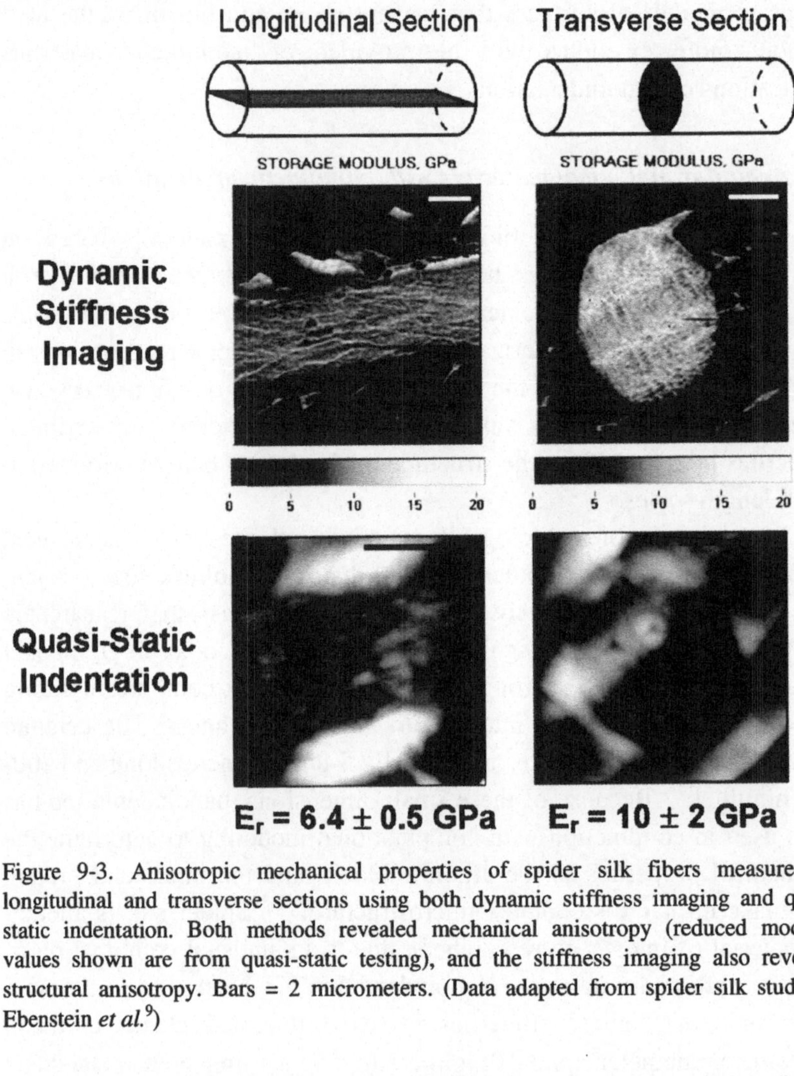

Figure 9-3. Anisotropic mechanical properties of spider silk fibers measured in longitudinal and transverse sections using both dynamic stiffness imaging and quasi-static indentation. Both methods revealed mechanical anisotropy (reduced modulus values shown are from quasi-static testing), and the stiffness imaging also revealed structural anisotropy. Bars = 2 micrometers. (Data adapted from spider silk study by Ebenstein *et al.*[9])

measurement of the mechanical properties of the veins of the cicada forewing,[46] and investigation of correlations between hardness and local chemical composition in polychaete jaws.[47] Because of the small dimensions of many of these materials, and the desire to map mechanical properties or study gradients in mechanical and chemical properties, nanoindentation was selected as the ideal mechanical characterization techniques for each of these studies. By coupling this mechanical data with spatially resolved chemical and structural information, engineers and materials scientists hope to learn to design new composite materials with enhanced materials properties at the macroscale.

3.4. *Summary*

The studies of biological materials outlined above highlight the advantages of submicrometer resolution mechanical testing capabilities in the characterization of natural materials, many of which have features with small dimensions or hierarchical structures of interest to researchers. The tissue characterization studies have implications for improved understanding of structure-property relationships in healthy, diseased, damaged, and repairing tissues. This information may aid in the development of diagnostic tools, disease treatments, and replacement tissues. Studies of plants and other biological materials reveal valuable information about structure-property relationships in a variety of natural composite materials, which may in turn help engineers design synthetic materials with comparable properties. While the nanoindentation studies described above have helped to answer many scientific questions, they have also served to uncover the many challenges associated with adapting the nanoindentation technique to the study of soft tissues and other biological materials. These challenges, as well as some of the methods currently being used to overcome those challenges, will be discussed in detail in the next section.

4. Challenges Associated with Nanoindentation of Biological Materials

While nanoindentation has been successfully utilized to characterize a wide variety of biological materials, many challenges have been faced in adapting a mechanical testing system designed for characterization of engineering materials like metals and ceramics to the study of compliant, hydrated biological materials. In addition, there are still many areas where improvement in indentation methods or instrumentation would greatly enhance the field of indentation of biological materials.

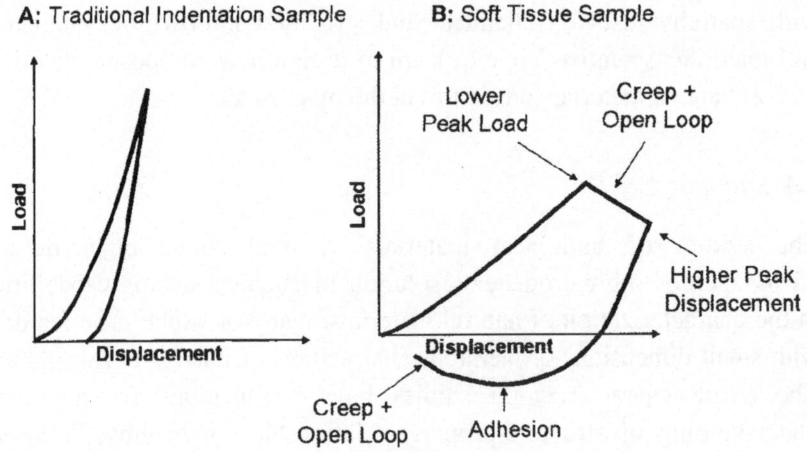

Figure 9-4. Comparison of idealized load-displacement behavior for indents in (A) traditional indentation samples using load function (a) in Chapter 5, Fig. 5-2, and (B) soft tissue samples using loading function (b) in Chapter 5, Fig. 5-2. Soft tissues tend to have larger displacements at lower applied loads, due to their lower elastic moduli ($E < 1$ MPa vs. $E >> 10$ GPa). In addition, soft tissues typically exhibit time-dependent behavior, which is often most easily visualized as a creep response, or a change in displacement with a constant applied load. Adhesion can also complicate data collection and analysis in soft tissues, and is most easily visualized as a measured negative load in a load-displacement curve upon unloading. Note that creep and adhesion interpretation are further complicated by data acquisition without closed-loop load- or displacement-control. Under open-loop conditions, even when a "constant" load is applied to the sample, part of that load goes into deforming the springs that support the indenter tip. At large displacements, or in the presence of large amounts of creep, this is observed as a change in load applied to the sample as a function of tip displacement during the hold period in a trapezoidal load function. These open-loop effects can be observed both during the hold period and after load is removed from the sample, and can also lead to a negative load during unloading.

To introduce the challenges associated with indentation of soft tissues using traditional indentation methods, consider the idealized load-displacement curves for indents in traditional indentation samples (*e.g.*, engineering materials with elastic moduli >> 10 GPa) and soft tissue samples (*e.g.*, compliant, hydrated samples with moduli < 1 MPa) using a trapezoidal load function presented in Fig. 9-4. Notice both the changes in the shape and the magnitude of the load-displacement curves. As illustrated in the figure, the sample compliance, adhesion and time-dependent properties (*e.g.*, creep behavior) account for some of the changes in load-displacement behavior. However, instrumentation issues, such as open-loop indentation *vs* closed-loop load- or displacement-control also influence the behavior of soft tissues. Hence, it is not surprising that adaptations to traditional nanoindentation techniques have focused on both data collection and instrumentation concerns.

This section will highlight some of the major concerns and identified challenges in sample and tip preparation, instrumentation, data acquisition and analysis, and data interpretation that have been identified as traditional nanoindentation techniques have been adapted to the study of soft tissues and other biological materials. Where relevant, examples of some common approaches to addressing these concerns will also be provided. Soft tissues will be the primary focus of this section, since soft tissue samples show the most extreme differences from traditional indentation substrates and hence represent the most challenging materials to test. However, subsets of these concerns and challenges apply to each type of biological material presented in this book.

Many of the techniques developed to adapt nanoindentation instrumentation and methodologies to the study of biological materials were derived from studies on polymers and hydrogels. Hydrogels in particular have similar properties to soft tissues, including high compliance, hydration, and time-dependent mechanical properties. However, they have the advantage of being more homogeneous than soft tissues, and can be produced in large, relatively uniform quantities, which makes them an optimal control material for development of nanoindentation protocols for soft tissues.

4.1. *Sample and tip preparation*

In preparing for a nanoindentation experiment, there are many challenges associated with the use of soft tissues as sample materials. These include both sample preparation issues, such as methods to maintain sample hydration and prepare the sample surface for indentation, and tip preparation issues, such as selecting an appropriate tip geometry and calibrating the tip area function. The challenges and concerns related to these aspects of sample and tip preparation, as well as approaches that have been used to overcome these challenges, are described in the next four sections.

4.1.1. *Sample hydration*

Most biological materials are hydrated in their native state, and many studies have demonstrated the importance of maintaining sample hydration in biological materials during nanoindentation.[28,41,48-53] The mechanical properties of mineralized tissues and plants, which have a low water content relative to soft tissues, have been shown to be strongly influenced by their degree of hydration.[41,48,49,51-53] The mechanical properties of soft tissues, which are made up predominantly of water, are even more heavily influenced by their water content. For example, a study of fully demineralized dentin showed a 4-fold increase in modulus after dehydration, with only partial recovery of mechanical properties after rehydration.[28] Hence, the ability to maintain hydration of soft tissue samples and other biological materials during both sample preparation and the nanoindentation process is critical for accurate mechanical characterization. The fluid used to hydrate the sample has also been shown to influence the mechanical properties of soft tissues,[19,27,48,54] so care must also be taken in selecting an appropriate hydration fluid for the experiment.

The most common method utilized to address this issue of sample hydration is to indent a submerged sample contained in a fluid cell.[26,28,48,49,51,53-56] However, submerging the sample complicates data acquisition and analysis due to the variable meniscus forces acting on the shaft of the indenter tip.[53,57] In addition, this method requires specialized tips with long shafts to protect the transducer from contact with the hydrating fluid and can also make surface detection difficult[57] (see Section 4.2.1). To avoid these and other challenges, some researchers have used other approaches to maintain sample hydration during indentation testing. These methods include applying several drops of fluid to a surface prior to starting the indent process,[58] removing fluid from the surface of submerged samples just prior to initiating the indent process,[20,27] and hydrating the samples from the edges using fluid-soaked gauze,[18] hydrating foam,[24,25,50] or specialized irrigations systems.[59,60]

However, adhesive forces between the tip and the sample are typically higher when the samples are tested moist rather than fully submerged, and the need to account for adhesion can also complicate the indent process (see Section 4.3.2).

4.1.2. *Surface preparation*

Another challenge associated with indentation testing of soft tissues is surface preparation. Studies have shown that surface roughness can heavily influence the modulus measured by nanoindentation.[61-63] For traditional indentation substrates, and even mineralized tissues such as bone and teeth, samples are typically microtomed or polished to present a "smooth" surface to the indenter tip.[64-66] Because of the complex, heterogeneous structure of many biological materials, sectioning of less stiff biological materials may also result in local damage on the surface, which has been a particular problem for indentation of wood.[8] To minimize the influence of surface damage or surface roughness on mechanical properties measured through indentation, deep indents should be performed.[67]

Researchers must also consider whether sectioning the tissue will disrupt the natural microstructure and change the properties that are of interest in the study. For example, sectioning has been avoided when surface properties are of interest, such as investigations of the waxy coatings on plants,[1,41] or when integrity of the surface is important in maintaining the structure and behavior of the tissue, such as investigations of cartilage.[18-21] Under these conditions, soft tissues are typically tested in their "native" state rather than undergoing sectioning and polishing. If surface roughness is a concern, or if the region of interest for indentation is not on the surface of the tissue, the best option for soft tissue sectioning is often cryomicrotoming since it does not require dehydration and paraffin embedding like other microtoming methods. Embedding samples in agarose or a different hydrogel of similar compliance to the tissue may also facilitate sectioning of small biological samples. Polishing is usually not an option for soft tissues and many other biological materials, so a sharp knife must be used to ensure a smooth surface.

4.1.3. *Tip selection*

When testing biological samples, the selection of the geometry and size of tip will depend on the length scale of interest and the mechanical properties being measured. For characterization of the mechanical properties in stiffer, small-diameter biological materials, such as sponge spicules and spider silk, sharp Berkovich (3-sided pyramidal) tips are commonly used to provide high spatial resolution.[9,45] For fracture testing of cartilage, flattened conical tips with small included angles and end diameters were used to ensure penetration into the samples.[21] However, for most soft tissue characterization, flat punch tips or large diameter (> 20 μm) spherical tips are typically preferred.[7,18,20,24-28,50] These blunter tips are preferred because the contact areas between the tip and the sample are large enough to sample tissue level properties (larger than individual cells and protein fibers),[7,22,50] and there is less danger of puncturing or fracturing the sample.[22] In addition, the larger contact area between the tip and the sample facilitates surface detection in compliant materials where initial contact stiffnesses are quite small. Only a few soft tissue studies have used Berkovich tips[19,27] due to concerns about contact sizes, puncturing the surface, and surface detection. For an illustration of contact sizes vs. cell sizes for different tip geometries, see Fig. 9-5.

The flat punch tip has the advantage of a known, constant contact area independent of depth. This is particularly helpful if load-relaxation or creep tests are being performed to measure time-dependent properties. However, for this contact area to be valid, full contact is needed with the surface. Hence, the flat punch is optimal only for very flat surfaces or at deep depths, so that full contact can be ensured. An additional concern associated with the use of the flat punch is that stress concentrations may arise at the perimeter of the punch. Spherical tips, on the other hand, do not generate stress concentrations and allow for some inaccuracy in the approach to the sample since they will contact the surface with an orb shape even if there is a slight tilt to the sample. In addition, many adhesion models and other indentation methods have been developed for the spherical tip, which can make this geometry advantageous. However, the solutions to some indentation problems are more difficult for

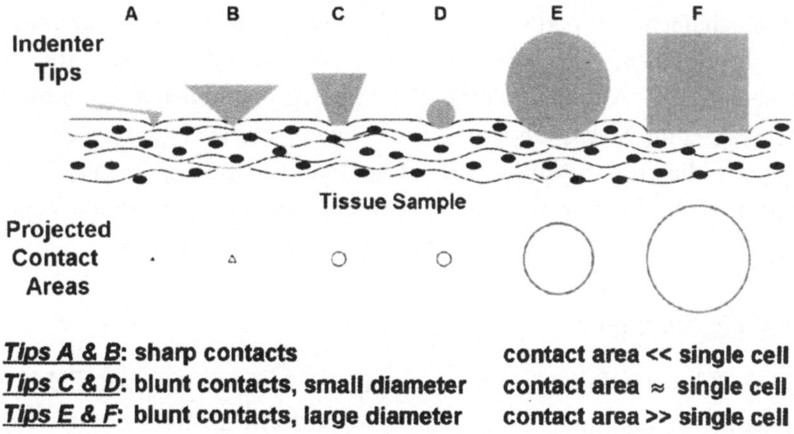

Figure 9-5. Schematic illustrating tip selection considerations for indentation of a tissue sample. Comparison of contact sizes during indentation using sharp tips (*e.g.*, AFM tip (A) or Berkovich tip (B)), small diameter blunt tips (*e.g.*, 10 μm diameter flat-ended conical tip (C) or 20 μm diameter spherical tip (D)), and large diameter blunt tips (*e.g.*, 100 μm diameter spherical tip (E) or 100 μm diameter cylindrical flat punch (F)). Note that projected contact areas will depend both on tip geometry and depth of penetration (except for flat-ended conical tips and flat punch tips, which have a constant contact area independent of penetration depth). Also, flat-ended conical tips, spherical tips, and flat punch tips come in a range of diameters ranging from micrometers to millimeters. The tips shown here are merely representative of geometries used for soft tissue indentation testing.

spherical tips than for flat punches (constant contact area) or Berkovich tips (self-similar shape as a function of depth).

Another concern that can affect tip selection and data interpretation is the possibility of substrate effect, or a stiff substrate underlying a soft material influencing the mechanical properties measured by nanoindentation. When indenting in soft tissues of finite thickness, such as cartilage on subchondral bone, the thickness of the sample can limit both the contact radius of the tip and the depth of penetration that should be used in the study. Studies of thin films of relatively homogeneous engineering materials have led to the rule of thumb that the indentation depth should not exceed 10% of the sample thickness.[68] Similar constraints also apply to the aspect ratio, the ratio of contact radius to sample thickness. For example, larger scale indentation studies in cartilage have also demonstrated that the aspect ratio influences the mechanical properties measured through indentation, leading to the development of correction factors that compensate for this influence.[12,14,17] Although rigorous studies of nanoindentation substrate effects in thin layers of soft tissues have not been performed, it is expected that the substrate will affect nanoindentation measurements when indenting soft tissues with large tips or deep penetration depths. Hence, this is an area that should be pursued in future studies.

4.1.4. *Tip calibration*

For traditional nanoindentation testing of engineering materials, the standard tip calibration procedure is to perform a series of indents at different loads in a fused silica sample of known modulus. Since fused silica is presumed to have a constant elastic modulus of 72 GPa and Poisson's ratio of 0.16–0.18 as a function of depth, analysis of the load-displacement curves can be coupled with knowledge of the mechanical properties of the fused silica sample to extract a tip area function as a function of penetration depth (based on elastic analysis via the compliance method[6]). At a peak load of 10 mN, the highest value for some commercial nanoindentation systems, the penetration depth of a sharp Berkovich tip is typically between 300 and 400 nm. When indenting in fused silica with a blunt tip, such as a large diameter spherical tip, the penetration depth will be much smaller. However, indents in compliant materials like soft tissues typically involve penetration depths on the order of micrometers, not nanometers. Hence, the traditional method of tip calibration does not calibrate the tip to sufficient penetration depths to be valid for soft tissue characterization, at least not using a sample as stiff as fused silica.

For this reason, researchers in the areas of polymer and soft tissue indentation have been seeking improved tip calibration materials and methods. Since soft tissues have moduli on the order of kPa to MPa, it has been difficult to identify a suitably compliant calibration material that does not have time-dependent or depth-dependent properties, and has repeatable behavior and minimal sample to sample variability. Polycarbonate ($E \sim 3$ GPa) has been used to calibrate 20 μm radius of curvature spherical tips for some stiff polymer studies,[63] while urethane rubbers (E ~ 4 MPa) have occasionally been used to calibrate larger diameter tips for applications in more compliant polymers and soft tissues.[21] However, there is currently no industry-accepted standard for calibrating tips to depths on the order of micrometers. A common alternative to tip calibration for blunt spherical tips is the use of an ideal shape function ($A(h) = \pi(2Rh - h^2)$, where h is the penetration depth and R is the radius of curvature of the tip) to infer contact areas.[18,20,50] This method has the advantage that it relies only on knowledge of the radius of the tip, which can be determined under a microscope or using the nominal value supplied by the manufacturer. However, it also assumes a perfect tip of known constant radius with no surface defects. For flat punch tips, which have a constant contact area with depth, no calibration is needed.

4.2. *Instrumentation limitations*

The challenges and concerns outlined in the previous section focused on issues relating to sample and tip preparation for indentation. The high compliance and time-dependent behavior of soft tissue samples leads to additional challenges in adapting commercial instrumentation to perform indents in these complex materials. The next few sections will discuss particular instrumentation issues that have complicated soft tissue indentation, including automated methods of surface detection, nonlinear spring stiffness in the transducer at large displacements, limited displacement ranges, open-loop load- or displacement-control, and thermal drift. In addition, the potential errors introduced to indentation data due to the use of the standard instrumentation will be discussed, as well as critical areas for future instrumentation development. Hardware and software vary from commercial instrument to commercial instrument, and hence instrumentation issues will vary from instrument to instrument. The critique in the next few sections was written based on experience with one particular indentation system, which uses a capacitive transducer for both load application and displacement monitoring. However, use of other commercial instruments may yield similar limitations so care should be taken to examine how the instrument functions.

4.2.1. *Surface detection and initial sink-in*

Detection of the sample surface typically relies on changes in load or contact stiffness sensed by the indenter systems.[69] For example, in some nanoindentation systems, the surface is detected by sensing a change in the load on the transducer equal to a predetermined setpoint value. The default setpoint value can be as high as 2 μN to minimize the likelihood of a false surface detection due to noise. In a stiff substrate a preload of 2 μN results in a negligible penetration depth, and does not have a significant impact on the mechanical properties measured by indentation. However, with a compliant substrate and a blunt tip there can be substantial penetration of the tip into the sample even at small preloads.[68-71] The additional unknown penetration depth due to this "sink-in" effect can lead to substantial errors in modulus calculation for compliant materials, as demonstrated by experiments in a polydimethylsiloxane (PDMS) elastomeric sample ($E \sim 1$ MPa) that showed that the apparent modulus value increases with increasing pre-load.[69] In addition, with a pre-load of 0.1 μN and a Berkovich tip, an error in displacement of 580 nm was observed for an indent in PDMS with a total displacement of approximately 1500 nm (based on comparison to an FEM model).[69] If the maximum depth of the indent is much greater than the displacement error due to sink-in, then the effect of the preload may be negligible. However, it is difficult to determine how much "sink-in" has occurred, since most commercial indentation systems zero out the displacement in the data files. Hence, there is an unknown amount of error associated with indents in compliant samples due to this sink-in effect.

In addition to the problem of sink-in, it can be difficult to even detect the surface of a compliant sample, a problem that is exacerbated when samples are submerged in a hydrating fluid during indentation. If the indent approach is started above the fluid surface, then the change in load when the tip penetrates the fluid layer may trigger the surface detection setpoint, resulting in an aborted indent. Even if the tip is already submerged before the sample approach is started, meniscus and capillary forces acting on the shaft of the tip may vary as the tip steps towards the sample, which can also lead to a false triggering of the setpoint. To avoid accidental triggering of the setpoint by environmental factors, it may be necessary to use a setpoint value even higher than the default, which will in turn lead to a more substantial sink-in and associated modulus calculation errors. Recently Kaufman *et al.*[97] have looked at the problems and potential solutions surrounding nanoindentation of very compliant materials. Methods to overcome the difficulty in surface detection under hydrated conditions include using manual methods of surface detection such as intentionally starting indents out of contact with the sample or using continuous stiffness measurements to more sensitively detect the surface.[69,72] However, to fully resolve the issue of surface detection and sink-in there is a clear need for a more sensitive method of surface detection that is more suitable for soft tissue applications.

4.2.2. *Large displacements*

Because soft tissues are so compliant (E ~ kPa or MPa), they exhibit large displacement under very low loads. To accommodate these large displacements, calibrations need to take into account any nonlinear behavior of the transducers at large displacements. This can typically be achieved by performing an indent in air (or the hydrating fluid) to the maximum displacement anticipated during the indent process, and then subtracting that control curve off the experimental curve.[25] Most commercial indentation systems have this type of calibration built into their software, though you need to be sure to do the calibration in the appropriate environment (*i.e.* air vs. fluid). When performing dynamic testing at large displacements, additional calibrations may also be needed to account for dynamic calibration constants that vary with transducer displacement.[72]

The impact of many challenges associated with indentation of soft tissues, such as surface roughness, sink-in effects and adhesion, can be minimized by performing deep indents in the material. In addition, the high compliance of soft tissues can mean that a high stroke length is necessary to observe the material's behavior at physiologically relevant levels of compression, as has been pointed out for cartilage studies.[22] Some commercial indentation systems have displacement limits as low as 5 micrometers, which is often insufficient to accurately observe the mechanical response of the soft tissue under physiologically relevant conditions, or in the presence of substantial adhesion. High load indenter systems have higher displacement limits (often exceeding 100 micrometers) which can solve this problem, but the tradeoff is decreased load resolution. Combining a high stroke length with high load resolution is an issue to be considered in the development of improved nanoindentation systems for soft tissue applications.

4.2.3. *Open-loop vs. closed-loop control*

Many indentation systems have both open- and closed-loop modes of operation. When operating a nanoindenter, a certain load (or displacement) profile is defined by the user, and that profile is then transferred to the transducer. In open-loop mode, there is no feedback from the actual load or displacement at the tip back to the system for corrections. Instead, it is assumed that the desired profile is transferred to the tip-sample interface without any losses. While this is a reasonable assumption for stiff engineering materials that exhibit only small amounts of displacement (< 500 nm) and no creep, for soft tissues which exhibit large displacements (typically > 1 micrometer) and substantial amounts of creep there is substantial loss of the applied load or displacement in the instrumentation before reaching the tip-sample interface. For example, in some commercial nanoindentation systems the tip is supported by leaf springs. When operated in open-loop load control, the amount of load that goes into deflecting the springs increases with the amount of displacement of the tip. This results in the characteristic linear sloped region of the load-displacement curve (Fig. 9-4) during the hold period of the trapezoidal load function, and upon removal of load.

To correct for these problems, most nanoindenters now offer closed-loop load- and displacement-control options. These methods provide feedback from the actual load or displacement of the tip back to the system to correct the applied load or displacement signal to the transducer to compensate for losses in the system. When used correctly, indents collected using a trapezoidal load function should show a perfectly horizontal load-displacement behavior during the hold period, while indents collected using a trapezoidal displacement profile should show a perfectly vertical load-displacement behavior during the hold period. Using closed-loop load- or displacement-control is essential for performing creep or load-relaxation studies, as it decouples system losses from displacement or load due to time-dependent properties. In addition, closed-loop control should eliminate negative loads during unloading in the absence of adhesion.

4.2.4. *Drift*

A common method used to monitor drift in indentation tests involves holding the tip on the sample surface with a small preload force and monitoring the change in displacement. It is assumed that any change in displacement is due to thermal drift. However, in compliant time-dependent materials the displacement observed while holding on the surface is a combination of thermal drift, material creep, and sink-in that cannot be easily separated. Automatic "drift correction" options on commercial instruments are therefore inaccurate when indenting time-dependent biological materials, and should not be used. It is likely that for experiments run over short periods of time (*i.e.* less than 1 minute) the displacement of the indenter tip due to creep will be much higher than the thermal drift rate, so thermal drift can be considered negligible.

Thermal drift may become a more significant issue when performing load-relaxation or creep tests to characterize time-dependent properties, as indent times can be much longer than 1 minute and thermal drift may no longer be negligible. In addition, some indentation systems use data collected from the thermal drift monitoring method described above to set up the feedback loop for displacement- or load-control, which can make it difficult to achieve accurate displacement-control in time-dependent biological materials. Clearly this is an issue that will need to be addressed in future instrumentation development to ensure accurate characterization of time-dependent biological materials.

4.3. *Data acquisition and analysis*

The challenges and concerns outlined in the previous sections focused on issues relating to sample and tip preparation for indentation, and areas for development of improved instrumentation. The unique properties of soft tissues, namely their time-dependent properties, the high amounts of adhesion sometimes observed between the tip and sample during indentation of soft tissues, and their anisotropy, are also leading to the development of new data acquisition and analysis methods for nanoindentation. Since the models behind time-dependent properties and adhesion are presented in other chapters, the following sections will focus on specific methods proposed or used for indentation of biological materials to address the issues of time-dependent properties, adhesion, and anisotropy.

4.3.1. *Time-dependent properties*

All soft tissues and most biological materials exhibit time-dependent properties due to their water content and protein-based structures. Time-dependent mechanical behavior of soft tissues can be analyzed using viscoelastic or biphasic poroelastic models in bulk testing. However, in the field of nanoindentation, the majority of tissue studies have not used rigorous viscoelastic or biphasic poroelastic analysis to characterize time-dependent behavior.[7] Instead, traditional quasi-static analysis methods are used except that a trapezoidal load function (load-hold-unload) replaces the typical triangular (load-unload) load function. The hold period allows time for creep to dissipate prior to unloading, and then the unloading curve can be analyzed using the traditional compliance method.[73] This approach has been utilized in the majority of indentation studies of biological materials, with hold periods ranging from 3 to 120 seconds.[18-20,22,24,27,50,53,58,67,74] Because thermal drift and other constraints may limit the practical hold period time that can be imposed during indentation, other researchers have developed *post-hoc* methods to correct trapezoidal indentation data to remove any remaining influence of creep.[75-78] These numerical correction methods have been applied to recent studies of bone[79,80] and teeth,[81] revealing substantial error in modulus calculations when the corrections were not used.

In recent years, more rigorous viscoelastic models have been developed that measure the time-dependent behavior rather than allowing it to dissipate.[82] Many of these methods are reviewed in Chapter 5. Applications of viscoelastic models to biological tissues include stress-relaxation analysis of arteries[26] and demineralized dentin[28], dynamic indentation of the stratus corneum layer of skin,[27] and mapping of dynamic mechanical properties in spider silk cross-sections.[9] More complex viscous-elastic-plastic models have been applied to the study of bone.[48,49,82] Biphasic poroelastic models that they take into account both fluid flow and the deformation of the solid matrix seem like the most promising future direction for characterization hydrated, compliant biological materials, and have recently been used for analysis of nanoindentation data.[98] Because biphasic poroelastic tests typically involve long hold periods to allow fluid flow to reach equilibrium and evaluate matrix behavior, the challenges of thermal drift correction will need to be overcome to develop reliable poroelastic analysis methods. However, since most biological materials exhibit time-dependent properties, the continued development and application of explicit viscoelastic or poroelastic analysis techniques is likely to be central to future nanoindentation studies of biological materials.

4.3.2. *Adhesion*

In addition to time-dependent properties, many biological materials have a high surface energy, and adhesion between the tip and the sample can interfere with measurements of modulus using the compliance method.[63,83-85] Adhesion can sometimes be observed as the load falling below zero during the unloading portion of the load-displacement curve, as shown in the cartilage indents in Fig. 9-3 and the idealized soft tissue load-displacement curve in Fig. 9-4. (Note that negative loads can also result from a combination of open-loop control and creep, as illustrated in Fig. 9-4 and described in Section 2.4.3. So care must be taken in interpreting negative loads observed in open-loop load-displacement curves, as one or both factors may be playing a role.) Recent studies of soft polymers (silicone elastomers) have demonstrated that the compliance method overestimates the sample modulus when there is significant tip-sample adhesion.[83,84] This is because one of the assumptions of the compliance method is negligible adhesion.[6] In the presence of significant adhesion, an appropriate adhesion model must be used to determine the modulus. While numerous models exist, contact mechanics methods based on the Johnson-Kendall-Roberts (JKR)[86] adhesion model (which applies for compliant samples, large surface energies) have been most frequently applied to the characterization of compliant polymers using spherical tips.[72,83,84,87,88]

Applications of the JKR model vary from applying JKR equations to the analysis of quasi-static load-displacement curves[83,84,87] to using a full force curve method.[72] In the force curve method, the indent process is started out of contact so that a full force curve is captured as the tip approaches, indents, and retracts from the sample.[72,88,89] A typical force curve against a polydimethylsiloxane (PDMS) sample is shown in Fig. 9-6 along with a quasi-static indent in the same material, illustrating the error in displacement measurement (and hence contact area calculation) that occurs when adhesion is not taken into consideration. Analyzing compliant polymer indentation data using force curves and JKR models has been shown to provide a more accurate measurement of sample modulus than using the compliance method.[72,83,84] More complex nanoindentation adhesion models have also been developed that combine JKR and viscoelastic analyses using a force curve approach.[72,88-90] While adhesion methods have not yet been applied to many nanoindentation studies of biological materials, force curve methods are widely used in the AFM community for characterization of cells and other biological materials.[91,92] For soft tissue studies, where fluid-mediated adhesion is likely playing a role in complicating data analysis, continued development and application of adhesion models will probably be essential for measuring accurate quantitative mechanical properties.

4.3.3. *Anisotropy*

In addition to elastic behavior and negligible adhesion, the contact mechanics theory upon which the compliance method is based also assumes isotropic, homogeneous sample materials. Indentation is a complex loading condition, with compressive load near the center of the

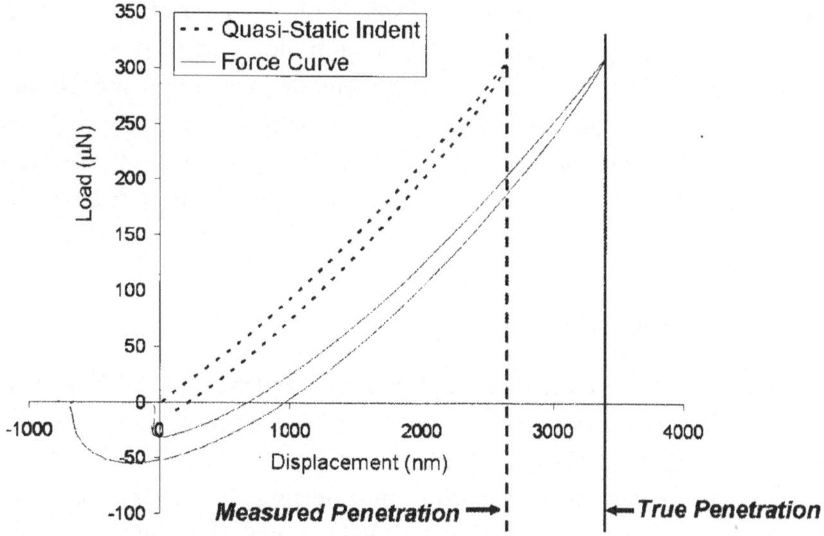

Figure 9-6. Comparison of a quasi-static indent (dashed line) and a force curve (solid line) in the same sample, a polydimethylsiloxane (PDMS) sample with a nominal elastic modulus of 3 MPa. Comparison of maximum displacements for the two types of indents reveals the large error in displacement calculation (and hence contact area calculation) that results from analyzing only the quasi-static indent and losing the information contained in the full force curve.

contact but tensile loading near the edges. When indenting anisotropic materials, it has been demonstrated that the indentation modulus represents a weighted average of moduli along the principal axes of the material, dominated by the modulus in the orientation of indentation.[93] Hence, anisotropic moduli measured by indentation are influenced by the moduli in the other orientations, and are not as different from one another as anisotropic moduli measured through uniaxial testing techniques. Vlassik and Nix[93] presented a method for calculating anisotropic moduli from indentation moduli in anisotropic materials, and this has been applied to measurements in bone samples.[94-96]

Although many soft tissues, such as cartilage, also exhibit structural anisotropy down to small length scales, rigorous anisotropic indentation analysis has not been applied to other biological materials due to the complexity of the current models. Moduli in multiple orientations have been measured in numerous biological materials using nanoindentation, including wood,[30,38,39] bamboo,[40] and spider silk,[9] but anisotropy methods than account for the influence of the moduli along other axes were not used in the analyses. Further development of anisotropic models and experimental methods specifically for soft tissues would greatly enhance the field of nanoindentation of biological materials by allowing researchers to more fully capture the complex mechanical behavior of these materials, rather than relying on simplifying assumptions.

4.4. *Data interpretation*

A final area of concern in applying nanoindentation to biological materials is how to interpret the resulting data. There is often a desire to compare mechanical parameters measured by nanoindentation with the same properties measured using bulk testing. However, due to the complexity in structure of many biological materials, and the local heterogeneities associated with their complex and hierarchical microstructure, it may not be realistic to assume that parameters measured at the nano- or micro-scale should correlate with bulk properties. In fact, a key advantage of nanoindentation is the ability to map local variations in mechanical properties within a tissue, which will not translate to bulk characterization. For validation of experimental methods, a homogeneous compliant material can (and should) be characterized, and results correlated with macroscale testing techniques.[21] However, care should be taken in extrapolating the results from nanoindentation studies of heterogeneous biological materials to macroscale properties.

An additional concern with data interpretation is whether modulus values or other parameters measured through indentation testing have quantitative meaning or only qualitative meaning. For example, cartilage studies demonstrated that the modulus measured by nanoindentation was dependent upon tip size,[22] and the authors suggested that only experimental data collected under indentical experimental conditions be compared. Considering all the sources of error in measuring mechanical properties of soft tissues using current commercial instrumentation and experimental techniques (*e.g.*, sink-in, adhesion), the lack of suitable control materials to validate experimental methods, and the expectation that biological materials will have different mechanical properties at different length scales due to their hierarchical structures, the interpretation of the quantitative values measured by indentation does appear daunting.

However, even though they are desirable, quantitative values that translate to macroscale properties may not be essential for all applications. Even as a qualitative tool, nanoindentation of biological materials has the ability to answer many research questions that are unachievable by traditional bulk mechanical testing techniques. For example, when comparing the effects of different clinical treatments on tissue properties,[18,19] investigating the effects of surface treatments on wood mechanics,[8,34-37] mapping mechanical properties in small samples,[9,20,45] or correlating changes in mechanical properties with variations in chemical composition in heterogeneous samples,[24,25,41,44,47] qualitative analysis by nanoindentation is sufficient and provides valuable insight into local structure-property relationships. Hence, it is always important to consider the question to be answered when determining the appropriate analysis to be used.

5. Conclusions and Future Directions

Nanoindentation has been used to measure quasi-static and viscoelastic mechanical properties in a broad range of biological materials, including soft tissues, plants, and spider silk. Because nanoindentation was developed for mechanical characterization of engineering materials such as metals and ceramics, adapting nanoindentation to the study of biological materials has posed many challenges to researchers. The study of soft tissues has posed the most challenges because soft tissues differ the most from traditional engineering materials: they are hydrated, compliant, have time-dependent properties, and have a complex hierarchical microstructure. Efforts to address these properties have led to the development of new nanoindentation sample preparation techniques and data acquisition and analysis methods. While these methods have furthered the field of soft tissue characterization using nanoindentation, there is still a need for continued developments and improvements in many areas in order to ensure that accurate, repeatable, quantitative mechanical properties will be measured in these complex materials using nanoindentation.

The breadth of data in the field shows that nanoindentation, at a minimum, provides a valuable method for characterizing mechanical behavior of biological materials at the nano- and microscales, mapping properties in small biological samples, and comparing different tissue regions after disease or repair. Whether the quantitative values reported in many of these studies are meaningful as absolute quantitative values is still uncertain, but the qualitative trends still answer many valuable research questions. Future developments in instrumentation and nanoindentation models tailored specifically to soft tissues and other classes of biological materials will lead to improved confidence in quantitative analyses of the mechanical properties of biological materials at small length scales.

Acknowledgements

Unpublished work presented in this manuscript was supported by a Denny Dickenson Orthopaedic Scholarship, a grant from the Orthopedic Research and Education Foundation, and a Giannini Family Foundation Medical Research Fellowship (Fig. 9-2) and the Air Force Office of Scientific Research (AFOSR), the Office of Naval Research (ONR), and a National Research Council Postdoctoral Associateship (Figs. 9-3 and 9-6). DME would like to thank Michelle Oyen for valuable comments on early drafts.

References

1. E. Gorb, K. Haas, A. Henrich, S. Enders, N. Barbakadze and S. Gorb, *J. Exp. Biol.* **208**, 4651 (2005).
2. G. Binnig, C. F. Quate and C. Gerber, *Phys. Rev. Lett.* **56**, 930 (1986).
3. N. A. Burnham and R. J. Colton, *J. Vac. Sci. Technol., A* **7**, 2906 (1989).
4. W. R. Bowen, R. W. Lovitt and C. J. Wright, *Biotechnol. Lett.* **22**, 893 (2000).
5. A. Vinckier and G. Semenza, *FEBS Lett.* **430**, 12 (1998).
6. W. C. Oliver and G. M. Pharr, *J. Mater. Res.* **7**, 1564 (1992).
7. D. M. Ebenstein and L. A. Pruitt, *Nano Today* **1**, 26 (2006).
8. R. J. Moon, C. R. Frihart and T. Wegner, *Forest Prod. J.* **56**, 4 (2006).

9. D. M. Ebenstein and K. J. Wahl, *J. Mater. Res.* **21**, 2035 (2006).
10. V. C. Mow, W. Y. GU and F. H. Chen, in *Basic Orthopaedic Biomechanics and Mechanobiology (3rd Edition)*, 2nd ed., eds. V. C. Mow and R. Huiskes, Lippincott Williams and Wilkins (2005).
11. V. C. Mow, A. Ratcliffe and A. R. Poole, *Biomaterials* **13**, 67 (1992).
12. W. C. Hayes, G. Herrmann, L. F. Mockros and L. M. Keer, *J. Biomech.* **5**, 541 (1972).
13. G. E. Kempson, M. A. R. Freeman and S. A. V. Swanson, *J. Biomech.* **4**, 239 (1971).
14. A. F. Mak, W. M. Lai and V. C. Mow, *J. Biomech.* **20**, 703 (1987).
15. V. C. Mow, M. C. Gibbs, W. M. Lai, W. B. Zhu and K. A. Athanasiou, *J. Biomech.* **22**, 853 (1989).
16. R. C. Appleyard, M. V. Swain, S. Khanna and G. A. C. Murrell, *Phys. Med. Biol.* **46**, 541 (2001).
17. T. Lyyra, J. Jurvelin, P. Pitkanen, U. Vaatainen and I. Kiviranta, *Med. Eng. Phys.* **17**, 395 (1995).
18. D. M. Ebenstein, A. Kuo, J. J. Rodrigo, A. H. Reddi, M. Ries and L. Pruitt, *J. Mater. Res.* **19**, 273 (2004).
19. O. Franke, K. Durst, V. Maier, M. Goken, T. Birkhoiz, H. Schneider, F. Hennig and K. Gelse, *Acta Biomat.* **3**, 873 (2007).
20. C. Li, L. Pruitt and K. King, *J. Biomed. Mater. Res. Part A* **78A**, 729 (2006).
21. N. K. Simha, C. S. Carlson and J. L. Lewis, *J. Mater. Sci. - Mater. Med.* **15**, 631 (2004).
22. N. K. Simha, H. Jin, M. L. Hall, S. Chiravarambath and J. L. Lewis, *J. Biomech. Eng.* **129**, 767 (2007).
23. D. Coughlin, Ph.D. Thesis, UCB/UCSF Bioengineering Graduate Group (University of California at Berkeley and University of California at San Francisco), 2004.
24. D. M. Ebenstein, D. Coughlin, J. Chapman, C. Li and L. A. Pruitt, *J. Biomed. Mater. Res.* **91A**, 1028 (2009)
25. D. M. Ebenstein, Ph.D. Thesis, UCB/UCSF Bioengineering Graduate Group (University of California at Berkeley and University of California at San Francisco), 2002.
26. A. Lundkvist, E. Lilleodden, W. Siekhaus, J. Kinney, L. Pruitt and M. Balooch, in *Thin Films: Stresses and Mechanical Properties VI (Mater. Res. Soc. Proc. 436)*, eds. W. W. Gerberich, H. Gao and J. E. Sundgren, Materials Research Society, Boston, MA, p. 353 (1997).
27. Y. H. Yuan and R. Verma, *Colloids Surf. B* **48**, 6 (2006).
28. M. Balooch, I. C. Wu-Magidi, A. Balazs, A. S. Lundkvist, S. J. Marshall, G. W. Marshall, W. J. Siekhaus and J. H. Kinney, *J. Biomed. Mater. Res.* **40**, 539 (1998).
29. M. Kohane, A. Daugela, H. Kutomi, L. Charlson, A. Wyrobek and J. Wyrobek, *J. Biomed. Mater. Res. Part A* **66A**, 633 (2003).

30. W. Gindl, H. S. Gupta, T. Schoberl, H. C. Lichtenegger and P. Fratzl, *App. Phys. A* **79**, 2069 (2004).
31. W. T. Y. Tze, S. Wang, T. G. Rials, G. M. Pharr and S. S. Kelley, *Composites Part A* **38**, 945 (2007).
32. R. Wimmer, B. N. Lucas, T. Y. Tsui and W. C. Oliver, *Wood Sci. Technol.* **31**, 131 (1997).
33. G. A. Zickler, T. Schoberl and O. Paris, *Philos. Mag.* **86**, 1373 (2006).
34. W. Gindl and H. S. Gupta, *Composites Part A* **33**, 1141 (2002).
35. W. Gindl, T. Schoberl and G. Jeronimidis, *Int. J. Adhes. Adhes.* **24**, 279 (2004).
36. J. Konnerth and W. Gindl, *Holzforschung* **60**, 429 (2006).
37. J. Konnerth, A. Valla and W. Gindl, *App. Phys. A* **88**, 371 (2007).
38. W. Gindl and T. Schoberl, *Composites Part A* **35**, 1345 (2004).
39. S. H. Lee, S. Q. Wang, G. M. Pharr, M. Kant and D. Penumadu, *Holzforschung* **61**, 254 (2007).
40. Y. Yu, B. Fei, B. Zhang and X. Yu, *Wood Fiber Sci.* **39**, 527 (2007).
41. S. Enders, N. Barbakadse, S. N. Gorb and E. Arzt, *J. Mater. Res.* **19**, 880 (2004).
42. D. R. Katti, K. S. Katti, J. M. Sopp and M. Sarikaya, *Comput. Theor. Polym. Sci.* **11**, 397 (2001).
43. B. J. F. Bruet, H. J. Qi, M. C. Boyce, R. Panas, K. Tai, L. Frick and C. Ortiz, *J. Mater. Res.* **20**, 2400 (2005).
44. N. Barbakadze, S. Enders, S. Gorb and E. Arzt, *J. Exp. Biol.* **209**, 722 (2006).
45. M. Sarikaya, H. Fong, N. Sunderland, B. D. Flinn, G. Mayer, A. Mescher and E. Gaino, *J. Mater. Res.* **16**, 1420 (2001).
46. F. Song, K. L. Lee, A. K. Soh, F. Zhu and Y. L. Bai, *J. Exp. Biol.* **207**, 3035 (2004).
47. J. H. Waite, H. C. Lichtenegger, G. D. Stucky and P. Hansma, *Biochemistry* **43**, 7653 (2004).
48. A. K. Bembey, M. L. Oyen, A. J. Bushby and A. Boyde, *Philos. Mag.* **86**, 5691 (2006).
49. A. J. Bushby, V. L. Ferguson and A. Boyde, *J. Mater. Res.* **19**, 249 (2004).
50. D. M. Ebenstein and L. A. Pruitt, *J. Biomed. Mater. Res. Part A* **69A**, 222 (2004).
51. S. Hengsberger, A. Kulik and P. Zysset, *Bone* **30**, 178 (2002).
52. C. E. Hoffler, X. E. Guo, P. K. Zysset and S. A. Goldstein, *J. Biomech. Eng.* **127**, 1046 (2005).
53. J. Y. Rho and G. M. Pharr, *J. Mater. Sci. - Mater. Med.* **10**, 485 (1999).
54. S. Habelitz, G. W. Marshall, M. Balooch and S. J. Marshall, *J. Biomech.* **35**, 995 (2002).
55. J. H. Kinney, M. Balooch, G. M. Marshall and S. J. Marshall, *Arch. Oral Biol.* **44**, 813 (1999).
56. G. W. Marshall, S. Habelitz, R. Gallagher, M. Balooch, G. Balooch and S. J. Marshall, *J. Dent. Res.* **80**, 1768 (2001).
57. A. B. Mann and J. B. Pethica, *Langmuir* **12**, 4583 (1996).
58. L. Angker, M. V. Swain and N. Kilpatrick, *J. Biomech.* **38**, 1535 (2005).

59. C. E. Hoffler, K. E. Moore, K. Kozloff, P. K. Zysset, M. B. Brown and S. A. Goldstein, *Bone* **26**, 603 (2000).
60. P. K. Zysset, X. E. Guo, C. E. Hoffler, K. E. Moore and S. A. Goldstein, *J. Biomech.* **32**, 1005 (1999).
61. E. Donnelly, S. P. Baker, A. L. Boskey and M. C. H. van der Meulen, *J. Biomed. Mater. Res. Part A* **77A**, 426 (2006).
62. W. W. Gerberich, W. Yu, D. Kramer, A. Strojny, D. Bahr, E. Lilleodden and J. Nelson, *J. Mater. Res.* **13**, 421 (1998).
63. C. Klapperich, K. Komvopoulos and L. Pruitt, *J. Tribol.* **123**, 624 (2001).
64. S. P. Ho, H. Goodis, M. Balooch, G. Nonomura, S. J. Marshall and G. Marshall, *Biomaterials* **25**, 4847 (2004).
65. M. E. Roy, J. Y. Rho, T. Y. Tsui, N. D. Evans and G. M. Pharr, *J. Biomed. Mater. Res.* **44**, 191 (1999).
66. J. Xu, J. Y. Rho, S. R. Mishra and Z. Fan, *J. Biomed. Mater. Res. Part A* **67A**, 719 (2003).
67. S. Habelitz, S. J. Marshall, G. W. Marshall and M. Balooch, *Arch. Oral Biol.* **46**, 173 (2001).
68. A. C. Fischer-Cripps, *Vacuum* **58**, 569 (2000).
69. J. Deuschle, S. Enders and E. Arzt, *J. Mater. Res.* **22**, 3107 (2007).
70. S. A. Hayes, A. A. Goruppa and F. R. Jones, *J. Mater. Res.* **19**, 3298 (2004).
71. M. R. VanLandingham, J. S. Villarrubia, W. F. Guthrie and G. F. Meyers, *Macromol. Symp.* **167**, 15 (2001).
72. D. M. Ebenstein and K. J. Wahl, *J. Colloid Interface Sci.* **298**, 652 (2006).
73. B. J. Briscoe, L. Fiori and E. Pelillo, *J. Phys. D-Appl. Phys.* **31**, 2395 (1998).
74. J. Y. Rho, M. E. Roy, T. Y. Tsui and G. M. Pharr, *J. Biomed. Mater. Res.* **45**, 48 (1999).
75. Y. T. Cheng, W. Y. Ni and C. M. Cheng, *J. Mater. Res.* **20**, 3061 (2005).
76. G. Feng and A. H. W. Ngan, *J. Mater. Res.* **17**, 660 (2002).
77. A. H. W. Ngan, H. T. Wang, B. Tang and K. Y. Sze, *Int. J. Solids Struct.* **42**, 1831 (2005).
78. B. Tang and A. H. W. Ngan, *J. Mater. Res.* **18**, 1141 (2003).
79. B. Tang, A. H. W. Ngan and W. W. Lu, *Philos. Mag.* **86**, 5653 (2006).
80. B. Tang, A. H. W. Ngan and W. W. Lu, *J. Mater. Sci. - Mater. Med.* **18**, 1875 (2007).
81. J. H. Kinney, S. J. Marshall and G. W. Marshall, *Crit. Rev. Oral Biol. Med.* **14**, 13 (2003).
82. M. L. Oyen and A. J. Bushby, *Int. J. Surf. Sci. Eng.* **1**, 180 (2007).
83. F. Carrillo, S. Gupta, M. Balooch, S. J. Marshall, G. W. Marshall, L. Pruitt and C. M. Puttlitz, *J. Mater. Res.* **20**, 2820 (2005).
84. F. Carrillo, S. Gupta, M. Balooch, S. J. Marshall, G. W. Marshall, L. Pruitt and C. M. Puttlitz, *J. Mater. Res.* **21**, 535 (2006).

85. J. C. Grunlan, X. Xinyun, D. Rowenhorst and W. W. Gerberich, *Rev. Sci. Instrum.* **72**, 2804 (2001).
86. K. L. Johnson, K. Kendall and A. D. Roberts, *Proc. R. Soc. London, A* **324**, 301 (1971).
87. S. Gupta, F. Carrillo, C. Li, L. Pruitt and C. Puttlitz, *Mater. Lett.* **61**, 448 (2007).
88. K. J. Wahl, S. A. S. Asif, J. A. Greenwood and K. L. Johnson, *J. Colloid Interface Sci.* **296**, 178 (2005).
89. M. Giri, D. B. Bousfield and W. N. Unertl, *Langmuir* **17**, 2973 (2001).
90. J. A. Greenwood and K. L. Johnson, *J. Colloid Interface Sci.* **296**, 284 (2005).
91. H. J. Butt, B. Cappella and M. Kappl, *Surf. Sci. Rep.* **59**, 1 (2005).
92. B. Cappella and G. Dietler, *Surf. Sci. Rep.* **34**, 1 (1999).
93. J. J. Vlassak and W. D. Nix, *J. Mech. Phys. Solids* **42**, 1223 (1994).
94. Z. Fan, J. Y. Rho and J. G. Swadener, *J. Mater. Res.* **19**, 114 (2004).
95. Z. Fan, J. G. Swadener, J. Y. Rho, M. E. Roy and G. M. Pharr, *J. Orthop. Res.* **20**, 806 (2002).
96. J. G. Swadener, J. Y. Rho and G. M. Pharr, *J. Biomed. Mater. Res.* **57**, 108 (2001).
97. J. Kaufman, G. Miller, E. Morgan and C. Klapperich, *J. Mater. Res.* **23**, 1472 (2008).
98. S. Gupta, J. Lin, P. Ashby and L. Pruitt, *J. Mech. Behav. Biomed. Mater.* **2**, 326 (2009).

CHAPTER 10

CELLS AND MEMBRANES

Kuo-Kang Liu[1] and Kai-Tak Wan[2]

[1]*School of Engineering, University of Warwick*
E-mail: I.K-K.Liu@warwick.ac.uk
[2]*Mechanical and Industrial Engineering, Northeastern University*
E-mail: ktwan@coe.neu.edu

Nanoindentation of biological cells and thin biomembranes has been demonstrated to be an effective means for measuring and understanding the mechanical behavior of these bio-entities with relatively small dimensions and unusual mechanical properties. Several new and variant indentation techniques have been developed recently. New solid mechanics models spanning multi-scale indentation deformation are constructed to analyze the force-displacement data so that the quantitative materials parameters can be found.

1. Introduction

Mechanical characterization of cells and membranes becomes increasingly important for the advancement of bioscience and bioengineering. For example, the changes in elasticity and viscoelasticity of cells which normally associate with their structure and molecular alterations are regarded as a prelude or even prime cause of pathogenesis. In tissue engineering applications, the engineered membrane is highly "process-dependent", that is, its mechanical (viscoelastic) properties vary with the physiochemical conditions and protocols used during the culture period. Intriguingly, the quantitative modulus obtained for tissue-engineered membrane equivalents are shown to be fairly time-dependent and the value is nearly three orders of magnitude lower than that of their natural counterparts when collagen is utilized as the scaffold.[1,2] Hence, a reliable and efficient method is necessary to characterize membrane properties for providing feedback to optimize the factors and procedures in tissue construction. Nanoindention has been regarded as one of the viable methods. It is obvious that such mechanical measurement is nontrivial and in fact quite challenging due to the small sample size and thickness and the unusual mechanical properties such as low modulus and viscoelasticity. However, recent advances in indentation instrumentation and the various modes of testing geometry (their details are addressed later) have facilitated the mechanical characterization of biological cells and biomembranes. Several new models which have recently been developed for quantitative data analysis are also presented in this chapter.

1.1. *Cell structure*

All living organisms are composed of cells. According to their structure, cells can be categorized as eukaryotic and prokaryotic cells. A eukaryotic cell is a cell with a nucleus which hosts the cell's chromosomes, while prokaryotic cells have no nuclei. Animals, plants, protists, and fungi are eukaryotes, while Eubacteria and Archaea are common prokaryotes. Generally eukaryotic cells comprise at least three major structural domains: (i) a plasma membrane to identify the boundaries of the cell and to retain its contents, (ii) membrane-bounded organelles, a nucleus to house its DNA, and (iii) a cytoplasm with its cytoskeleton of tubules and filaments. In addition, plant cells usually possess a rigid cell wall, while animal cells are encapsulated by a compliant plasma membrane coated with glycolproteins.[3] Most biologists agree that cell walls are essential to maintain the cell integrity. In fact, the membrane is not simply a lipid bilayer studded with protein as many biologists suggest, but "a material with a very small thickness, in comparison with its radii of curvature, which separates two adjacent liquid-like domains and supports the stress created by the embedding medium".[4]

1.2. *Soft tissue membrane*

Soft tissue normally provides connections and supports for structures or organs in the body. For example, skin, blood vessel and cornea are a few among many tissue membranes. In general, they are anisotropic and inhomogeneous materials exercising mechanical non-linearity and viscoelasticity. Most soft tissues resemble nonlinear stiff gels with significant internal loss when subject to mechanical shear, but become fluidic under compression.[5]

1.3. *Hydrogel membrane*

A hydrogel is a polymeric chain network that is insoluble in an aqueous environment, yet contains very high water content (up to 99%). Resembling natural tissues with excellent biocompatibility and processing amiability, these biomimetic gels are widely used in biomedical applications such as prosthetic tissues and artificial skins for wound healing. Among limited types of hydrogels successfully used in tissue engineering or drug delivery, agarose and alginate are most popular due to their excellent availability and biocompatibility. Agarose is natural polymer derived from agar, while alginate is derived from seaweed. However, hydrogel-based tissues in general possess poor mechanical properties compared with their natural counterparts (commonly 100 times lower). Various methods have been developed to improve their mechanical properties, *e.g.*, UV cross-linking.[6] In order to gauge and to validate their performance, there is an urgent need to develop synergistic mechanical characterization techniques.

2. Experimental Setup

There are a few experimental techniques to characterize the mechanical properties of biological cells and membranes. To align with the main theme of this book, only indentation-based methods and instruments are discussed in this chapter.

2.1. *Cell poking*

The "Cell-poking" technique developed by Daily *et al.*[7] and Zahalak *et al.*[8] can be regarded as an extended depth-sensing indentation method. Figure 10-1 is the schematic view of the instrumental set-up. The setup mainly comprises a micro- and nano-size punch pressing a sample cell against a rigid substrate. Applied force is measured as a function of punch displacement during indentation loading. However, some serious drawbacks are identified. For instance, there is a potential detachment of cytoplasm from the thin lipid bilayer. The local mechanical properties of the sample cell can be determined, provided the indentation depth is less than roughly 10% of the cell wall thickness in order to avoid a "substrate backing" effect.[9] Measurements with such fine displacement control are technically difficult to control and to yield feasible data is challenging.

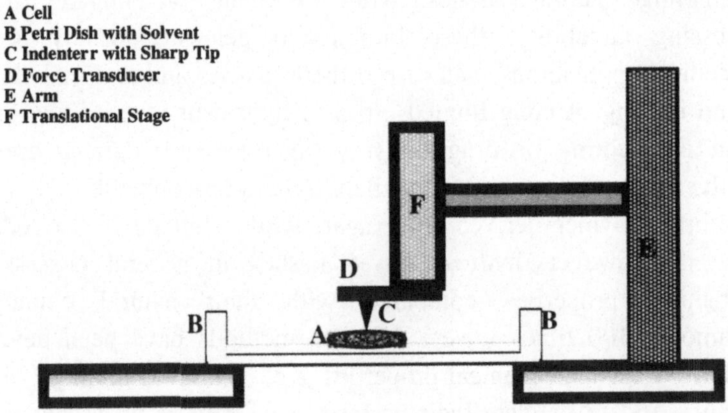

Figure 10-1. Schematic of experimental setup of "cell-poking" test.

2.2. *Atomic force microscope (AFM)*

AFM, originally designed for surface topography mapping, is now widely used for force measurement of cells. The instrument comprises a sharp tip for scanning the sample surface with atomic resolution and an ultra-sensitive cantilever beam for sensing topographical changes and tip-surface interactions. Figure 10-2 shows a schematic of AFM indentation of biomembranes (testing geometry for single cell is shown later in Fig. 10-5). One remarkable and unique feature is the simultaneity of surface scanning and indentation of biomembranes. In the present context, the instrument is also capable of the now-celebrated "Relative Microelastic Mapping"[10], *i.e.* to resolve the local mechanical compliance and material properties of a single biological cell. AFM-based indentation recently becomes an attractive method to investigate cellular nano-biomechanics. Properties such as elastic modulus of cell membranes,[11] turgor pressure, and surface properties of a single bacterium[12] can be quantified to a certain accuracy surpassing many conventional methods.

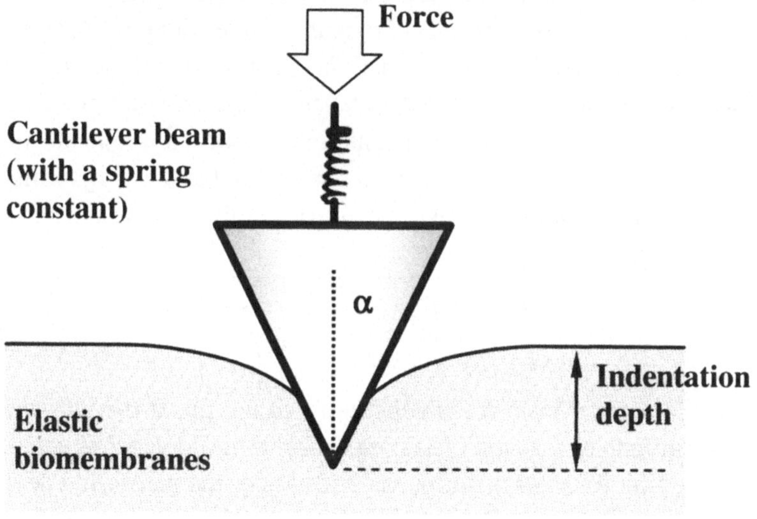

Figure 10-2. Schematic of an AFM applied for nanoindentation test.

2.3. *Nanoindentation*

Micro- and nano-indentation have been widely used to characterize a wide range of hard materials such as ceramics, metals and hard biomaterials (*e.g.* natural and prosthetic bones, hydroxyapatite). External load is applied to the sample surface via the hard indenters of steel, diamond or silicon nitride in the shapes of sphere, Vickers (square based pyramid), Knoop (rectangular based pyramid) and Berkovich (triangular based pyramid), leaving a permanent dent. Hardness of the sample is defined to be the ratio of applied load to the indenter-sample contact area or the projected area at the pre-indented plane. Complications such as the hysteretic mechanical response during loading-unloading cycle, elastic recovery at the indent based upon unloading, piling-up or sinking-in at the contact edge, interplay of elastic and plastic deformation, and viscoelastic behavior upon cyclic loading are accounted for by a number of standard models in the literature. The complex stress fields at the indent can be derived as variant forms of the classical Hertz contact model that is based on deformation of an elastic half-space by a spherical indenter.

Unlike the typical hard surfaces of metals or ceramics, soft biological tissues do not leave a permanent dent on the sample surface upon external load removal in most situations because of elastic recovery of the shallow indentation depth. Nevertheless, there are voluminous studies in the literature reporting indentation using a conical tip on biomembranes laying flat on a rigid substrate. Ignoring the aforementioned complications, the most straightforward mechanical response is given by

$$h = \frac{F}{k_c} + \sqrt{\left(\frac{\pi}{2\tan\alpha}\right)\left(\frac{1-\nu^2}{E_t}\right)F} \tag{10-1}$$

with F the applied load, h the penetration at the tip, α the half-angle of the conical indenter, E_t the elastic modulus in the traverse direction (*i.e.* across the film thickness), and k_c the stiffness of the measuring device. It is obvious that the measurement only yields E_t, rather than the more relevant two-dimensional planar film area. As noted earlier, it is the "large area" of cell membrane of individual cells that makes adhesive contact at the cell-cell and cell-substrate interfaces. Indentation confined to volumes small compared to the sample thickness is therefore quite irrelevant to the overall deformation of cells, especially the cell membrane that encloses the internal cytoplasmic fluid. In fact, the classical Hertz contact assumption of an elastic half space breaks down completely in the case of indentation of a single cell with an encapsulating membrane. The external point load acting locally causes the membrane to deform, the internal pressure to rise, and ultimately the entire cell to swell globally.

Depth-sensing nanoindentation techniques have recently been extended to characterize soft films and membranes such as polymer or biopolymer membranes.[13] However, the limitations of the conventional nanoindentation method still prompt for a need to develop new techniques to characterize multi-scale freestanding membranes from the nano-scale of cell membrane to macro-scale of tissues.

2.4. *Shaft-loaded blister test*

Conventional indentation is not suitable for measuring material behavior of thin soft membranes because it is normally associated with a "substrate backing" effect. To avoid such complication, the sample film is clamped by two rings to form a circular diaphragm, shown as Fig. 10-3. Such a configuration also helps to prevent a delicate ultrathin membrane from being damaged by the conventional mechanical grips. An external load is then applied to the center via either a ball bearing or a spherical capped cylindrical shaft, causing the film to deform to a conical cusp. Simultaneous monitoring of the load, central displacement, and the deformed profile yields the material parameters of the sample. The testing configuration can be used to characterize membranes with dimensions in the centimeter range. The smallest possible radius can be down to submicron if standard lithography is to drill a through-hole in a rigid plate while the film is clamped between two such identical plates.

The mechanical response of the sample can span multiple length-scales depending on the ratio of the membrane thickness to radius. In a typical loading process, as the external load increases incrementally from null, film deformation is confined to a local volume negligibly small comparable to the film thickness according to the Hertz indentation contact model. Depending on the indenter dimension, the micro-scale indentation range represents the response of single cells or even cell membrane at the surface layer. Progressive increase in external load subjects the entire sample to deform by plate-bending, mixed bending-stretching transition, and ultimately pure membrane stretching, manifesting the meso- to macro- scale behavior. While bending subjects the film to shearing perpendicular to the sample plane, stretching puts a

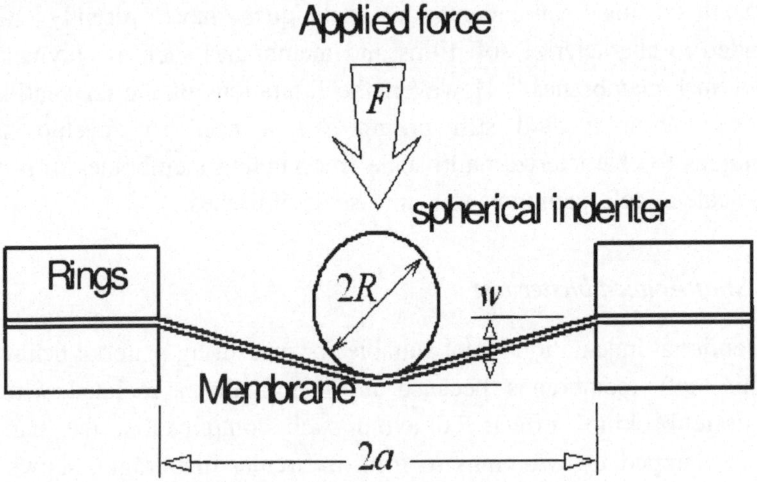

Figure 10-3. Schematic of experimental setup of shaft-loaded blister test.

tensile force on the film in the radial and circumferential directions. The bending-stretching range can be used to characterize from multi-cell up to tissue level. Extreme loading leads to large deformation governed by rubber elasticity.

It is worthwhile to note that the new indentation method for freestanding membrane, though similar, is significantly different from conventional nanoindentation. Either configuration requires the applied load to be in the sub-mN range, but the former might require the shaft displacement to be in the millimeter range depending on the membrane radius and thickness. The large actuation displacement requirement might exceed the maximum extent that most nanoindentation systems can handle, which is less than 1 mm.

Besides the capability of measuring materials properties in a typical loading-unloading cycle, the new method is useful for fixed-load measurements, *e.g.* creep test.[14] A ball bearing of fixed weight can be dropped onto the film center, thus allowing the load to be unchanged in either a dry or aqueous environment for a prolonged duration in terms of weeks, if not months or years (Fig. 10-4). There is virtually zero thermal and mechanical drift as in all force sensing devices. This is by far the

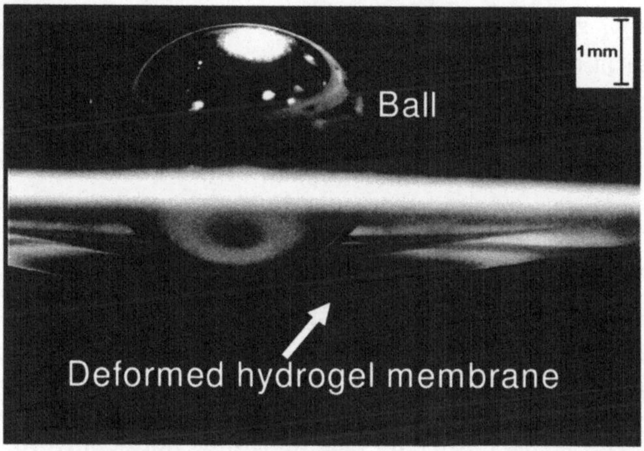

Figure 10-4. A novel shaft-loaded (ball-bearing) blister test for characterizing creeping deformation of a hydrogel membrane (adopted from Ahearne *et al.*[14]).

only known long term creep test for an ultra-thin polymer membrane. Creep was measured in natural and prosthetic cornea constructs for 6 weeks using this method.

2.5. *Other methods*

Fairly recently, a novel micro- or nanoindention technique based on Optical Coherence Tomography (OCT) has been developed technique for characterizing mechanical properties of biomaterials, particularly for soft biomimetic membranes.[15] OCT is an interferometric imaging method that utilizes the short coherence lengths of a broadband light source. This method provides an on-line, non-destructive and real-time way to characterize soft biomimetic or biological membranes. The equipment is capable of in-situ monitoring of a creep test on soft hydrogel membranes.

A Micro-Electro-Mechanical-System (MEMS) has been developed for the indentation of single biological cells.[16] The MEMS-based capacitive sensor-actuator provides high accuracy in calibrating force and displacement in 0.01 μN and 1 μm, respectively. More importantly, the device can simultaneously aspire the cell as a function of pipette actuation. The method has been applied to indent and pierce through the outer porous glycoprotein shell of mouse oocytes and measure the force-displacement curves of the deformed egg. In combination with a simple membrane model (stretching deformation only), the quantitative elastic modulus and piercing of the egg wall before and after fertilization is reported the first time in the literature.

3. Theories

3.1. *Mechanical models for single cells*

3.1.1. *Balloon model*

A typical cell is taken to be an incompressible liquid enclosed by a thin spherical elastic membrane, similar to a balloon. The bending inertia of the cell membrane is usually ignored so that membrane stretching serves as the only deformation mode. Several theoretical works are based on the same assumption, though various forms are reported in the literature. For example, Yao *et al.*[17] developed a model to account for the turgor pressure within a bacterium under AFM-indentation (Fig. 10-5). In their model, a uniform tensile deformation of the cell membrane is assumed, and therefore, with no surprise, the calculation shows a linear force-displacement relation,

$$k = \frac{F}{w_t} = \frac{2\pi pR}{\log\left(\frac{R+c}{c}\right) + \frac{R+2c}{R+c}} \tag{10-2}$$

where F is the force applied via an AFM tip with radius c, p is turgor pressure, and R is the radius of the cell. The total indentation depth at the pole, w_t, is given by

$$w_t = w_0 + 2w_c \approx w_0 + 2\left(\frac{F}{2\pi pR}\right) \tag{10-3}$$

where w_0 is the depth of the indentation dimple and w_c is the height of the collapsed cap at one pole.

Sen *et al.*[18] developed a similar theoretical model which also ignores bending effects. Adhesion between the cell membrane and substrate is,

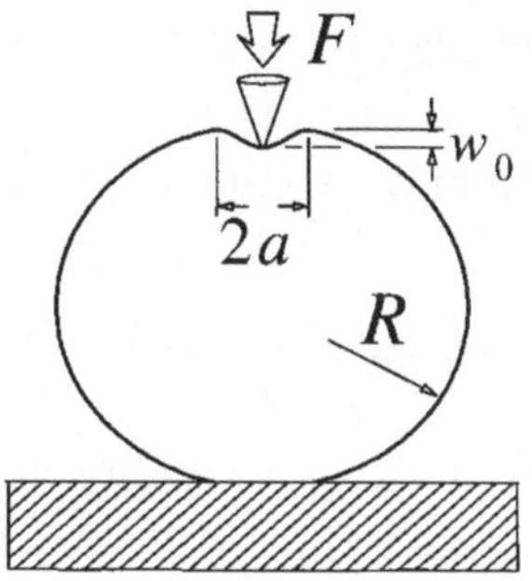

Figure 10-5. Schematic of an AFM indentation of a thin-walled microcapsule.

however, taken into account for the deviation of force-displacement relation from a non-adhesive surface indented by an AFM-tip.

3.1.2. *Shell model*

Contrasting the balloon model, Boulbitch *et al.*[19] derived a shell model that ignores membrane stretching but accounts for the out-of-plane bending stresses only. The indented profile w is found to be

$$w = w_t K_0(\xi)/K_0(\zeta) \tag{10-4}$$

and a linear spring constant k as

$$k = \pi R p \zeta \; [K_1(\zeta)/K_0(\zeta)] \tag{10-5}$$

where $\xi = r/a$, $\zeta = c/a$, and K_i is the ith order of second kind modified Bessel function. For the model to be valid, the indentation depth must be negligibly small compared to the cell membrane thickness.

3.1.3. *Model including both bending and stretching*

To circumvent the shortcomings of the two aforementioned modeling approaches, Wan *et al.*[20] developed a model which takes account of both bending and stretching stresses. Here a spherical capsule with wall thickness, b, elastic modulus, E, Poisson's ratio, ν, and original membrane pressure, $\sigma_0 = pR/2b$, is indented via an AFM tip. The loading process breaks down into 2 consecutive steps: (i) the capsule is compressed by 2 parallel plates forming 2 planar contact circles at the polar regions with radius, a; (ii) the top plate is then replaced by the AFM tip, resulting in a dimple of radius, a. The dimple profile can be found by linear elasticity[21]

$$\kappa \nabla^4 w - \sigma b \nabla^2 w = -p + F\,\delta(r) \tag{10-6}$$

with $\kappa = Eb^3 / 12(1-\nu^2)$ the flexural rigidity, σ the apparent membrane stress within the dimple, and $\delta(r)$ the delta function. A few assumptions are made: (i) the cell wall deformation is dominated by membrane stretching of the wall with bending moment a perturbation; (ii) the deformation is assumed small and local such that increase in p and R are negligible; (iii) equilibrium requires $F = (\pi a^2)\,p$; and (iv) σ is uniform within the dimple. The external force is supported by the local membrane stress and bending moment within the dimple. The external load induces a local concomitant stress σ_m superimposing on the intrinsic membrane stress, such that $\sigma = \sigma_0 + \sigma_m$, where

$$\sigma_m = \left[\frac{E}{2(1-\nu^2)}\right] \int_c^a \frac{1}{2}\left(\frac{dw}{dr}\right)^2 r\,dr \Bigg/ \int_c^a r\,dr \tag{10-7}$$

The global geometry of a truncated sphere resumes beyond the dimple ($r > a$). The dimple profile is found to be

$$\omega = \frac{1}{\beta}\{C_1[I_0(\beta\xi) - I_0(\beta)] - C_2[K_0(\beta\xi) - K_0(\beta)]\}$$
$$- \frac{\varphi}{\beta^2}\log\xi - \frac{\rho}{2\beta^2}(1-\xi^2) \tag{10-8}$$

with I_i the ith order of the first kind modified Bessel functions, and C_1 and C_2 some functions of φ, $\omega = w/b$, $\rho = pa^4/2\kappa b$, $\varphi = Fa^2/2p\kappa b$, $\beta = (\sigma ba^2/\kappa)^{1/2}$ and $\beta^2 = \beta_0^{\,2} + \beta_m^{\,2}$. The dimple depth is given by $\omega_0 = \omega(\xi{=}0)$. It can be shown that $\beta_m^{\,2} = 6\,[f(\beta)\text{-}f(\beta\xi)]/(1-\zeta^2)\beta^2$ with $f(x)$ some mathematically involved functions given earlier.[20] The relation $\varphi(\omega_0)$ can now be found analytically by eliminating β_m. In case of pure stretching ($\kappa = 0$), the constitutive relation $F(w_0)$ is shown in Fig. 10-5, which is linear at small w_0 and cubic otherwise. In the linear region, the total displacement traveled by the AFM tip is given by

$$w_t = \frac{F}{2\pi pR}\left[\log\left(\frac{a^2}{c^2}\right) + \frac{c^2 + a^2}{a^2}\right] \tag{10-9}$$

which is linear ($\varphi \propto \omega_0$). It is interesting to compare the model with the existing theories. Yao *et al.*[17] assumes a pure stretching model, ignores the σ_0 contribution and derives Eq. 10-2. Comparing with the present model, the following are noted: (i) R and c are used in Eq. 10-2 and both are linear, but a and c are squared in Eq. 10-9; (ii) σ_m is ignored and thus a non-cubic region of $F(w_t)$. It is also worthy of pointing out that the following terms are ignored by Boulbitch *et al.*[18] in the Eqs. 10-4 and 10-5: (i) $I_0(\beta\xi)$ in $\omega(\xi)$; (ii) $\log(\zeta)$ in ω_0; (iii) σ_m; (iv) the dependence of a upon F.

3.1.4. *Viscoelastic model*

In general, most cells under external load behave like a viscoelastic body rather than a purely elastic entity. Hence it is sensible to construct a theoretical model of cell indention based on viscoelasticity. Several research groups have developed such viscoelastic models for describing cells under indentation.

Darling *et al.*[22] have a simple model which incorporates viscoelasticity of a standard linear solid into the Hertz equation to account for small deformation of an isotropic, incompressible solid sphere indented by a hard, spherical indenter. The model has been successfully applied for characterizing the viscoelastic properties of zonal articular chondrocytes. The concise form of the final constitutive relation of the cell is given by

$$F(t) = \frac{4R^{1/2}h^{2/3}E_R}{3(1-\nu)}\left(1 + \frac{\tau_\sigma - \tau_\varepsilon}{\tau_\varepsilon}\exp\left(\frac{-t}{\tau_\varepsilon}\right)\right) \tag{10-10}$$

where $F(t)$ is the time-dependent indentation force (*i.e.*, stress-relaxation), h is the indentation depth, E_R is the relaxation elastic modulus, τ_σ and τ_ε are the relaxation times under constant load and constant strain, respectively.[23]

Roca-Cusachs *et al.*[24] have developed an alternative viscoelastic model for AFM-indentation of neutrophils. Based on a cell cortex exerting a force on the spherical tip, they developed force-displacement relationships based on several viscoelastic models such as Newtonian liquid, Maxwell liquid, and Kelvin body. In particular, the final constitutive relation based on the Kelvin body is listed as following:

$$F = \frac{4E_i}{3(1-\nu^2)} R^{1/2} h^{3/2} + \frac{2T}{r_t} \pi \delta R \tag{10-11}$$

where E_i is the Young's modulus of the interior of the cell, T is cortical tension, and r_t is the radius of the AFM tip. Viscoelasticity is then incorporated into the model through converting the shear modulus $G=E/2(1+\nu)$ into the frequency domain as G^*_{NM}

$$G^*_{NM}(\omega) = G^*_i(\omega) + \kappa_t \tag{10-12}$$

with

$$\kappa_T = \frac{\pi(1-\nu)}{2r_t}\left(\frac{R}{h_0}\right)^{1/2} T \tag{10-13}$$

$$G^*_i(\omega) = \frac{1-\nu}{4R^{1/2}h_o^{1/2}}\left(\frac{F(\omega)}{\delta(\omega)} - i\omega b^*(0)\right) \tag{10-14}$$

where $\omega = 2\pi f$ is the angular frequency, h_0 is approximate depth around an indentation point and $i\,\omega\, b^*(0)$ is the correction for the viscous drag force exerted by the liquid medium on the AFM cantilever.

3.2. *Mechanical models for membranes*

3.2.1. *Shaft-loaded deformation*

Most bio-membranes exhibit large and time-dependent deformation pertaining to the mathematically involved rubber elasticity. We will begin with classical linear elasticity. A thin film with planar elastic modulus, E, and Poisson's ratio, ν, bending rigidity $\kappa = Eb^3/12(1-\nu^2)$, thickness, b, and radius, a, being clamped at the periphery forms a freestanding diaphragm. An external force, F, applied to the film center via a ball bearing or a spherical capped shaft with radius, R, results in a central displacement, w_0, film-shaft contact circle with radius, c, and the associated membrane stress, σ. The deformed profile is governed by

$$\kappa \nabla^4 w - \sigma b \nabla^2 w = F\,\delta(r) \tag{10-15}$$

similar to Eq. 10-6. An exact solution to Eq. 10-15 can be derived and can be found elsewhere. Here we present only the asymptotic solution as a guide to the experimentalists. The overall mechanical response is a simple superposition of indentation, plate-bending, mixed bending-stretching, and membrane-stretching, such that

$$w_0 = \underbrace{\left(\frac{9}{16R}\right)^{1/3}\left(\frac{1-\nu^2}{E_t}\right)^{2/3} F^{2/3}}_{\text{Indentation}} + \underbrace{\frac{3a^2}{4\pi\, b^3}\left(\frac{1-\nu^2}{E}\right) F}_{\text{Plate Bending}} + \underbrace{\left(\frac{4a^2}{\pi\, b}\right)^{1/3}\left(\frac{1-\nu^2}{E}\right)^{1/3} F^{1/3}}_{\text{Membrane Stretching}}$$

(10-16)

Figure 10-6 shows a schematic of the mechanical response. The three deformation modes take consecutive turns to dominate the membrane deformation in different length scales. When the shaft comes to initial contact with the film surface, the micro-scale ($w_0 << b$) response is dominated by indentation with $F \propto w_0^{3/2}$, followed by the meso-scale ($w_0 \sim b$) deformation governed by plate-bending with $F \propto w_0$, and ultimately macro-scale ($w_0 > b$) deformation approaches $F \propto w_0^3$. Note that Eq. 10-16 shows only the asymptotic solutions and does not describe the transition from one deformation mode to the next. The transition from indentation to plate bending depends on the dimension and geometry of the indenter and is difficult to derive exactly. The deformation mode transition can be characterized by a crude gauging factor

$$\lambda = \sqrt{12(1-\nu^2)}\;\frac{F\,a^2}{\kappa\, b} \tag{10-17}$$

Plate-bending dominates when $\lambda < 85$, the gradual bending-stretching transition occurs in the intermediate range $85 < \lambda < 3 \times 10^4$, and stretching dominates when $\lambda > 3 \times 10^4$. Note that the intersection of the pure bending and pure stretching asymptotes occurs at $\lambda \approx 350$ and $w_0 \sim 2b$. The asymptotes can be adopted as the constitutive relation of

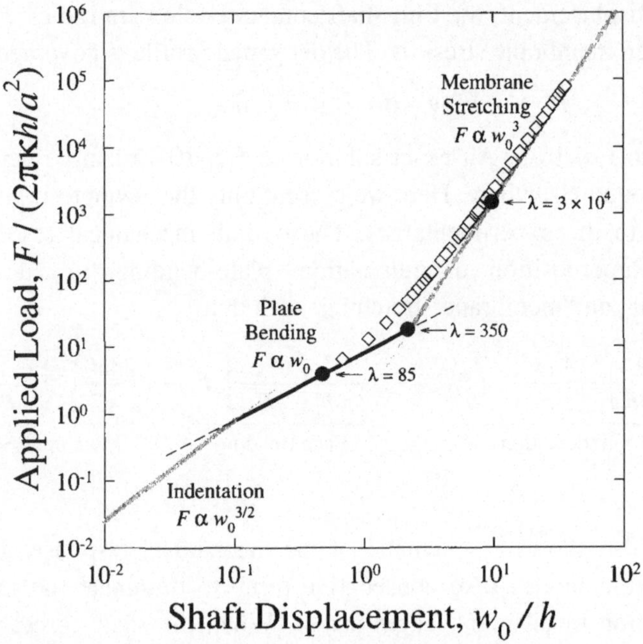

Figure 10-6. The multi-scale behavior of the shaft-loaded blister test. Loading is dominated consecutively by indentation, plate bending, mixed bending-stretching, and membrane-stretching. The data points (diamonds) are obtained by finite element analysis.

the sample in the first approximation for the empiricists, provided the applied load falls in the λ – range indicated above.

Residual membrane stress and viscoelasticity can also be incorporated into the solid-mechanics model. The biaxial tensile pre-stress, σ_0, causes the compliance of the sample membrane to decrease, similar to strain-hardening. A rigorous derivation shows the mechanical response to be

$$w_0 = \frac{F}{2\pi\sigma_0 b}\left\{\frac{1-\beta_0 I_0(\beta_0)-\beta_0 K_1(\beta_0)+\beta_0^2 I_0(\beta_0)K_1(\beta_0)}{\beta_0 I_1(\beta_0)}+K_0(\beta_0)-\log\left(\frac{2}{\beta_0}\right)+\gamma\right\} \quad (10\text{-}18)$$

where $\beta_0 = (\sigma_0\, b\, a^2/\kappa)^{1/2}$ and $\gamma = 0.577216$ is the Euler-Mascheroni constant. The effect is that the bending asymptote shifts upwards reminiscent of classical strain hardening. The stretching asymptote, on the other hand, remains unchanged, because the increasing concomitant stress due to deformation will ultimately supersede the residual stress. The λ-value for the bending-stretching transition must be redefined based on Eq. 10-18. The viscoelastic behavior of the sample can be investigated by cyclic loading of the central load, and the materials parameters can be obtained by expressing the elastic modulus in the complex form, $E = E' + j E''$, in the above model, assuming the simplest standard model of linear viscoelasticity.

3.2.2. *Large-deformation model*

When an ultra-thin biomembrane is investigated using the shaft-loaded blister test, a large deformation model is needed in terms of rubber elasticity. Without loss of generality, the Mooney-Rivilin model is adopted for analysis, where the strain-energy function is given by

$$W = C_1(I-3) + C_2(II-3) \qquad (10\text{-}19)$$

where $I = \lambda_1^2 + \lambda_2^2 + (\lambda_1\lambda_2)^{-2}$ and $II = \lambda_1^{-2} + \lambda_2^{-2} + (\lambda_1\lambda_2)^2$ are the strain invariants, λ_1 and λ_2 are the meridian and circumferential stretch ratios for the deformed membrane, and C_1 and C_2 are the material parameters of the membrane. For a central load applied via a ball bearing, the membrane deformation is governed by the Young-Laplace equation

$$\frac{dT_1}{d\rho} + \frac{1}{\rho}(T_1 - T_2) = 0 \qquad (10\text{-}20)$$

$$K_1T_1 + K_2T_2 = 0 \qquad (10\text{-}21)$$

with $\rho = \rho(r)$ the radial coordinate parallel to the undeformed plane, T_1 and T_2 the corresponding tensile stress, and K_1 and K_2 the principal curvatures. The equilibrium deformed profile gradient is governed by

$$\frac{d\rho}{d\xi} = \lambda_a \left\{ 1 - \frac{F^2}{[2\pi a \lambda_p T_1(\lambda_a, \lambda_p)]^2} \right\}^{1/2} \qquad (10\text{-}22)$$

with $\lambda_p = d\xi/dr$ the meridian gradient along the deformed membrane profile and $\lambda_a = \rho/r$. A numerical iterative method can be devised to compute the deformed profile in a self-consistent manner and thus the constitutive relation.

4. Adhesion of Cells and Tissues

The many-faceted cell adhesion has from time to time serious consequences in biomedical manifestations and applications. On the negative side, life-threatening diseases develop due to inability to express cell adhesion such as leukocyte adhesion deficiency-I (LAD-I), or excessive adhesion such as atherosclerosis. Conversely, formation of natural-prosthetic tissues-organs, morphogenesis and developmental embryology also depend on adhesion of apposing cells. In fact, there are situations when the mechanical properties of macroscopic tissues depend on the hierarchical constitutive components. The multi-scale behavior of the nano-scale lipid bilayer membrane (~100 Å), the micro-scale cell with an encapsulated viscoelastic cytoplasm (~10 μm), meso-scale multi-cell aggregates (~100 μm), and macro-scale biomembranes and organs (>1 mm) can be expressed by a simple phenomenological equation for elastic modulus,

$$E_{macro} = \Theta_{meso} \times \left(\sum_{j=1}^{m} \Theta_{micro,j} \times \left(\sum_{i=1}^{n} \Theta_{nano,i} \times E_{nano,i} + C_{nano,i} \right)_{micro,j} + C_{micro,j} \right)_{meso} + C_{meso} \tag{10-23}$$

where Θ is the structural factor governing the geometrical arrangement, and C is the mechanical properties of the interconnecting or interlocking mechanism (*e.g.* surface forces, extra-cellular matrix).

Cell adhesion and detachment are apparently complex interplay of biochemistry and mechanical properties of the cell membrane and substrate. Rather than tackling the comprehensive cell adhesion theory and phenomena, we focus only on the mechanical aspects of intersurface physical and chemical forces that lead to adhesion interfaces. The origin of surface forces can be roughly divided into two main categories. The non-specific forces arise from electrostatics, van der Waals interaction, configurational forces (*e.g.* intercalating adsorption layers at the adhesion interfaces), DLVO double layer interaction. The specific forces involve ligand-receptor interactions and are mainly responsible for cell signaling. The actual surface force law or potential is difficult to measure experimentally. Most biomedical and engineering literature assumes adhesion to be effective only for adherends in intimate contact with zero intersurface separation, though this is not true in general since the force range usually spans the nm-mm range, yet the cell dimension is in micrometers and intercellular separation is in nanometers.

Notwithstanding the many qualitative techniques for thin film adhesion, a number of quantitative methods are documented in the literature, *e.g.* statistical distribution of cells remaining in adhesion contact when subject to a liquid flow, micro-pipette aspiration, and reflection interference contrast microscopy (RICM). Some of these techniques are also applicable in macroscopic tissue level.

In this chapter, we focus on a modified version of nanoindentation similar to the shaft-loaded blister test where the deformation is confined to a local area comparable to the probe dimension. A general force potential is devised, which takes the form of a step function or a Heaviside function with a finite magnitude, F_s, and range, w_s, corresponding to finite adhesion energy, γ. We will consider adhesion of isolated thin films in general before proceeding to single cells.

4.1. *Membrane adhesion*

Figure 10-7 shows a sketch of the adhesion test configuration. Rather than applying a compressive force on the sample membrane via a ball bearing, a tensile force is applied to a flat punch (*e.g.* AFM tip with a planar or blunt tip) which is in adhesion contact with the membrane. A circular delamination front is driven into the probe-membrane interface. Simultaneous measurement of the force, probe displacement and contact radius yields the adhesion energy. The governing elastic equation is the same as Eq. 10-15 and the delamination trajectory can be formulated using a thermodynamic energy balance between the energy stored in the overhanging elastic annulus and energy expenditure on making new surfaces.

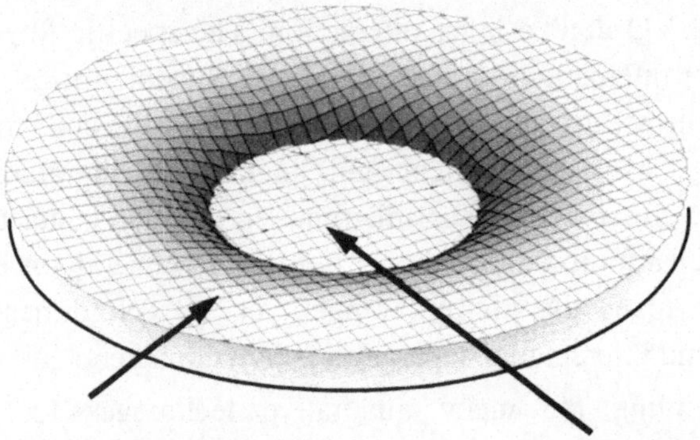

Figure 10-7. Sketch of the punch test to measure membrane adhesion.

The limiting case of finite adhesion energy with zero range and infinite magnitude was derived in our earlier work. The experimentally verified delamination trajectory for pre-stressed membrane is found to be

$$F=\frac{4\pi}{2^{1/2}}(\sigma_0 b)^{1/2}\gamma^{1/2}\left\{\frac{I_1(\beta_0)K_1(\beta_0\zeta)-I_1(\beta_0\zeta)K_1(\beta_0)}{I_1(\beta_0)K_0(\beta_0\zeta)+I_0(\beta_0\zeta)K_1(\beta_0)-1/\beta_0}\zeta\right\} \quad (10\text{-}24)$$

with $\zeta = c/a$. Figure 10-8 shows the delamination trajectory OABC for a punch radius smaller than the clamped film. As the punch moves in a displacement controlled manner, the overhanging annulus deforms with no delamination until point A. An increase in punch displacement shrinks the contact circle and reduces the applied load along path ABC. At point C, the contact radius reduces to a critical non-zero value of c^*, where $(\partial w_0/\partial F)_{r=c^*} = 0$. An incremental increase in w_0 leads to "pull-off" where the film spontaneously detaches from the punch (denoted by * hereafter). The present model shows $0.1758 \le \zeta^* \le e^{-1}$ (= 0.3679), where the lower and upper limits refer to zero and infinite residual membrane stress, respectively. One can characterize the mechanical and adhesion behavior of the film and film-substrate interface using F^* and w^*. Based on thermodynamic reversibility, "pull-in" occurs when the film-punch gap narrows to w^*.

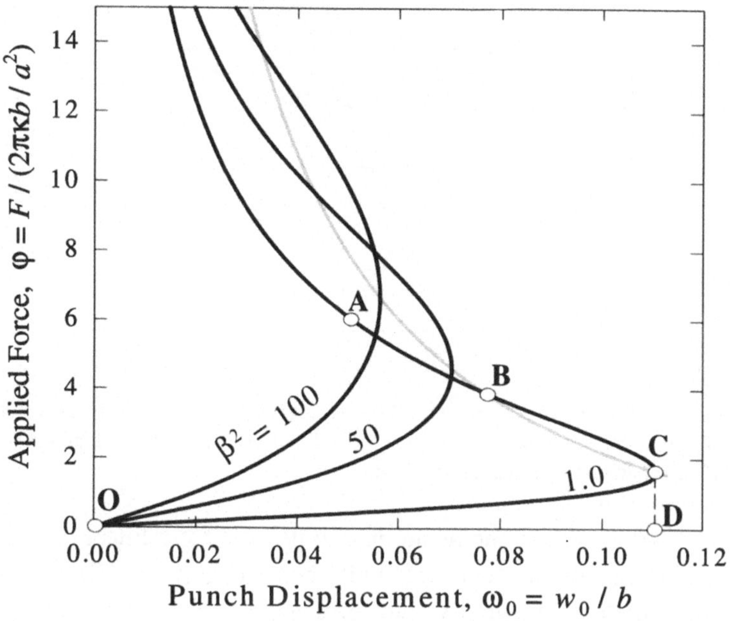

Figure 10-8. Delamination trajectory ABCD with "pull-off" at C. The curves show different pre-stress on the sample membrane.

Generalized intersurface forces with finite and non-zero magnitude and range can be incorporated into the adhesion model in the form of a square well using the Barenblatt-Dugdale-Maugis cohesive zone theory. The short range force discussed above corresponds to the Johnson-Kendall-Roberts (JKR) limit, while the situation with infinite range but vanishing magnitude is referred to the Derjaguin-Muller-Toporov (DMT) limit. The lengthy and mathematically involved JKR to DMT transition for membrane adhesion is not given here but only the qualitative discussion.[25]

At mechanical equilibrium with a non-zero load and fully developed cohesive zone, the membrane profile can be divided into three regions (Fig. 10-9): (i) within the contact circle ($r \leq c$), the membrane is in intimate contact with the substrate; (ii) the inner cohesive annulus ($c < r \leq c_s$), the uniform intersurface force, F_s, acts as the membrane-substrate separation falls within the force range ($w \leq w_s$); and (iii) the outer annulus ($c_s < r < a$), intersurface interaction vanishes.

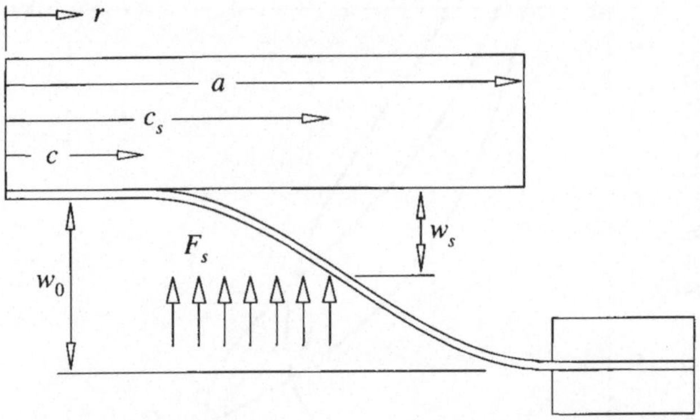

Figure 10-9. Membrane delamination from a substrate surface. Long range surface forces, F_s, are present immediately outside the intimate contact edge, and cut off once the intersurface separation exceeds the force range, w_s.

Figure 10-10 shows the delamination process for a linear elastic, thin and flexible membrane with negligible bending rigidity. The interfacial adhesion energy, essentially the product of the intersurface force magnitude (normalized force, ρ_s) and range (normalized range, ω_s), is made to be a constant here (*i.e.* $\rho_s\,\omega_s = 1$), and the combination of force magnitude is allowed to vary over a range. The maximum displacement the probe can travel is identified as the unstable "pull-off". The dashed curve traces the "pull-off" trajectory in Fig. 10-10.

It is worthwhile to note the changing constitutive relation as the interface character gradually transits from JKR to DMT. As mentioned above, the JKR limit requires the contact radius to shrink as the external diminishes until "pull-off" occurs with a non-zero radius. In the other extreme, when the force range is large approaching the DMT limit, the entire overhanging annulus from the contact edge to the clamped periphery is always under the influence of the surface forces until the membrane completely detaches from the substrate. Here the applied load is an increasing function of the probe displacement, contrasting the JKR limit. It is therefore obvious that the two segments of the "pull-off" trajectory separated by the local maximum can be identified with the JKR (the hook-like envelop) or DMT (monotonic increasing) traits. Experimental determination of the delamination trajectory allows one

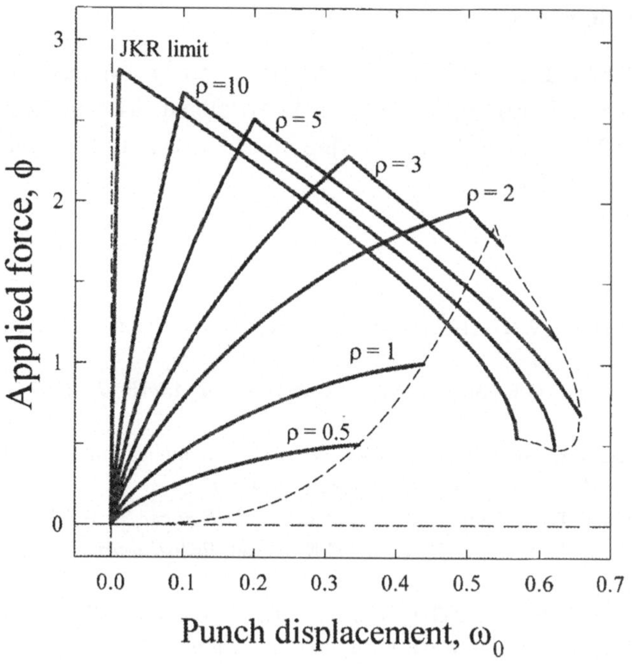

Figure 10-10. Delamination trajectory for a fixed adhesion energy but with a range of magnitude and range of surface forces.

to deduce the intersurface force range and magnitude. Further complications due to non-zero flexural rigidity, tensile pre-stress, large deformations due to rubber elasticity, and viscoelasticity can be incorporated into the model, but are beyond the scope of this chapter.

5. Conclusions and Outlook

Various modes of nanoindentation have been extensively applied for mechanically characterizing biological cells and membranes. Non-invasive, in-situ and experimentalist-friendly indentation methods are important for the instrumental developments. Multiscale linear elastic models for indentation, bending and stretching mechanical stresses for small to large deformation for data analysis are presented. Nonlinear large deformation models are highly desirable for analyzing the force-displacement relation of most biological materials.

Acknowledgments

Support from BBSRC/EPSRC (BB/D014786/1) for KKL and NSF (CMMI # 0757140) for KTW is acknowledged. The authors are also grateful to Mr. Scott Julien at Northeastern University for generating the data for Fig. 10-10.

References

1. M. L. Borene, V. H. Barocas and A. Hubel, *Annals Biomedical Engineering* 274 (2004).
2. E. J. Orwin, M. L. Borene and A. Hubel, *Trans. ASME- J .Biomech. Eng.* 439 (2003).
3. W. M. Becker, *The World of the Cell*, Benjamin/Cummings Publishing (1986).
4. L. Fisher, *J. Chem. Soc. Faraday Trans.* 2567 (1993).
5. C. W. Gadd, W. A. Lange and F. J. Peterson, *ASME Paper No-WA/HUF-8* (1965).
6. M. Ahearne, Y. Yang, K. Y. Then and K. K. Liu, *British Journal of Ophthalmology*, in Press.
7. B. Daily, E. L. Elson and G. I. Zahalak, *Biophysical Journal* 671 (1984).
8. G. I. Zahalak, W. B. McConnaughey and E. L. Elson, *J. Biomechanical Eng. ASME* 283 (1990).
9. F. Yang, *Mat. Sci. Eng.* 226 (2003).
10. E. A-Hassan, W. F. Heinz, M. D. Antonik, N. P. D'Costa, S. Nageswaran, C. Schocenenberger and J. H. Hoh, *Biophys. J.* 1564 (1998).
11. A. Vinckier and G. Semenza, *FEBS Letters* 12 (1998).
12. E. Miller, T. Garcia, S. Hultgren and A. F. Oberhauser, *Biophys. J* 3848 (2006).
13. B. J. Briscoe, L. Fiori and E. Pelillo, *J. Phys. D: Appl. Phys.* 2395 (1997).
14. M. Ahearne, Y. Yang, A. El Haj, K. Then and K. K. Liu, *J. Roy. Soc. Interface* 455 (2005).
15. Y. Yang, P. O. Bagnaninchi, M. Ahearne, R. K. Wang and K. K. Liu, *J. Roy. Soc. Interface* 1169 (2007).
16. Y. Sun, K. T. Wan, K. Robert, J. Bischof and B. J. Nelšon, *IEEE Transaction on NanoBioScience* 279 (2003).
17. X. Yao, J. Walter, S. Burke, S. Stewart, M. H. Jericho, D. Pink, R. Hunter and T. J. Beveridge, *Colloids and Surfaces B: Biointerfaces* 213 (2002).
18. S. Sen, S. Subramanian and D. E. Discher, *Biophysical Journal* 3203 (2005).
19. M. Arnoldi, M. Fritz, E. Baeuerlein, M. Radmacher, E. Sackmann and A. Boulbitch, *Phys. Rev. E* 1034 (2000).
20. K. T. Wan, V. Chan and A. Dillard *Colloid & Surface B: Biointerface* 241 (2002).
21. K. T. Wan, *J. Applied Mech. ASME* 110 (2002).
22. E. M. Darling, S. Zauscher and F. Guilak, *Osteoarthritis and Cartilage* 571 (2006).

23. Y. C. Fung, *Biomechanics*, 2nd edition, Springer (1996).
24. P. Roca-Cusachs, I Almendros, R. Sunyer, N. Gavara, R. Farre and D. Navajas, *Biophysical J.* 3508 (2006).
25. J. A. Greenwood *Proceedings of Royal Society of London* **A453**, 1277 (1997).

CHAPTER 11

CONCLUSIONS AND OUTLOOK

Michelle L. Oyen

Department of Engineering, Cambridge University
E-mail: mlo29@cam.ac.uk

This volume has considered the application of nanoindentation testing techniques for the evaluation of the mechanical properties of biological materials. The book began with an exploration of the fundamental contact mechanics principles upon which nanoindentation methods are based, including elastic, plastic, viscous and viscoelastic deformation modes, and adhesion between the indenter and sample. The book then considered specific applications of nanoindentation testing to biological tissues such as bone, tooth tissues, soft tissues and other biological materials such as plants, woods and seashells. The indentation of membranes and structures, applicable to the indentation of isolated biological cells, was also considered.

There are several reasons for considering this subject. Biological materials can have intrinsically mechanical functions, and understanding the underlying structure-property relationships can promote general knowledge. There are at least two possible applications for the information gained by understanding biomechanics in this context. First, the knowledge can be used to aid in biomedicine in the context of repairing injuries or replacing extracted tissues. Tissue engineering is a large and growing area of investigation, but more attention has been paid to the composition and biological function of engineered tissues than to their mechanical function. Second, there are also significant opportunities for future developments in the area of biomimetics, where natural principles are exploited to design novel materials and devices.

Finally, and not unrelated to the other two applications, there is a great need in the medical device industry to make implants with mechanical properties which are not incompatible with the natural tissues. This field—man-made biomaterials and medical implants—has not been considered extensively in this volume to this point, but represents an economic faction of interest in the discussions to follow.

The earliest applications of nanoindentation to biological tissues over a decade ago examined bones and tooth-tissues—materials that were comparable in mechanical stiffness to the engineering materials and systems for which commercial nanoindentation instruments were originally developed. The stiffness of bones and tooth tissues ranges from around 10 GPa to over 100 GPa, while the elastic modulus of two common calibration materials—fused silica (sometimes "fused quartz") and single crystal aluminum—is around 70 GPa for both, such that the working range is a mN-nm load-displacement (P-h) framework for either mineralized tissues or engineering metals and ceramics. Metals and ceramics exhibit little to no time-dependent deformation (with the exception of extremely soft metals such as single crystal indium or soft ionic crystals such as single crystal sodium chloride). Bones and tooth tissues exhibit minimal time-dependence when dried, although they can exhibit a greater degree of time-dependent deformation when in a physiologically hydrated state. Regardless, the stiffness and the dominance of elastic or elastic-plastic deformation in bones and teeth meant that the existing nanoindentation instrumentation and data analysis techniques were directly applicable to mineralized tissues.

There were significantly greater challenges when the first experiments were performed on soft and hydrated biological materials, such as articular cartilage. Cartilage is approximately 75% water, which gives rise to dramatic viscoelastic and poroelastic deformation. The base elastic modulus ranges from less than 1 MPa to several MPa. Instead of working in a mN-nm load-displacement (P-h) framework, we move to a mN-μm framework. For transducers with a maximum displacement range of several micrometers, the forces developed in soft tissue indentation may challenge the load resolution of transducer. Of course, softer materials can be indented using an atomic force microscope instead of an instrumented indenter. In terms of environmental control,

AFMs are far advanced compared with commercially available indenters. However, the cantilever-based loading mechanism introduces challenges for time-dependent characterization, due to a lack of direct closed-loop control of the load or position of the the indenter tip.

Instrumentation challenges are only one factor preventing the widespread adoption of nanoindentation techniques in the context of soft tissues. Instrument manufacturers appear reluctant to introduce options for testing and data analysis into the software driving current commercial nanoindentation systems. Clearly the widespread adoption of nanoindentation techniques for analysis of engineering materials was driven by the incorporation of Oliver-Pharr elastic-plastic analysis into indenter software. Mechanical testing underwent a change almost overnight, as samples could be inserted into what was almost a "black box" mechanical property meter, where mechanical tests could be performed and data analyzed with just a few clicks of a mouse. This has resulted in the widespread reporting of the "hardness-and-modulus" as if these were the relevant mechanical properties for all materials. Such an approach assumes that there is a one-size-fits-all testing method for nanoindentation and that an elastic-plastic analysis is appropriate in all cases.

We therefore encounter the first significant challenge that must be addressed prior to the routine implementation of nanoindentation testing for mechanical analysis of *all* biological materials: one or more commercial instruments must be developed, including both hardware and software modifications compared with current commercial systems, meant to address the needs of the biological materials community. It is useful to note at this point that there are compliant nonbiological materials for which similar testing capabilities would be useful, such as polymer foams, elastomers and hydrogels (Fig. 1-11), and the market for such instruments would be larger than just the biological materials community. As emerging areas of biomedical research develop, the need will emerge for both researchers and medical device developers to mechanically test materials with compliances comparable to biological soft tissues, such as hydrogel coatings and tissue-engineered implants.

The development of a compliant material-optimized nanoindentation system should driven by instrument manufacturers. For this to happen, companies must be convinced of the economic need for such a system; the thin films and microelectronic devices industries motivated development of today's nanoindenters. Currently there are a number of commercial nanoindentation systems on the market, and thus a number of different manufacturers. Novel developments in the past decade or more have included coupled instrumented indentation and AFM devices and indenters that function inside the chamber of an electron microscope. The developments continue to be driven towards advanced applications for elastic-plastic materials, and not yet for biological systems.

It is hoped that this collection of chapters has demonstrated the research potential for nanoindentation of biological materials, while alluding to other related applications in biomedicine that would justify the infrastructure and development costs of optimized system development. The instrumentation requirements for an optimized instrument were discussed in Chapter 3 of this volume, with particular emphasis on load and displacement resolution and range, and the critical need for environmental control. Data analysis techniques for materials with responses more complex than elastic-plastic were summarized in Chapters 4 and 5. Given the large body of literature that has supported the widespread adoption of commercial indenters for elastic-plastic materials, a similar research effort will be required in support of new approaches to instrumentation and data analysis, to add to the relatively small body of current literature reviewed in this volume. It is worth noting at this point that the mechanical properties of biological materials are more complicated than the simple linearly viscoelastic or poroelastic approaches considered to date. Ongoing research into indentation of materials exhibiting responses that are nonlinear viscoelastic, hyperelastic, multiphasic and exhibiting advanced poroelastic behavior is still needed.

In conjunction with the development of hardware and software optimized to handle soft, compliant and time-dependent materials, there are additional considerations related to standards for the calibration of indenter tips and validation of system performance. Material standards development clearly presents a different need than instrumentation or data analysis, and would likely be led by a standards organization, research group or materials testing laboratory.

Without the intervention of one or more instrument manufacturers, research on nanoindentation of biological materials will continue, but perhaps at a slower pace than would be found were a novel and optimized instrument available. Current nanoindentation studies on soft and hydrated materials utilize existing hardware—such that researchers are left to "rig up" modifications to accommodate the limitations of current instruments—with off-line data analysis requiring the efforts of individual researchers to write custom data analysis routines. As such, nanoindentation characterization of biological materials is different from one laboratory to another, and retains a "research" feel instead of the more uniform testing and analysis for elastic-plastic materials since the introduction of the Oliver-Pharr method and commercialization of dedicated nanoindentation instruments. The biological materials nanoindentation community is thus fragmented and the impact of the work minimized due to a lack of standardization.

An intermediate possibility, if no manufacturer customized an instrument for both testing and analysis of nanoindentation data on biological materials, would be the development of a stand-alone data analysis program. With such a model, the data (however obtained) from any current nanoindenter, AFM, or a custom- or home-built indentation system could be analyzed in a uniform framework. There are two obvious options for the development of such a system, either commercially-funded or supported by research organizations. There are also two (or more) options for a stand-alone data analysis package: open-source software or a commercial product. The factors associated with determining which model would prevail are intrinsically coupled to the funding mechanism used to develop the software package, and the projected commercial revenue associated with the sale of such a product. A third model could prove viable, with a simple version available open-source and an advanced version only available by fee-paying license. This author certainly hopes that the arguments presented here help motivate the community to form such a consortium (with financial support as well) if an instrument manufacturer-led project does not arise in the near future.

In conclusion, nanoindentation of biological materials represents a fast-growing and fast-developing area of research and development inquiry. The potential for gaining understanding of the natural world is immense. The extensions to the development of biomedical implants, tissue engineered replacement body parts and biomimetic materials for non-medical applications represent an extremely large opportunity for the research community and corresponding commercial entities. A model was presented in this chapter for expanding the impact of this research, either by intervention on the part of one or more instrument manufacturers to develop a new integrated hardware and software solution, or by the formation of a consortium of researchers for developing a stand-alone software solution. Even without commercial developments in either hardware or software, this area of inquiry is only going to grow and develop, with advances in human understanding likely to result in improvements in quality of life.

Index